Inference and Representation

Inference and Representation

A Study in Modeling Science

MAURICIO SUÁREZ

The University of Chicago Press
Chicago and London

The University of Chicago Press, Chicago 60637
The University of Chicago Press, Ltd., London

Published 2024
Printed in the United States of America

33 32 31 30 29 28 27 26 25 24   1 2 3 4 5

ISBN-13: 978-0-226-83002-5 (cloth)
ISBN-13: 978-0-226-83004-9 (paper)
ISBN-13: 978-0-226-83003-2 (e-book)
DOI: https://doi.org/10.7208/chicago/9780226830032.001.0001

Library of Congress Cataloging-in-Publication Data

Names: Suárez, Mauricio, author.
Title: Inference and representation : a study in modeling science / Mauricio Suárez.
Description: Chicago ; London : The University of Chicago Press, 2024. |
Includes bibliographical references and index.
Identifiers: LCCN 2023026892 | ISBN 9780226830025 (cloth) |
ISBN 9780226830049 (paperback) | ISBN 9780226830032 (ebook)
Subjects: LCSH: Science—Philosophy. | Science—Methodology. |
Representation (Philosophy) | Inference. | Knowledge, Theory of.
Classification: LCC Q175 .S9345 2024 | DDC 501—dc23/eng20230919
LC record available at https://lccn.loc.gov/2023026892

♾ This paper meets the requirements of ANSI/NISO Z39.48-1992 (Permanence of Paper).

Para Samuel Hugo, hijo pródigo

Contents

Preface and Acknowledgments ix

1 **Introducing Scientific Representation** 1

PART I **Modeling**

2 **The Modeling Attitude: A Genealogy** 19

3 **Models and Their Uses** 44

PART II **Representation**

4 **Theories of Representation** 83

5 **Against Substance** 103

6 **Scientific Theories and Deflationary Representation** 132

7 **Representation as Inference** 155

PART III **Implications**

8 **Lessons from the Philosophy of Art** 189

9 **Scientific Epistemology Transformed** 224

Notes 261

References 289

Index 311

Preface and Acknowledgments

This book has been long in the making. As I explain in the biographical notes in chapter 1, I became interested in scientific representation through my work on the practice of modeling in science. Most of my initial work on modeling took place while I was a PhD student at the London School of Economics during the years 1992–96. I came across the topic of scientific representation during my last year there, while I was writing up my PhD thesis on modeling in quantum mechanics. R. I. G. Hughes was a visiting professor that year, and I had the chance to talk to him extensively while he was developing the pioneering paper on representation that eventually appeared as Hughes (1997). This book certainly owes much to Hughes and those conversations I had with him. He is possibly the one person about whom I can say that in truth this book would not exist without him. In memoriam, and thank you, R. I. G.

There are others who have been decisive along the way in making the book become what it is. A memorable exchange with Margaret Morrison and Arthur Fine after a seminar that the former delivered in March 1998 at Northwestern University, where I was spending the year as a postdoctoral research fellow, was a catalyst for much of my own thinking on the topic. Back in Britain, the award of a Leverhulme Trust Fellowship during 2001–2 finally allowed me to get on with the project, and it led to the publication of the two papers of mine on the topic with the highest citation count to date (see Suárez 2003, 2004). The intense response that followed provided a veritable mine for philosophical argument but delayed the project further. A symposium that I organized on the topic at the Philosophy of Science Association conference in Milwaukee in 2002, with Bas van Fraassen, Ronald Giere, Mary Morgan, and Andrea Woody as cospeakers, also provided much impetus, while Bas van Fraassen's Locke Lectures at Oxford that year were a source of inspiration.

(The symposium papers were eventually published as Mitchell 2004, while Bas's lectures eventually gave rise to van Fraassen 2008.) Research stays at the Department of History and Philosophy of Science and the Centre for Time at Sydney University in 2003 allowed me to study the importance of inferential semantics and pragmatics in general. I want to thank Rachel Ankeny and Huw Price for hosting me and for their feedback. Another research visit at the Centre for Philosophy of Natural and Social Science at the London School of Economics in 2005 allowed me to discuss representation in relation to fictions and the work of Nelson Goodman. Thanks to Stephan Hartmann, Nancy Cartwright, and Mary Morgan for making that visit possible. Later still, three consecutive stays at Harvard University (in 2007, 2009, and 2011) allowed me to link my then developing ideas on representation with several pragmatist themes in some detail as well as to come to terms with some of Hilary Putnam's works on closely related topics regarding cognitive representation, truth, and reference. I thank Catherine Elgin, Peter Godfrey-Smith, Ned Hall, Sean Kelly, and, especially, Hilary Putnam (twice) for sponsoring those visits and making them possible as well as for their comments and suggestions.

The project was then shelved while I got on with a different project on probability from 2011. I returned to representational matters in 2018, partly through an ongoing honorary position at the Department of Science and Technology Studies, University College London, for which I thank Chiara Ambrosio, Phyllis McKay Illari, and Julia Sánchez-Dorado. It was clear by then that a lot of new literature had emerged that made the prospect of writing a comprehensive guide to the topic unviable. Wise counsel—not least from my tireless University of Chicago Press editor—advised turning the book into a leaner but I hope crisper presentation of an inferential approach to scientific representation simpliciter. A summer school at the Institute Vienna Circle in the summer of 2018 provided one last impetus, and it led to a series of research stays at the IVC from January 2020 on for the completion of the project. I am grateful to Jim Brown, Katarina Kinzel, Martin Kusch, John Norton, Sabine Koch, and Fritz Stadler.

More people than I can recount have offered comments and advice as the project slowly progressed over years. They certainly include, besides those already mentioned, Brandon Boesch, Alisa Bokulich, Maria Caamaño, Jimena Canales, Natalia Carrillo-Escalera, Elena Castellani, Anjan Chakravartty, Hasok Chang, Gabriele Contessa, Henk de Regt, José Díez, Mauro Dorato, Roman Frigg, Mathias Frisch, Peter Galison, Peter Godfrey-Smith, Carl Hoefer, Martin Jones, Philip Kitcher, Tarja Knuuttila, Uskali Mäki, Thomas Mormann, Francesca Pero, Isabelle Peschard, Chris Pincock, Pedro Sánchez-Gomez, Carmen Sánchez-Ovcharov, Michael Stoelzner, Paul Teller, Marion

Vorms, Eric Winsberg, and Rasmus Winther. Many thanks also to my colleagues and students at Complutense, London, Vienna, and elsewhere over the years.

Many of the ideas in the book have been rehearsed at different conferences and venues, and I thank audiences for their questions and reactions, particularly those at the Society for Philosophy of Science in Practice, Models and Simulation, and integrated History and Philosophy of Science meetings. I also thank five referees for the University of Chicago Press who reported on the book, and I apologize in advance to those who contributed yet are not mentioned. By now they know who they are better than I possibly could.

The inordinate length of this book project overlaps almost entirely with my son's lifetime, a period over which I have seen him develop into a mature and admirable young man. At several difficult times, extending over months or years, "Sam S-R" was my sole source of encouragement and the only reason I could find to go on. I owe it to him to have found the strength and resilience to complete this project. Two other people and places stand out for special mention. The first full draft of the book was completed in San Lorenzo de El Escorial in November 2021. The final revised manuscript was committed to the Press in Vienna in July 2022. It turns out that both places share a good deal of their history, architecture, and spirit. Their cafés, views, and cobbled streets unlocked my resolve in equal measure. Karen Merikangas Darling, executive editor at the University of Chicago Press, showed me unlimited patience, appreciation, and understanding over many years, and the unfailingly kind Press editorial associate Fabiola Enríquez Flores was critically helpful in the last few stages.

I am indebted for financial support to the Leverhulme Trust, the Centre for Time at the University of Sydney, the Dissent and Contingency research project at the London School of Economics, the Institute Vienna Circle, and the Spanish government both for funding my stays abroad (at the London School of Economics and Harvard) and through its research projects (PGC2018-099423-BI00 and PID2021-126416NB-I00). Chapters 1–3 and 8–9 are entirely new, while chapters 4–7 are extended, emended, or heavily redacted versions of published works of mine as follows. Chapters 4 and 5 are partly based on material in Suárez (2003), with permission from Taylor and Francis. Chapter 6 relies in different ways on claims made in Suárez (2005, 2015) and Suárez and Pero (2019). Chapter 7 is a heavily redacted, revised, and expanded version of Suárez (2004), with permission to reprint a few extracts from the University of Chicago Press.

Vienna, July 2022

1

Introducing Scientific Representation

The concept of representation is central to many of the concerns of philosophers of science, but until quite recently it was surprisingly one of the least explored topics in the philosophy of science. Within the philosophy of science of the last century, it has been only within the last two decades that attention has been paid to the nature as well as the philosophical and practical implications of scientific representation. These decades have seen an upsurge of interest in the notion of representation, with many symposia, conferences, and workshops organized on the topic and a large number of articles published. Why has it taken us so long to focus on this central issue? Answering this question will help us reveal the concerns and objectives that reflection on representation is supposed to accomplish. It will allow us to discern the aims of a philosophical theory or account of scientific representation.

It is now close to a mantra among those interested in scientific representation that an important cause of recent interest in the topic is the replacement of the old syntactic conception of scientific theories with a new semantic conception. Or conversely, it is presumed that a main reason for the lack of previous interest in this topic is to be found in the dominance of the received or *syntactic* conception of scientific theories well into the 1970s.[1] On the syntactic conception, scientific theories are essentially linguistic entities, and their relation to the world is necessarily mediated by the linguistic properties of reference, denotation, designation, etc. Now, it has been noted by many authors (e.g., Giere 1988, 1999a) that the terms *syntactic* and *semantic* are misleading in important respects. Any *linguistic* view worth its salt is also necessarily *semantic*—it declares the terms in scientific language interpretable and meaningful in given ways. Regardless of the terminology, however, representation can play a role in elucidating the relation of theories to the world on a syntactic, or linguistic, conception only by being intimately linked to—even identified with—descriptive notions. Not surprisingly, according to this "official history," interest in representation was essentially suppressed by the logical empiricist exclusive focus on the language of science, and the logical empiricists' disregard for models and modeling as heuristic tools (Bailer-Jones 2003) is essentially of a piece with the dismissal of representation. In other words, Carnap's *semantic ascent* (really a linguistic turn) is supposed to have driven any interest into the nature and practice of representation underground for decades within the philosophy of science.

According to this official history, then, interest in representation emerges only when the linguistic, syntactic view is abandoned in favor of the so-called semantic conception of scientific theories. This is in truth a nonlinguistic conception of theories that comes in two versions. In a first, structuralist rendition (advocated and defended originally by Patrick Suppes in a set of pathbreaking papers published in the 1950s and 1960s), theories are best conceived of as structures, and the relation of theories to the world is consequently best conceived in structuralist terms. The change is supposed to signal interest turning away from the linguistic category of description and toward the more general category of representation, and in particular its subspecies structural representation.[2] If so, the focus on scientific representation is driven by an attempt to explain and illuminate the relation between our theorizing about the world and the world itself once the strictures imposed by the logical empiricists' syntactic account are done away with.

I shall be addressing the nonlinguistic or semantic view in chapter 6, but it is important to emphasize at the outset that the starting point and presupposition of this book is rather different. I now beg to differ with this official history, and I locate the philosophical interest in representation elsewhere. True, the topic of scientific representation has become a very lively one indeed in recent philosophy of science. But it used to be very lively prior to the advent of logical empiricism and its syntactic conception of scientific theories. It appears prominently in fin de siècle disputes and throughout the early years of the philosophy of science in Britain, France, Germany, and elsewhere. It figures in ways that are altogether contemporary in the work of Hermann von Helmholtz, Ludwig Boltzmann, Heinrich Hertz, James Clerk Maxwell, Lord Kelvin, and Norman Campbell. These writers predate the syntactic conception yet treat models and modeling extensively—from a philosophical as well as a scientific point of view—and do not link modeling to theorizing or the question of the structure of theories, never mind its linguistic formalization. And, while the interest in recent decades is intense, it was not absent during the high tide years of logical positivism. Max Black (1962), Mary Hesse (1963/1966), and Stephen Toulmin (1963) were all major contributors to a continuing literature on models and modeling in the English-speaking world. And, although at present there are all the signs of an emerging philosophical wave or movement, it is by no means the case that all authors interested in representation are in addition defenders of the semantic conception—at least certainly not in its dominant structural version. In other words, the outpouring of interest in representation is too recent to have been directly caused by the emergence of the semantic conception of scientific theories in the 1970s. While the semantic view may be consistent with this upsurge of interest (and, in chap. 6, I argue that a nonstructural version of the semantic view is also hospitable to the inferential conception of representation defended in this book), it has not been the primary engine behind it.

On the contrary, many of us got interested in representation through an interest in models and modeling practices in science. Arguably, no account of theory is required in order fully to understand inferential modeling practice,[3] and historians and sociologists were already paying handsome attention to such practices without recourse to theory.[4] From a biographical point of view, my interest in modeling was kindled by my participation in the so-called mediating models movement, whose members were, if anything, critical of the semantic conception.[5] This movement attempted to direct attention away from questions of essence such as, What is a scientific theory? and refocus the debate on the more pragmatic questions of use and application of theories and models. During 1995–96, R. I. G. Hughes was a visiting professor at the London School of Economics (LSE) while I was finishing my PhD thesis on modeling in quantum mechanics there, and I had the chance to talk to him extensively while he was working out his views on representation.[6] This paper is a pioneering effort in many ways, defending the application of Goodman's views to scientific representation. The conversations explicitly brought out for me a connection between the largely descriptive literature on physics modeling that we were reading in the LSE's Research Group in Models in Physics and Economics and the much more prescriptive philosophical literature on the nature of representation. But what finally triggered me to start working on this topic was a memorable exchange with Margaret Morrison and Arthur Fine after a seminar at Northwestern University, where I was spending the year as a postdoctoral research fellow, that the former delivered in March 1998 on the topic of modeling. Roughly, Morrison defended the representational function of models from the point of view of scientific realism, while Fine attacked it from an instrumentalist point of view. I was left with a distinct feeling that this was a pseudo dispute generated by unduly strong assumptions regarding the nature of representation. The description of modeling practice was common ground, and it seemed as if the only differences hinged on two opposing attitudes to an underlying substantive approach to representation. If one could show how to deflate the concept of representation, I thought, one could get these two friends to agree.

My conjecture is then that the recent interest in representational matters is mainly a result of the renewed interest in the topic of modeling among philosophers of science during the last twenty years or so. The resurgence of representation might appear to be a consequence of the emergence of the semantic conception only because interest in modeling has been to some extent coextensive with interest in the semantic conception of theories. If this is correct, the driving issue behind recent work on representation is not metaphysical and is unrelated to anxieties about the referential relations between our theories and the world. What drives inquiry into representation is rather an attempt to understand modeling practices in science. And this is significant since it means that a theory of representation must be judged primarily by its ability to answer pragmatic questions about modeling practice, not analytic questions regarding the metaphysics of the relation between our thought and the world.

To my knowledge, the outlook advanced in this book makes it a pioneering effort in this area, at least in its explicitness. I will assume from the start that the aim of any account of scientific representation is to understand modeling practice. My minimal use of analytic distinctions and categories in metaphysics is driven by an attempt to understand and accommodate the practice of model building. The pragmatist tradition has not cared very deeply in its heart about questions of reference, denotation, or the connection between our thoughts and the world. I doubt that there is an essence to the relation of representation and that one will ever be found; but, in any case, this would be a poor basis against which to judge the present book since it is not its aim to locate or describe such essence. On the contrary, I tend to be in sympathy with those who think that these issues in metaphysics have, if anything, occasionally led philosophers of science astray. My main concern throughout will be to develop a conception of representation that provides us with a fair account of modeling practice. The inferential conception of scientific representation that I defend in this book is, I submit, the best account of representation *for the purposes of understanding the practice of modeling in science*. I do not claim for it any virtues as an analysis of the metaphysics of the relation between thought and the world. On the contrary, on such matters of analytic metaphysics the inferential conception is meant to remain silent.[7]

The Uses of Modeling

The project hence aims to derive lessons regarding the nature of representation from the practice of scientific modeling. Thus, part 1 of the book is entitled "Modeling" and comprises two chapters devoted to the history, nature, and methodology of model building. Chapter 2 is a historical review of the emergence of what I call the *modeling attitude*, a nineteenth-century and fin de siècle development that bears witness to the long history of modeling in scientific practice and the enduring attraction of philosophical reflection on the nature of representation involved in its practice. Chapter 3, by contrast, puts the focus more narrowly on several prominent cases of models—both historical and contemporary—that will be employed as benchmarks for the different accounts of representation discussed throughout the book. Together, they constitute the material grounds, historical and descriptive, on which philosophical arguments regarding the nature of representation will proceed. Not surprisingly, then, given the philosophical use to which they will be put, I make no excuses for presenting both the history and the details of the case studies in a form that is already suited to my subsequent purposes and arguments. As in much case study work typical of contemporary philosophy of scientific practice as well as the philosophically informed historical work typical of the integrated history and philosophy of science, the hope is that the presentation will strike the right balance between historical and descriptive accuracy, on the one hand, and philosophical relevance, on the other.

Chapter 2 charts a course through a fascinating and complex history, particularly in relation to nineteenth-century physics; I do simplify somewhat for the sake of extracting the elements that bear on the discussion of representation and particularly those that bear out the inferential conception that I defend in the book. But the simplification is, I hope, sufficiently faithful to the record. It is certainly informed by both primary sources and much secondary literature in the interpretation of those sources within their historical context. It would be tempting to attempt an encyclopedic treatment, and certainly such treatments already exist and are appropriately referenced in the text. But I have overall and in full conscience avoided any attempt at a comprehensive historical account. An account exquisitely fair to the letter in every dimension would more likely induce confusion in the reader, and it would serve little philosophical purpose.[8] Throughout the narrative, the emphasis is squarely on the thesis of the *relativity of knowledge* that informs, I find, most of the explicitly philosophical discussions about the nature of representation. As a necessary warning, *relativity* here has little to do with current forms of philosophical relativism and rather points to the fact that all scientific modeling is relative to prior or antecedent representations of some features, aspects, forms, or types of the systems involved. This ensures a sort of virtuous circularity of representation, and a putative task of a successful account of representation will be to explain how and why such representational circularity can amount to virtue.

The historical account focuses on the relativity thesis as it figures in the thought and practice of prominent defenders of the emergent modeling attitude, including Kelvin, Maxwell, Hertz, and Boltzmann. While the historical facts are hardly new or unknown, the tapestry of concerns regarding representation that is described adds, I hope, new elements. I do not argue that Kelvin, Maxwell, Hertz, Boltzmann, and other "modelers" shared any substantial philosophical thesis regarding representation. Instead, I ascribe to them a shared *attitude*: a far less committed stance, one requiring no significant theoretical backup but instead best revealed in practical decisions and judgments in their modeling work.[9]

Chapter 3 shifts away from history and focuses on the representational uses of scientific models. Scientific modeling putatively serves several functions in inquiry related to the aims of explanation, prediction, and control, and representing phenomena is a prominent vehicle for most of them.[10] A number of case studies are presented as illustrations of the diverse representational uses of modeling, namely: (i) an architect or engineer's diagrammatic scale model of a building or a bridge; (ii) concrete toy physical models of physical systems, such as planetary models of the atom or mechanical models of the ether; (iii) conceptual analogies such as the billiard ball model of gases; and (iv) mathematically sophisticated models such as the Lotka-Volterra model in evolutionary ecology, models of stellar structure in astrophysics, or the quantum state diffusion model for quantum measurement.

A few key distinctions are then introduced regarding these case studies. The convention is applied throughout to refer to the object doing the representational work as the *source* and the object getting represented as the *target*. And, most importantly, a distinction is drawn in this chapter between the *means* and the *constituent* of scientific representation.[11] A rough-and-ready definition of these terms will suffice for now as follows. At any given time, the means of a representation is the relation between source and target employed by scientists (more generally, by any inquiring agent) in their modeling practice. The constituent, on the other hand, is whatever relation is necessary and sufficient for the source to represent the target. The distinction has not been appropriately drawn in the literature often enough, an omission that has led to unnecessary confusion. Much of the debate over representation so far has attempted to settle issues regarding the constituent of representation via a discussion of the means of representation. Yet these can be utterly different relations. Certainly, in any case of successful modeling, there will be some relation between the source and the target that is employed actively by an inquiring agent to carry out the representational work. But this need not be the relation—if any—by virtue of which the source *is* a representation of the target.[12]

The Elements of Representation

The distinction between means and constituents is crucial for the purposes of this book for two main reasons, which I explore in greater detail in chapters 4–7, the main theoretical part of the book. First, in chapter 4, this distinction grounds a clear division of labor between epistemic and conceptual dimensions of representation. This leads naturally to a further distinction between *successful* representation, on the one hand, and the more general category of representation, on the other.[13] The epistemic norms built into representing practices derive from the former; that is, they derive from the aim to seek accuracy or truth. By contrast, the conceptual and philosophical issues discussed in this book mainly address the more general category; that is, they deal with the prior conceptual question of how a representational relation is set in the first place.

Second, distinguishing means and constituents helps us appreciate the limitations of accounts that aim to reduce representation to some relation between source and target that pertains only to the properties of the objects that play those roles (such as isomorphism or similarity). I will refer to such theories as *reductive naturalist* since there is a sense in which they aim to naturalize representation by identifying it entirely with the set of facts about the properties of the relata, thus avoiding any reference to human values and in particular the interests, desires, and purposes of the inquirers. In other words, a reductive naturalist account defines representation entirely in terms of the properties of the typical objects of scientific research. This is arguably a particularly strong form of naturalism about representation. It is not my aim in this book to argue that representation cannot be naturalized *tout court* since there are weaker forms of naturalism that are not reductive in this way. However, I maintain that no naturalization of representation will take the form of reductive naturalism. As is often the case in analytic philosophy, a preliminary defense of a conception of representation must first carry out a full critical study of any possible or potential alternatives, and reductive naturalist theories that appeal to similarity and isomorphism are among the only-half-cooked alternatives for a theory of scientific representation at present. Chapter 4 introduces all the relevant terminology, and it develops the proposals in detail. It thus provides a succinct and self-standing introduction to scientific representation. Chapter 5 turns to a critical analysis of reductive naturalist accounts. It develops fully five arguments against them, arguments that I have elsewhere called the *variety*, the *logical*, the *misrepresentation*, the *nonsufficiency*, and the *nonnecessity* arguments. These arguments show that reductive naturalist theories are untenable and consequently pave the way for alternative nonreductive views, which are first outlined in chapter 6 and then fully embraced in a particular form in chapter 7. Throughout the book, these arguments are employed as the benchmarks against which theories of representation must be judged.

Thus, the four chapters in part 2 have a logical order. Chapter 4 establishes the grounds that any discussion regarding the nature of scientific (or, more generally, cognitive) representation must take. Chapter 5 shows why the main reductive naturalist approaches to representation fail and extracts lessons from these failures. Chapter 6 urges a deflationary attitude and develops the more general outlines of a deflationary account. Finally, in chapter 7, I develop and defend my own account of scientific representation against the background of the definitions and distinctions drawn earlier in the book, particularly the means/constituent distinction. Indeed, the bulk of this book is in essence a sustained attempt to argue for pluralism regarding the means of representation and minimalism regarding its constituent. In other words, the aim of the book is to argue that there is a plurality of different relations between sources and targets that scientists legitimately employ in their representational practice—and that help them assess the accuracy of their representations—while arguing that there is nothing metaphysically deep about the constituent relation that encompasses all these diverse means. This combination of minimalism and pluralism at the heart of my conception of representation turns it into a complex position but not, I claim, an unstable one. Rather, the two requirements that I minimally impose on representation stand in a sort of reflective equilibrium vis-à-vis each other, thus providing the required stability.

These features can be explored through helpful analogies with debates on theories of truth involving similar combinations of pluralism and minimalism,[14] and this is indeed the purpose of some central parts of chapters 4 and 6. But the crucial test of the stability and coherence of this conception of representation will be in its detailed application to case studies from the modeling literature. It does not at all depend on whether deflationism about truth is viable. That is just a helpful analogy—a philosophical model of just the sort studied in this book. Hence, while chapter 7 is a detailed exposition and defense of the inferential conception, it also involves taking stock a bit in terms of the case studies provided.

In chapter 7, I explain how the inferential conception of representation is a two-vector concept: two conditions or requirements must be fulfilled for any source *A* to represent a target *B*. The first condition involves what I call *representational force* or *force simpliciter*. This is an oriented, merely formal relation between *A* and *B* such that the source *A* can be said to point toward its target *B*. It does not imply a metaphysically stronger relation that would require the existence of either *A* or *B*. Nonetheless, wherever this relation obtains, we might also say that *B* is the force of *A*. I take *force* to be an unanalyzable primitive, in line with the commitment to minimalism. That is, I assume that it cannot be conceptually reduced any further, but there is nonetheless much to say about its workings, application, and inner logical structure. For instance, representational force is precisely what is shared between all representational signs in Peirce's original classification of symbols, indexes, and icons. It is neither reflexive nor symmetrical nor transitive. The force of a source can be set up in the first instance by arbitrary convention, but it is afterward maintained in a community by normative practices that are certainly not conventional for any of the individuals who make up the community. A source's representational force can be ambiguous between different legitimate targets, and vice versa: different sources might have the same force. Thus, force is certainly not a one-to-one relation. Finally, the representational force of a source will typically be opaque to the uninitiated; an agent must possess a type of practical knowledge, including social competences and skills, to elucidate or comprehend it. Kuhn's methodology of instruction by means of *exemplars* (Kuhn 1962/1996, postscript) is a particularly good illustration of these features of representational force.

The other condition that must be fulfilled for *A* to represent *B* cognitively is what I call *inferential capacity*. The source *A* must be of the right kind to allow informative inference regarding the target *B*. The condition does not require the inferences to be infallible or to be true conclusions about *B*, but there is an important clause that requires them to reveal aspects of *B* that do not follow from the mere existence of a representational relation. There must be other informative inferences about *B* that can be drawn from *A* for "*A* represents *B*" to be true. The condition might appear to be a cumbersome addition, but it is in fact essential to rule out cases of mere arbitrary denotation, and it appropriately distinguishes the inferential conception from Goodman's (1968/1976) theory of representation as denotation and any of its derivatives, such as R. I. G. Hughes's (1997) DDI (denotation-demonstration-interpretation) or Frigg and Nguyen's (2020) DEKI (denotation-exemplification-keying upimputation) accounts. I then go on to explain how the combination of these two conditions successfully meets four of the benchmark arguments that I advanced in chapter 5, the logical, misrepresentation, nonsufficiency, and nonnecessity arguments.

The last section in chapter 7 applies the inferential conception to some of the case studies in some detail, and it argues that information transmission is a prime aim of modeling in every case. The main lesson is that the inferential conception fits well with the full range of model-building practices and often illuminates them. For instance, the inferential conception explains well the dynamic tension that directs and propels model-building activity—as the tension between the two main necessary conditions for representation. Once a model's target is determined, inferences are carried out by inquirers that might in turn result in a change of target. This is particularly clear in cases of mathematical models of physical processes or phenomena. For instance, in astrophysics, where the models of stellar structure are constantly refined in view of empirical data from different astronomical objects. Or in foundations of quantum mechanics, where many crucial parameters are adjusted progressively as the targets are refined—for instance, in decoherence models of the interaction of the systems with the environment. Or, elsewhere, in population ecological models, such as the Lotka-Volterra model, where additions to the equations result in modified descriptions of the target. It goes without saying that this happens too in blueprints, such as those for the Forth Rail Bridge, studied in chapter 3, where the target is in fact constructed to fit with the blueprint. In sum, chapter 7 shows that the inferential conception defeats the argument from variety since it can accommodate any *means* of representation, including cases in which the means are isomorphism or similarity. This is because, although neither representational force nor inferential capacity is committed to any means of representation, both can nonetheless accommodate them all.

Applications in Aesthetics and Epistemology

The third part of the book studies the broad implications of the inferential conception outside scientific modeling itself and in particular concerning aesthetics and epistemology. Chapter 8 explores analogies with debates on artistic representation over the last thirty years or so. I find myself defending intuitions about representation that have been common among philosophers of art for decades. Moreover, some aspects of the debates regarding scientific representation mirror some aspects of the debates regarding artistic representation; one can even find versions of the means/constituent distinction implicit in some key texts in the philosophy of art literature. Thus, although chapter 8 is in some ways a detour from the main line of argument (and can be skipped by the reader interested only in the philosophy of science), the similarities and analogies between the debates on representation in both these areas are enormously revealing and suggestive.

There is yet another important reason to pursue the analogy with art, which is connected to the integrity of the concept of representation. If, as scholars and students of science, we are to use the term *representation* legitimately to describe some of the functions that scientific models are designed to carry out, we had better make sure that our usage is in line with that of philosophers elsewhere—and indeed that of practicing scientists and artists themselves. We must explain what it is about the function of modeling that enables us to describe it as a representational function. So we need a benchmark, a philosophically well-studied paradigm case of representational human activity against which we can compare our account. Art is a paradigm case—if not *the* paradigm case—of representational practice. Hence, our use of the term *representation* ought to be in line—at least in its most relevant respects—with that of historians and philosophers of art. I develop fully some case studies from the history of art (painting) to which I appeal in more elliptical ways earlier in the book (as well as in past published work) to argue comprehensively against resemblance theories of artistic representation—the equivalent to reductive naturalist accounts applied to art. I then develop the equivalent conditions in the inferential conception as applied to art via first a discussion of Goodman's and Elgin's *exemplification* and then, mainly, through Wollheim's *seeing-in* theory. Finally, I show how the transposition of the inferential account over to the case of art further strengthens the argument and provides an alternative and interesting outlook on artistic representation.

Chapter 9 explores the consequences for epistemology—and in particular scientific epistemology—of the advocated switch from a reductive naturalist to an inferential conception of representation. A large part of the epistemological literature has presupposed, without much argument, some type of reductive naturalist conception of scientific representation. The mere fact that any such conception is implausible already sheds doubt on this key presupposition underlying many current debates within epistemology and, hence, opens new vistas. If, in addition, the arguments of this book in favor of a broadly inferential conception are correct, then the consequences are even more radical, for the presuppositions will have to be replaced by very different premises.

I argue that reductive naturalist conceptions of representation have often been weighted heavily in favor of realism and against instrumentalism. In these views, realism has the upper hand as soon as it is accepted that a main aim of science is to represent the world—since, to put it crudely, *to represent* turns out to be synonymous with *to represent accurately*. The inferential conception, on the other hand, distinguishes *representing* from *accurately representing*, and it is, hence, neutral between realism and instrumentalism. It thus becomes legitimate for an instrumentalist to accept that science aims at (among other things) representing the world; and the realism/antirealism debate almost instantly gains new life. I illustrate this general thesis by looking at and briefly taking issue with the epistemology of some of the most widely discussed philosophers in the literature: Nancy Cartwright, Ian Hacking, Bas van Fraassen, Philip Kitcher, Helen Longino, and Arthur Fine.

Ian Hacking (1982, 1983, 1989) famously argued for a sharp contrast between representation and intervention and for the importance of the latter over the former in a proper philosophical understanding of science.[15] The revolt against representation has been a constant of neopragmatist writing in the last forty years or so, including not just Hacking but Rorty and others too (see, e.g., Rorty 1980; and Rouse 1987). For these authors, representation is a metaphysical relation between mind and world whose apprehension necessitates no action on the part of the agent. It follows that some type of active intervention is needed for a full understanding of the scientific enterprise, for only that can provide the element of pragmatic action required for any epistemology of science that does not just turn into an instance of what Dewey rightly derided as "the spectator theory of knowledge" (see Dewey 1920, 1943). The need to consider intervention thus presupposes a reductive naturalist conception of representation that assumes the purposes, desires, and values of inquirers to be irrelevant in establishing the requisite relation. By contrast, on the inferential conception, representation turns out to be an activity of its own involving the normative practices of inference making and the establishing of force. So an epistemology that puts representation—understood in accordance with the inferential conception—at its core is neither a spectator theory of knowledge nor deficient from a pragmatist point of view.

Van Fraassen's constructive empiricism allows us to restrict belief to the substructures of our theories that correspond to observable phenomena only while remaining agnostic about the part of the theoretical structures that can have correlates only in the unobservable part of the world. Thus, constructive empiricism embraces a reductive naturalist conception of representation with a type of restricted isomorphism ("embedding," i.e., isomorphism to a substructure) at its core. If this assumption is replaced by the inferential conception, then a distinct form of constructive empiricism emerges, one that is, I argue, closer to Dewey's original instrumentalism and turns out to be more robust and defensible than the original view. The substitutions of acceptance for belief as the appropriate attitude toward theories and of empirical adequacy for truth as the appropriate aim of science turn out to have no implications regarding the relationship between theoretical and phenomenological structures and thus allow the instrumentalist to get around some powerful critical objections to constructive empiricism.[16]

The third section in chapter 9 elaborates more positively on the sort of epistemology that a deflationary account of representation, such as the inferential conception, would recommend. A natural aim of science is to represent the world, including its unobservable part, because it is already the aim of everyday cognition and a main element of our cognitive makeup from infancy. But, on the view defended, it remains a suitably minimal aim. It merely entails the building of symbol systems that might result in informative models of putatively real-world systems of interest. On the inferential conception of representation, this requires neither truth nor empirical adequacy as the aim of science since accurate, true, and empirically adequate representation is not to be conflated with representation per se. This resonates well with Arthur Fine's *natural ontological attitude* (NOA) at several levels, not least the emphasis on a *natural modeling attitude*.

Mutatis mutandis, but conversely, as applied to scientific realism, the notion of representation plays a key role in Philip Kitcher's "real realism" (Kitcher 2001a). The crucial question is how stable this form of realism is under a range of substitutions of different notions of representation. I argue that it is acutely unstable: Kitcher's argument produces the desired outcome (scientific realism) only under a reductive naturalist conception of representation. An application of the inferential conception yields neither realism nor antirealism. However, I argue that this is the most elegant and defensible form of the "Galilean strategy" recommended by Kitcher. The inferential conception in effect strengthens the core of Kitcher's epistemology, and it is more in agreement with his newly found pragmatism (Kitcher 2012a).

Helen Longino (1990, 2002) has for decades been advocating a key role for social values in science and in particular in theory choice. While agreeing with her general vision—she has been attempting to locate normatively charged values, or, in the terminology of Bernard Williams (1985), *thick* concepts—I dispute the underlying assumption regarding representation in the claims of underdetermination that ground her epistemology. Instead, in line with the inferential conception, all kinds of normative choices and judgments are already involved in the claim that a theory represents data or that the data support the representation provided by a theory. There is, on this account, no initially neutral way in which to cash out such relations, to which values are then necessarily added at a later stage. There are rather thick layers of normativity already involved at every stage of establishing and maintaining any representational relation between theory and data. In other words, sensitive choices must be made earlier than Longino supposes. Yet my analysis again, if anything, strengthens Longino's claims regarding the role of values and judgments in science while adding a valuable institutional dimension.

The final section of chapter 9 briefly addresses another topic in the philosophy of science literature that has received widespread recent interest: scientific understanding. My thesis is that scientific explanation—on most traditional philosophical accounts, such as the deductive-nomological and the causal-ontological accounts—is implicitly, if not explicitly, linked to a reductive naturalist conception of representation. At any rate, explanation is often understood to carry stringent criteria of truth or success for our representations. By contrast, the notion of scientific understanding is more typically associated with bare representation per se. On the inferential conception, the only commitment is to the cogency of our representations, which allows competent and informed agents to draw inferences regarding the systems represented efficiently. The types of familiarity that underlie the possibility of cogency and the practical skills required for successful inference have interesting overlaps with the sort of familiarity that is often thought to be required for scientific understanding, as opposed to the more formal notion of explanation.[17]

Inference, Reference, and Truth

The book is intended as a *manifesto* in favor of a pragmatist view of representation and a detailed defense of a specifically inferential conception of cognitive and scientific representation. This view is minimalist with respect to the constituent of the representational relation but pluralist regarding the means actively employed in the practice of representing via the building of models. The arguments presented in its favor are many and diverse, ranging from considerations of logical form, to analogies with art, and, in particular, to comparisons with scientific practice through as careful and detailed an analysis of case studies as possible. One form of argument that will, however, be singularly and perhaps conspicuously absent is metaphysical in nature. The reader will find no detailed discussion in this book of the well-known disputes within the philosophy of language surrounding the nature of reference. My contention is that the debate over scientific representation is largely independent of these disputes.[18]

In recent decades, there has been a notorious and influential attempt on the part of Robert Brandom (1994, 2000) to provide a sophisticated inferential account of such metaphysical notions as reference and truth.[19] The conception of representation defended in this book shares some of the general features of Brandom's inferentialism, but it is also markedly different in some regards. Both are attempts to turn the traditional dichotomies between reasoning and reference or denotation upside down. The traditional account takes reference to ground inference, providing the meaning of the elementary units (words, proper nouns, sentences, or, more generally, symbols in a symbol system; lines on a diagram; states of a phase space diagram) that make up the whole meaningful complexes with which reasoned inference operates. Thus, the norms of correct reasoning and inference naturally spring out of the proper account of the primitive meaning of the elementary units. On such an account, reference is a primitive brute metaphysical fact, while reasoned inference is derivative and can be explained by an appeal to such primitive facts. By contrast, on a broadly inferential account, this picture is reversed. The norms of correct inference are primitive brute social normative facts, while reference is epiphenomenal, that is, derived from the correct norms of inference. The referential account of inference is the traditional metaphysical approach that Brandom's inferential account of reference is intended to oppose. The conception of scientific representation defended in this book is certainly closer to the latter in spirit since it aims to ground representation on the inferential practices of informed and competent agents.

An example is the analysis, provided in part 2 of the book, of how the term *true* can be applied to representational sources. On the inferential conception, a true representation is defined as a representation of a target *B* by some source *A* that allows a competent and informed agent to draw surrogate inferences from *A* to only true conclusions about *B*. Since the kind and degree of competence as well as the required level of information are essentially dependent not just on the source and the target and their properties but also on the context of inquiry, the expression *true model* is revealed to be a mere *façon de parler*: a model (a scientific representation in general), by itself, cannot be true or false. It can be said to be so only in a derivative sense, one that depends on its context of use and application.

Nevertheless, it must be stressed that, despite the overall similarity in spirit, the inferential conception of scientific representation is not committed to any of the core elements of Brandom's inferentialism. It is not committed to inferentialist semantics for all our linguistic terms and predicates; it is in fact not committed to any semantics, whether inferential or referential. The key to appreciating this lies in the other condition that, on the conception defended in this book, is essential for representation, namely, that the source's representational force points toward its target. In chapters 5 and 6, I show that the notion of representational force is open to interpretation, in terms amenable to either a traditional referential or inferential semantics. Defenders of a traditional referential semantics of linguistic units such as proper nouns, sentences, or propositions involving the term *representation* will have no difficulty adopting the conception of scientific and cognitive representation defended in this book. They just need to pack their referential semantics into the notion of representational force.

This is precisely the sense in which I claim neutrality on behalf of the inferential conception in matters metaphysical and referential. "Representation without metaphysics" is a just and appropriate slogan. Metaphysical questions related to reference are gently put to one side, not so much in a spirit of rebuttal, confrontation, or refutation, but rather as not directly relevant to the purposes of my study. This includes both referential and inferential accounts of the semantics of proper terms and nouns and derivative attempts to ground the meaning of more complex expressions on such semantical notions. The deflationary approach adopted in this book is aimed instead at scientific representation via modeling. It does advance and defend a deflationary account of the use of models in practice. It does not, however, advance a thesis about semantics, truth, or any of the categories typical of analytic philosophy of language.

PART I

Modeling

2

The Modeling Attitude: A Genealogy

Science as we know it was born in the nineteenth century.[1] That is when it was socially institutionalized in the form that is common today—mainly in Germany, France, and Britain.[2] Its birth as a social institution coincided with the heyday of what I will call the *modeling attitude*. Of course, natural philosophers had employed models before and had reflected on the nature of those models. In fact, before science and philosophy parted ways decisively, at some point in the late nineteenth century, building models and reflecting philosophically on their nature often went together.[3]

Today, we have a clearer division of labor, one that we owe in part to our nineteenth-century predecessors. The scientists who build models now rarely reflect in depth or in detail about their methodology or epistemology, never mind relating their ideas to those of their reflective predecessors. The philosophers of science, on the other hand, have by now at their disposal an illustrious genealogy of past philosophers who have reflected long and hard on the topic. The modeling tradition in the philosophy of science is in fact over a century old. But philosophers' understanding of the nature and working of models is often secondhand and more akin to the spectator's view of a game than the fully engaged know-how of the participants themselves.

It is unclear that this contemporary division of labor is healthy. It is certainly expedient in the short term since it allows all concerned to get on with what they know how to do best. But what is expedient in the short term is not always enlightening or even efficient in the long term. Scientists are typically too invested in their own modeling activity to be able to judge it dispassionately or to frame it in a broader historical and methodological context. The philosophers' view from the distance may be less partisan and more objective, but it can also lead to misperceptions regarding the essentially practical judgments required to engage fully in the actual modeling game. Much of the tacit knowledge that goes into modeling is rarely made explicit, and it may remain invisible from the distance of the philosopher's gaze.[4]

An important motive of this book is to urge a reconciliation of the scientific and the philosophical outlooks on modeling. If we want to understand properly what goes on in modeling and hope to be able to employ that understanding in a productive manner to improve both on our theoretical knowledge and on our practical engagement, we had better combine the theoretical bird's-eye view of the philosopher with the know-how and practical engagement on the ground of the scientist.[5] In composing such an integrated and holistic view, we can do no better than to look back at the last generation of scientists-philosophers who did combine philosophical depth and technical skill and accuracy. The modelers of the late nineteenth century, exemplified by James Clerk Maxwell, William Thomson, Heinrich Hertz, and Ludwig Boltzmann, were simultaneously the first generation of scientists and the last generation of natural philosophers. They combined philosophical and technical training, and they also reflected deep and hard on the nature of modeling. On rereading their work, one is surprised by how little of substance philosophers of science have in fact added to the topic over the last century.[6]

In this chapter, I develop a genealogy of this modeling attitude[7] that will, I hope, provide a helpful historical background for the various philosophical approaches to the notion of representation that will be discussed later in the book. I focus particularly on the nineteenth-century modelers and summarize their insights and contributions, which, I claim, remain essentially unsurpassed. In subsequent chapters, I develop a philosophical framework—the inferential conception of representation—that captures, I claim, the main ingredients of the modeling attitude promoted by the late nineteenth-century modelers.

The Origins of the Modeling Attitude

Systematic philosophical reflection on the nature of models begins in the nineteenth century in connection with models of classic electrodynamics and the ether.[8] There are two sources or broadly defined schools. Although they are historically related and reach similar conclusions, they are essentially distinct and are rooted in different contexts and traditions, and they are practiced and instructed mainly in different languages. There is a British or English-speaking school, led by James Clerk Maxwell and William Thomson (Lord Kelvin), but also encompassing many of the other celebrated British physicists of the nineteenth century, such as Oliver Lodge, George F.

FitzGerald, Oliver Heaviside, John Poynting, and Joseph Larmor. Maxwell and Kelvin advanced several methodological considerations in their many attempts over the years to model the ether as a concrete physical medium of vortices. These considerations mainly bear on the importance of the modeling attitude to their ongoing development and understanding of electric and magnetic phenomena. There is then a somewhat later German-speaking school represented mainly by Heinrich Hertz and Ludwig Boltzmann and heavily indebted to Helmholtz's methodology of physics. Partly under the influence of the English-speaking school, but mainly because of an ongoing process that begins with Helmholtz's research into the nature of perception, the German and Austrian-Hungarian theoretical physicists gradually develop a more abstract and theoretical sort of modeling and provide a cogent philosophical defense for it. The defenders of the *Bildtheorie* (theory of images) have a philosophical agenda, nuanced and sophisticated even in contemporary terms (an agenda that in Boltzmann's case, at least, was linked to a defense of the tenability of the atomic hypothesis). Thus, the modeling attitude is born in Britain, but it grows of age and acquires the mature form that launches it into the twentieth century—and that in essence endures to the present day—in the hands of the skillful German-speaking school.

The English-Speaking School

The modeling attitude in Britain has a distinctive Scottish origin.[9] James Clerk Maxwell (1831–79) was born and educated in Edinburgh, while William Thomson (Lord Kelvin, 1824–1907) grew up in Glasgow and was linked to the city throughout his life. Their various Scottish instructors and teachers introduced them to the uses and importance of reasoning by analogy. Thomson was taught by his father, the reformist mathematician James Thomson (1786–1849), and, at Glasgow University, by the new radical professor of astronomy John Pringle Nichol (1804–59), who had in turn been trained at Aberdeen's King's College—all habitual locales in the history of Scottish commonsense philosophy.[10] Maxwell was educated at Edinburgh University, where he was mentored by James David Forbes (1809–68) and taught by Sir William Hamilton (1788–1856), both members of prominent and affluent Scottish merchant classes. Edinburgh, the capital of the Scottish Enlightenment—also dubbed the Athens of the North—had developed a robust form of commonsense philosophy,[11] and Maxwell was intensely exposed to it during his formative years.[12] It was typical in this environment to oppose "atomistic" theories of knowledge, that is, those views according to which knowledge is absolutely and exclusively of its own object. On the contrary, the

Scottish commonsense tradition emphasized the relativity of knowledge: the idea that knowledge of an object always emerges from a comparison of that object with something else. Comparison, likeness, resemblance, and analogy were all means to achieve knowledge and in fact the only means through which genuine empirical knowledge of the world could possibly come about.

In addition, Forbes, Hamilton, and Nichol also saw themselves as part of the abstractive tradition of metaphysical Scottish mathematics, inaugurated in the eighteenth century by the likes of Robert Simson (1687–1768) and Colin MacLaurin (1698–1746). The simultaneous development of the metaphysical Scottish school in mathematics and commonsense Scottish philosophy is not a coincidence.[13] Both are attempts to construct a middle way between rationalism and empiricism—between purely a priori and purely a posteriori forms of knowledge. Abstractive mathematicians argued that mathematical knowledge required a sense of "comparative experience," at least when it came to geometry, while commonsense philosophers argued that inductive empirical knowledge is grounded on the unreflective presuppositions of our shared commonsense forms of ordinary cognition. Geometry figures prominently among those presuppositions of ordinary cognition, and it is also the prime vehicle for mathematical abstraction. This makes it easy to understand the traditional Scottish preference for geometric over algebraic methods in the investigation of empirical reality.

Forbes, Nichol, and, particularly, Hamilton were all self-conscious and willing partakers in this tradition. They were knowledgeable and practicing users of mathematical abstraction and strong defenders of the primacy of geometry and geometric methods that characterized Scottish science and mathematics throughout the seventeenth and eighteenth centuries.[14] In Euclidean geometry, the method of mathematical abstraction was employed primarily to derive the concepts of surface, line, and point from the perceptual concept of a solid. The relevant point for our purposes is that, as will shortly be explained, mathematical abstraction required the comparison of an experience of a solid with an imaginary model of the experience of the solid under different circumstances. The shared content of such experiences was then extracted in the form of an analogy. Thus, in embracing the metaphysical Scottish school of mathematics, the early British modelers were consistently assuming the very comparative method of analogy that they would go on to employ in their application of models to physical phenomena.[15]

The principal textbook in the Scottish metaphysical school of mathematics, one that in many ways shaped the tradition, was Robert Simson's (1756/1762) commentary on Euclid's *Elements*. It went through ten editions between 1756 and 1799 and continued to be printed, particularly in the United States, to the end of the nineteenth century. In a notorious passage in the book,[16] Simson develops the concept of a surface, or a plane, by abstraction. He first invites us to consider a solid geometric object in three-dimensional space, such as a perfect rectangular-shaped block, and then asks us to imagine our experience of it under changed circumstances. We imagine the solid block divided in two perfect halves, right down the middle. We are then asked to compare our experience of the surface that both halves share in the initial situation with that which we would have *in a series of imaginary situations after the division*. Had the surface thickness, it would belong to either half. Yet it cannot be part of either half because, if we imagine that half being removed, the surface will still exist in the remaining half. Therefore, by reductio, the surface has no thickness, only length and width. It is not part of any of the halves. It is rather an abstraction that we can apprehend only when considering the comparison of the real block with the split block in two different imaginary situations.

In other words, the method of abstraction allows us to derive a result about a real object and its properties on the basis of a piece of reasoning that we carry out on an imaginary situation—described by means of what we would now call a *model*. The comparison allows us to consider the object in its relation to different circumstances and further objects, and the analogies between both situations yield conclusions regarding the object's abstract properties. The principle of abstraction was, of course, intended for the abstract objects of mathematics in the first place. So its repeated application to the abstract concept of surface allows first for the derivation of the concept of a line as what two nonoverlapping surfaces have in common and then for the derivation of the concept of a point as what two intersecting lines have in common. It is not hard to see that the modeling attitude inaugurated by Thomson and Maxwell is in essence the application of a similar principle to the concrete objects of empirical reality described in physics.

The method of abstraction is not a mere formal procedure but a philosophically subtle method for deriving or even constructing abstract geometric properties. It in turn requires a few ancient or medieval scholastic notions and concepts, including of course the idea that properties may be abstract, an idea often understood among Scottish metaphysical mathematicians along the lines of universals. It also requires a principle referred to by MacLaurin as Archimedes's *method of exhaustions* (see Davie 1961, 139–46). This is a method designed to deal conceptually with infinite limits from a broad empiricist standpoint. It has a relatively straightforward interpretation in geometric terms. For instance, a Euclidean line can be understood as that part of space that two series of converging curves (one upward and one downward) going through the same two points will converge on in their limit. The method of abstraction is (in a very general sense that ignores empiricist concerns about mathematics as a priori knowledge of abstract relations) an application of the method of exhaustions to the concrete objects of the physical world.

Thomson, Maxwell, and all their mentors had been trained at Scottish universities within the system of wide liberal humanistic education bestowed by the Scottish Enlightenment. As a result, their formative influences were far from narrowly technical, being instead based on a variety of disciplines, and included very prominently both the history of philosophy and the study of philosophical reasoning and argument. James Clerk Maxwell, most prominently, was imbued with many of the traditional themes and theses of this tradition, including Thomas Reid's emphasis on the unavoidability, however regrettable (in Reid's view) of analogical thinking and Dugald Stewart's exhortation of the virtues of hypothetical reasoning and the useful role that the imagination plays in discovery in general (see Olson 1975, chaps. 2–3).

Later, Maxwell would go on to develop his own philosophical views about these matters. In a paper delivered to the Apostles in Cambridge following his appointment as fellow of Trinity College (Maxwell 1856/1990), he makes explicit his philosophical debt to both Hamilton and William Whewell. The paper is a disquisition on the epistemology of *analogy*, which at that time had a significantly more general meaning than it does today, like our contemporary meaning of *model*. Maxwell's central question in this paper is whether analogies are in the mind or in nature. The answer he provides is not straightforward at all, and the discussion is somewhat dense and murky. But some elements seem clear enough and provide some insight into his modeling attitude and by extension that of his contemporaries and disciples. Maxwell is clearly of the view that there are genuine existing objects endowed with properties and holding an array of relations to each other. There are natural laws stipulating such relations, and there is natural necessity. Yet, when it comes to discussing the nature of analogy, Maxwell resorts to a different kind of necessity that operates among thoughts. These laws among thoughts can then be said to reflect natural laws only under some comparative statement that requires some act of will or perception on the part of the agent. Most intriguingly, Maxwell writes: "Whenever they [men] see a relation between two things they know well, and think they see there must be a similar relation between things less known, they reason from the one to the other" (1856/1990, 382). This is an important passage on the structure of analogy according to Maxwell, and it is revealing of the modeling attitude in the Victorian era. Mechanical models of the ether throughout the nineteenth century fit well this view of modeling as reasoning via perceived shared relations among distinct systems of objects. The Victorian physicists used mechanical models to infer conclusions regarding what they perceived to be properties of the ether in just this fashion.

Maxwell also fills in his concept of reason as follows: "A reason or argument is a conductor by which the mind is led from a proposition to a necessary consequence of that proposition" (1856/1990, 379). In the next chapters in the book, I shall argue that, in its bare elements, and given appropriate caveats,[17] this account of reasoning by analogy, which in a way represents the most sophisticated version of the English-speaking school, is significantly close to what I call the *inferential conception* of representation. Maxwell's account of reason also provides an appropriate point of entry into the related modeling attitude of the German-speaking school. There is a form of surrogative reasoning via models that is a shared feature of both schools.

Despite their respectful intellectual allegiance to the Scottish tradition, some of Maxwell's and Thomson's mentors (notably Forbes and Thomson Sr.) were nonetheless radical reformers. They were set on introducing some of the new Continental style of mathematical physics (particularly as practiced at Cambridge at the time) into the Scottish educational curriculum. Forbes and Thomson Sr. railed against intellectual nepotism and parochialism, and they made great efforts to attract eminent English mathematicians to Scottish universities. For instance, Forbes was instrumental in the appointment of Philip Kelland (1808–79), from Queens' College, Cambridge, against several local candidates. Once installed at Edinburgh, Kelland went on to become yet another one of Maxwell's mentors. It was also at the insistence of Thomson Sr. and Forbes and thanks to their contacts in Cambridge that both William Thomson and James Clerk Maxwell moved to Cambridge and enrolled in advanced courses of study there. At Cambridge, they were able to master the latest developments in mathematical physics, particularly those associated with the analytic treatment of heat in Fourier's (1822) *Théorie analytique de la chaleur*, a book that was hugely influential in Britain and played a particularly important role in the foundation of the Cambridge tradition in mathematical physics (see Harman 1985b; and esp. A. Wilson 1985).[18]

Cambridge was the source of another important influence on both Thomson and Maxwell. It was the exposure to Cambridge that made them generally inclined toward what we can refer to as *realistic modeling*, particularly throughout the 1860s.[19] Thomson became convinced that the underlying theory for any good model was classic mechanics. Thomson's allegiance to mechanical models was principled yet also firmly rooted in his conception of practical reality, which in turn had been shaped by his experience in industry and in particular the leading role he played in the development of the submarine telegraph cable in the second half of the nineteenth century (Smith and Wise 1989). A model on this view represents a phenomenon as the outcome of an underlying mechanical process, and the only theory involved in explaining the workings of the model is the classical Newtonian mechanics of discrete point particles and continuous fluids.

This signals an important change in the modeling attitude, which essentially begins with an instrumental notion of model, or analogy, as merely a tool for thought but progressively develops into a device for probing the nature of reality. The Scottish commonsense tradition in which it originates is inherently skeptical regarding the nature of reality. In this tradition, analogy is a means for illustration and heuristic guidance in the building of useful representations. Reid, Stewart, Hamilton, and Forbes all claim that it is a central and unavoidable tool of thought owing to the essentially relational nature of knowledge. Yet, on their account, analogy is by no means to be reified, and their writings (particularly Reid's) are plagued with severe warnings against taking an overly realistic view of it. In other words, in the Scottish commonsense tradition, analogies are useful tools in inquiry and can play a role in the development of theories, but they can mislead if they are taken at face value. However, as analogy travels south toward Cambridge in the 1850s, in Thomson's and Maxwell's hands it becomes something else. It develops into a magnifying glass, a window on the underlying laws of a whole range of apparently distinct and detached phenomena. Under the influence mainly of William Whewell's "consilience of inductions," first Thomson and then Maxwell embraced analogy as the means for the discovery of the mechanical laws of nature.[20]

The transition is best carried out in two of Maxwell's most famous papers: "On Faraday's Lines of Force" (Maxwell 1856–57/1890) and "On Physical Lines of Force" (Maxwell 1861–62/1890). The driving analogy in the earlier paper is that between electric and magnetic phenomena and the flow of an incompressible fluid through a porous medium. Maxwell's attitude toward this analogy is typically Scottish. It is used as a provisional template for investigating electric and magnetic phenomena, but it is not to be taken seriously as a realistic representation at all. Maxwell even claims that the incompressible fluid is not merely hypothetical but to be taken to be imaginary (Maxwell 1856–57/1890).[21] The flowing incompressible fluid model must, on this account, be taken as a mere illustration of how things might hang together to generate Faraday's magnetic induction phenomena. It does not, however, describe how things actually hang together.

Maxwell's attitude changes in the years leading up to 1861, partly as the result of the influence of both Thomson and Whewell, but also out of his progressive success in conceptualizing induction as a unified electromagnetic phenomenon. By the time he publishes "On Physical Lines of Force," analogy has for him become more than a merely heuristic device. The analogy itself has also changed. Rather than modeling the induction in currents as a flow, we now model it in terms of the famous vortices and idle-wheels model. This mechanical model (which originates with Thomson and which Helmholtz demonstrated to be energetically stable) conceptualizes the ether as molecular vortices in rotational motion.[22] Some tiny counterrotating "idle wheels" are introduced to account coherently for such rotational motion (see fig. 2.1). The model contains some elements that Maxwell continued to think of as helpful heuristically for the drawing of inferences, namely, the idle wheels inserted among the vortices whose rotation counterclockwise is required to preserve the consistency of the mechanical model. Yet, alongside them, there existed also what Maxwell called *real* analogies, describing objects that he took to exist in nature—namely, the molecular vortices themselves.

The genesis of this model—now evidently discarded—and its ulterior development in the hands of the so-called Maxwellians[23] bring home some lessons regarding modeling in general that will be of use in this book, particularly

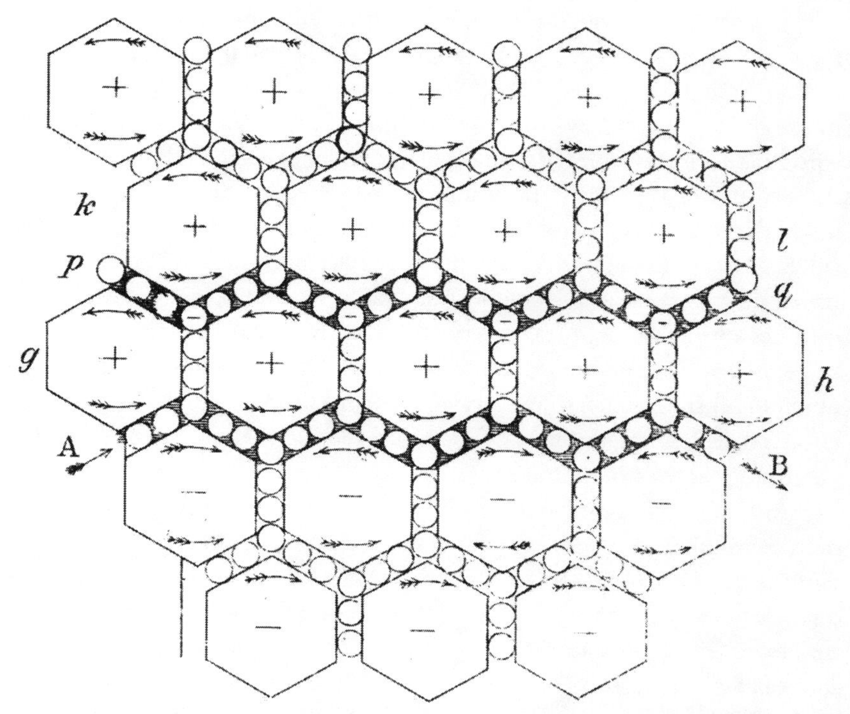

FIGURE 2.1. Maxwell's representation of electromagnetic induction as a vortices and idle-wheels mechanism (Maxwell 1861–62/1890, vol. 1, pl. 9, fig. 2).

in the development of the inferential conception of representation in part 2. This and other mechanical models of the ether illustrate the powerful avenues for inquiry opened by illustrative theoretical analogue models, which are explanatory in their own terms.[24] Maxwell was inspired by Thomson's (1847/2011) analogy between the linear and rotational strains in an elastic solid and the distribution of electric and magnetic forces in conductors in motion (Harman 1991, 71–80; Siegel 1991, chap. 2 and pp. 58ff.). Thomson in turn was attempting to make theoretical sense of Michael Faraday's experimental discoveries of electromagnetic induction and the very physical representation of magnetic forces as geometric lines around a magnet. The main theoretical source in both cases is to be found in William Rankine's (1855) molecular vortices theory of elasticity,[25] which provides the template for the sort of elastic mechanical solids that the English-speaking modelers took to underpin both thermodynamic and electromagnetic phenomena (see Harman 1991, 93–94; Hunt 1991, chap. 4; and Siegel 1991, 34–35, 60). The relativity of knowledge drives both Thomson's and Maxwell's attempts to illustrate electric and magnetic phenomena by means of mechanical models of elastic solids or fluids.

Maxwell first provided a kinematic representation of Faraday's lines of forces (see Maxwell 1856–57/1890) and then proceeded to provide a dynamic analysis (see Maxwell 1861–62/1890), putting Rankine's molecular vortices to use in a comprehensive theory of magnetoelectric phenomena. The idle-wheel model is the critical driving analogy, and it helps investigate Maxwell's detailed use of it in his inquiry into the properties of the electromagnetic medium. This is instructive because the conclusions of Maxwell's inferences still hold today, in the form of the theory of electromagnetism that carries his name, yet his premises are notably not what we would accept today. This illustrates a pervasive feature of the modeling attitude that ought to take center stage in any contemporary account of scientific representation, namely, the ubiquitous use of model sources as vehicles for fortunate and insightful inferences, regardless of how faithful the sources themselves may be when literally taken as depictions of their targets.

Thus, Maxwell took the mechanical description of molecular vortices to be paramount, and it was his insistence on developing his model in accordance with the laws of mechanics that drove his discovery of both the displacement current and eventually the theory of electromagnetism that carries his name.[26] First, as regards the displacement current, this is now presented as the result of generalizing Ampère's law: curl $\boldsymbol{H} = 4\pi \boldsymbol{J}$, where $\boldsymbol{H}$ stands for the magnetic field, and $\boldsymbol{J}$ stands for the current density. Supposedly, Maxwell extended this law to apply it to dielectric materials and open capacitor circuits:

curl $\boldsymbol{H} = 4\pi\boldsymbol{J} + (1/c^2)(d\boldsymbol{E}/dt)$, where $\boldsymbol{E}$ stands for the electric field, and the extra term $(1/c^2)(d\boldsymbol{E}/dt)$ corresponds to the so-called displacement current. However, it is sometimes overlooked that Maxwell's reasoning in arriving at the precise form of the extension is explicitly reliant on the molecular vortices model. More particularly, the displacement current—and its mathematical formulation in the expression above for the modified version of Ampère's law—was a consequence of the introduction by Maxwell of the idle wheels into his model (fig. 2.1). As we saw, these tiny wheels were introduced by Maxwell to maintain the mechanical consistency of the molecular vortices model. It is impossible otherwise for all the vortices to rotate in the same direction with the same angular velocity. Yet, once introduced, in the context of the analogies presented, they lead to the supposition that any distortion of the molecular vortices (which by 1861 Maxwell was conceiving along the lines of Rankine's conception as elastic solid vortices) would generate a displacement of idle wheels in a particular direction. Within the analogy, the idle wheels stand for electric current (which, again critically, Maxwell was thinking of, not in terms of being composed by charged particles, as we do today, but rather in terms of the coalescence of Faraday's fields), thus yielding the conclusion that a displacement current would be generated.[27]

Thus, we can begin to see how the electromagnetic theory that eventually emerges (out of the work of mainly Oliver Heaviside and Heinrich Hertz) is the outcome of an ingenious mechanical model devised with specific properties (vortices, idle wheels, rotations, elastic deformations, and displacements). We would now not countenance or accept this as in any way real, but it genuinely led to the correct inferences regarding the mathematical form of the electromagnetic relations. This is entirely in keeping with the thesis of the relativity of knowledge and the analogical modeling techniques inaugurated by the likes of Thomson and Maxwell. Recall that, for Maxwell, an inference is a conductor by which the mind is led from a proposition to its necessary consequences. In the case of Maxwell's mechanical model, the model is the conductor that leads from the electromagnetic induction phenomena discovered by Faraday to the existence of a displacement current and beyond that, eventually, to a comprehensive electromagnetic theory of light. This capacity of a model to act as a conductor of reasoning—a sort of communication channel—will become a central motif in the rest of this book.

The modeling attitude inaugurated by William Thomson and James Clerk Maxwell is thus the outcome of a convergence of two formative elements. The basic element is the traditional Scottish commonsense emphasis on analogical reasoning and abstractive principles that they were educated into. Thomson and Maxwell added to the Scottish tradition a distinctive layer of formidable analytic skills in mathematical physics, which they acquired in Cambridge. It was there, under the influence of William Whewell (1794–1866) and George Gabriel Stokes (1819–1903), that Maxwell developed an inclination and a skill for mathematical and formal analogy.[28] The result of this convergence of Scottish and Cambridge formative elements is the basic template for mathematical modeling that has been characteristic of the physical sciences ever since and that is the main subject of this book.

Certainly, like most nineteenth-century physicists, Thomson and, to a lesser extent, Maxwell were in addition committed to the mechanical worldview. This was after all the period that Martin Klein (1972, 73) has described as the "high Baroque era of the mechanical world-view." After Maxwell's premature death, Thomson eventually went as far as to reject Maxwellian electrodynamics altogether, including the displacement current, on account of its being too abstrusely mathematical and divorced from "practical reality." By this, Thomson essentially meant that it was impossible to reconcile Maxwellian electrodynamics, particularly as developed posthumously by Maxwell's disciples in the late nineteenth century, with the mechanical worldview. In particular, the displacement current had no practical reality, according to Thomson, unless it could be interpreted as a genuinely real displacement of matter (Wise and Smith 1987, 342).[29] More generally, Thomson continued to insist on a flowing fluid model of electric current, which caused the so-called Maxwellians—and FitzGerald in particular—to disparage him (Hunt 1991, 84–86, 131; Smith and Wise 1989, 459). Yet Thomson was prescient. The history of the development of Maxwell's theory at the hands of the so-called Maxwellians is in essence the history of the emancipation of electrodynamics from mechanics. And, as we shall shortly see, this emancipation trend is taken further forward by the German-speaking school and, of course, given its fullest extent yet at the hands of Albert Einstein in his theory of relativity.

History makes clear, moreover, that the mechanical worldview is essentially independent of the modeling attitude. For the modeling attitude did not disappear with the demise of classical mechanics but instead prospered a good deal and in fact arguably achieved its greatest extent and degree of sophistication once the mechanical worldview was left well behind. Thomson's rejection of Maxwell's theory cannot therefore be taken to imply a rejection of the modeling attitude. Instead, it is the result of Thomson's additional recalcitrant commitment to a strong form of mechanical explanation. And it was precisely this additional commitment that was ultimately superseded by the followers of Maxwell and that lies deep in the heart of Thomson's irresolvable conflict with the likes of FitzGerald and Hertz over the interpretation of Maxwellian electrodynamics (Buchwald 1985; Wise and Smith 1987).[30] The principles of abstraction embodied in the method of analogy were crucially not in dispute, I claim, and the Maxwellians (both in England and across the Channel) were as intensely committed to the use of analogical modeling techniques to describe physical nature as was Thomson himself, if not more intensely so.[31]

Thus, in its modern form, which will be studied in this book, the modeling attitude in the physical sciences can be said to have begun with the method of analogy first implemented by Thomson, Maxwell, and other British physicists during the nineteenth century. It is certainly the case that, in the work of these physicists, particularly Thomson's, the modeling attitude is intertwined with a mechanical worldview aiming to understand all physical phenomena in terms of a particular version of Newtonian mechanics. Yet, far from perishing with the demise of the mechanical worldview and the advent of the new physics in the early years of the twentieth century, this attitude is arguably what led to the new physics. At any rate, it certainly remained unabated and can be regarded as the main methodological legacy of nineteenth-century Victorian science. It appeared not only in physics but throughout the natural and social sciences, including biology, chemistry, geology, economics, and sociology, where it remains dominant to the present day. The sort of abstract mathematical physics that Thomson so strongly objected to and that has become characteristic of twentieth-century physics is, far from being opposed to the modeling attitude, arguably its offspring. As we will see in the next section, and as will be argued throughout this book, there is nothing in abstract mathematical modeling that departs from the analogical and comparative methods that inform the modeling attitude as captured by Maxwell and Thomson.

The German-Speaking School

Simultaneously with the development of the English-speaking school of modeling, Hermann von Helmholtz (1821–94) set up his laboratory in Berlin in 1871 and, in so doing, set into motion a different modeling tradition. The progressive evolution of the Berlin school of physics, led by Helmholtz, eventually gave rise, in the hands of Helmholtz's most outstanding pupil, Heinrich Hertz (1857–94), to a fully developed and distinct German-speaking modeling school. Its scientific and philosophical roots are not just in Berlin, however, but also in Vienna, for Ludwig Boltzmann (1844–1906) independently developed a very similar modeling attitude of his own, first in Graz, and then at the end of his career in Vienna. Hertz's view developed because of his interaction with Helmholtz, and his version of *Bildtheorie* is heavily indebted to his mentor's neo-Kantianism. By contrast, Boltzmann developed his views only indirectly, under the remote influence of Helmholtz. His views owe at least as much to his own Austrian philosophical upbringing and are mainly the result of his own close reading and interpretation of Maxwell's writings.[32] Moreover, Boltzmann's reflections on modeling became focused and central to his preoccupations only during his last phase in Vienna. In constant confrontation with the positivism of his colleague Ernst Mach (1838–1916), he there turned most definitely toward methodological and philosophical problems. As a result of this very different path, his version of *Bildtheorie* leans much more toward empiricism than toward neo-Kantianism (D'Agostino 1990; de Regt 1999; Wilson 1989). We may even properly speak of two variants of the German school of modeling: a neo-Kantian variant, which is dominant in Hertz's writings, and an empiricist variant, which can be found in Boltzmann's later methodological writings. Although their scientific sources are similar (and they are both explicitly indebted to Maxwell and the English-speaking school), their philosophical orientations are different and mark out different concerns.

This is not to say that Hertz and Boltzmann did not know each other's work or that they failed to appreciate it. On the contrary, they were keen on and appreciative of each other's scientific contributions. Thus, in the introduction to *Electric Waves*, the collection of his most important and celebrated papers on electromagnetism, Hertz refers in very positive terms to Boltzmann's systematic interpretation of Maxwell's theory, appropriately linking it to his own (see Hertz 1893/1962, 27). For his part, Boltzmann was, like almost everyone else at the time, most complimentary in his description of Hertz's experimental work leading up to the discovery of electromagnetic waves.[33]

As for their more philosophical views on modeling and representation, Boltzmann and Hertz probably also influenced each other reciprocally, even though only Hertz's influence on Boltzmann is explicitly recorded.[34] After Hertz's untimely death in 1894, Boltzmann saw it as one of his missions to develop Hertz's views further. He enthusiastically took up the torch of the modeling attitude, becoming a main propagandist of Hertz's views as well as his own and adumbrating in this manner a genuinely unified *Bildtheorie*. Thus, if William Thomson's (Kelvin's) Baltimore lectures in 1884 (Kargon and Achinstein 1987) signal the culmination of the English-speaking school of modeling, the publication of Ludwig Boltzmann's *Populäre Schriften* (1905/1974) may well be said to signal the culmination of the German-speaking school.[35] It thus becomes possible to speak of a unique German-speaking school despite the different philosophical backgrounds of and the diverse paths followed by Boltzmann and Hertz in developing their positions.[36]

Hermann von Helmholtz moved from Bonn to Berlin in 1871, and he founded his laboratory there. That laboratory was devoted to an empirical examination of electrodynamics and related phenomena. The theoretical framework initially pursued was the form of action-at-a-distance electrodynamics developed by Wilhelm Eduard Weber (1804–91) and Franz Ernst Neumann (1798–1895), but eventually Helmholtz came around to accept a version of the Faraday-Maxwell field theory.[37] The two key moments in the transition are Helmholtz's (1870) proof that action-at-a-distance theories entail Maxwell's equations and Hertz's (1888) epoch-making experimental discovery of electromagnetic waves. Buchwald (1993) makes it clear that the dialectics was not a mere choice between the Neumann-Weber action-at-a-distance theory and the Faraday-Maxwell field theory. It would be too simplistic to suppose that Hertz simply represented the pinnacle of a German conversion to British field-theoretical electrodynamics exerted under the guidance of Helmholtz. Rather, Helmholtz had already developed his own understanding of electrodynamics as a curious mixture of action-at-a-distance and field-theoretical elements, and it is within this Helmholtzian tradition that Hertz's discoveries must be primarily understood. Hertz inherited a schooling in electrodynamics from Helmholtz, but he was also deeply wedded to his mentor's philosophical attitudes. His astonishingly deep and philosophical *Principles of Mechanics* (Hertz 1894/1956) is in great part the result of the exposure to the dialectics between alternative theories of electromagnetism that he and everyone else who worked in Helmholtz's academic environment and laboratory experienced.

According to Buchwald (1993), Helmholtzianism consisted in an open-ended set of methodological maxims for the practice of experimental science that emphasized active interventions on the part of experimenters to alter the conditions of experimental setups in order to obtain anomalous results or effects. Underpinning this methodology was a commitment to a particular ontology that diverged from both field theories and the standard form of action-at-a-distance theory proposed by Weber. In fact, according to Buchwald, there are significant differences between Weber's and Neumann's electrodynamics (see Buchwald 1994, chap. 2). The former—following a hypothesis proposed by Gustav Theodor Fechner (1801–87)—assumes that electric currents consist in the equal and opposite flow of two kinds of particles (positive and negative) in a conductor. The particles that make up charges, and whose motion constitutes currents exert forces on each other that depend on the distances between the particles. Hence, the strength of a current in a circuit is proportional to the forces driving the particles. The Fechner-Weber approach therefore postulates forces acting at a distance among the particles that it takes charges and currents to consist of. By contrast, Neumann's theory does not presuppose any account of charges or currents, nor does it specify forces acting at a distance among them. Instead it postulates that between any two charges a potential function obtains whose shape depends on their distance. This potential function determines the energy of the system of charges so that the interaction forces between any two circuits can be obtained as the gradient of such a potential function. There is no assumption regarding the nature of the system of charges itself and the forces operating, other than that the system can be ascribed to a "state" that may figure in a potential "function" that fully describes its interaction properties.

Soon after arriving in Berlin, Helmholtz generalized Neumann's form of electrodynamics and applied it to all of physics.[38] He understood physics in terms of the ascription of states to systems, together with functions representing potential interactions. Everything else was redundant or derivative, including forces. So, although the Neumann-Helmholtz approach to electrodynamics condones action at a distance, it does not seem right to refer to it as an action-at-a-distance theory of electrodynamics. Helmholtzianism as an approach to physics is rather essentially linked to the thought that any subtle change in the conditions of an experiment generates changes in the state that represents the system and the potential function that represents the possible interactions of that system. This is independent of whether the changes are mediated by fields (as in the Faraday-Maxwell view) or by actions at a distance (as in the Fechner-Weber view). Helmholtz does not assume any ontology of action. Hence, the progressive adoption of field theory by Helmholtz's school and the abandonment of Weberian action-at-a-distance theory do not signify any deep break with Helmholtzianism but are instead of a piece with the ontological flexibility afforded by the Helmholtzian methodology.[39] Helmholtz's (1870) proof that both theories entail the same experimental predictions—which so impressed his contemporaries, including Boltzmann, Hertz, and Thomson himself—trades on the adoption of a Helmholtzian methodology, and its most profound philosophical lesson (namely, that both theories are alternative schemata that provide the same empirical results) makes sense only within such a methodology.[40]

This is important for our purposes because a commitment to Helmholtzianism, understood in this generic methodological sense, turns out to be at the heart of what distinguishes the German-speaking school of modeling. The principal lesson that both Boltzmann and Hertz learned at Helmholtz's laboratory is that nature is dynamic and that the most appropriate representation of it must abstract from the concrete material details and concentrate instead on dynamic states and potential functions. Once the appropriate mode of representation in terms of potentials and dynamics is adopted, ontological disputes prove to be beside the point. Whether there really are forces or just masses or whether atoms exist or instead just packets of energy—these are ontological disputes that are beyond the purview of scientific representation per se. Science itself cannot determine the answer to such questions, for it aims at the building of dynamic representations, and these always by definition leave open such ontological issues regarding their interpretation.

The impact of Helmholtzianism on both Hertz and Boltzmann is sometimes obscured by superficial accounts of their opposition to action-at-a-distance forces and energentism, respectively. Hertz's *Principles of Mechanics* (1894/1956) has sometimes been considered an attempt to eradicate the notion of force from mechanics altogether, and Boltzmann's famous and public quarrels with Wilhelm Ostwald (1853–1932) over Ostwald's energetics are often understood to be the result of his profound commitment to the existence of atoms. It is true that the thought that both Hertz and Boltzmann were deeply invested in disputes for or against a particular ontology of mechanics seems incompatible with methodological Helmholtzianism, which promotes a neutral stance on such matters. But then such thought is not in accord with an appropriate rendition of Hertz's and Boltzmann's views, as recent scholarship makes clear.[41] Instead, their most profound commitment was to a Helmholtz-inspired methodological "abstractionism," and this is of a piece with pluralism regarding both scientific representation and its ontology.

Helmholtzianism thus appears to be the origin of *Bildtheorie*. But what then are the origins of Helmholtzianism itself? They seem to lie at the crossroads of four distinct constitutive elements, which I consider in turn. I have already discussed Helmholtz's methodological practice, which provides further clues regarding his philosophical views in general. We should, then, next consider Helmholtz's own philosophical upbringing and education. There is also Helmholtz's early work on the physiology of perception, on which his scientific reputation was built, and which made him well-known worldwide. It too reveals a great deal regarding his philosophical outlook. Finally, there is Helmholtz's own work on electrodynamics and his employment of analogies, in particular, the analogy with fluid dynamics.[42]

Helmholtzianism as a methodological practice seems to emerge out of Helmholtz's initial commitment to a style of causal realism in line with his original commitment to a general Kantian epistemology (Heidelberger 1998, 10ff.). In this early phase, Helmholtz presupposed that matter was endowed with force and that the changes in matter generate effects, which in turn result in our perceptions. However, we do not perceive matter or forces as causes directly. They are not given to us in our direct experience; rather, they are essentially metaphysical entities that are moreover necessary for science. Later in life, as Helmholtz moved away from Kantianism toward a more empiricist framework, avowedly as a result of his acquaintance with Faraday's work, the causal realism was filled in differently. He moved away from the presumption that there must be any necessary ontology for science and came to reject that the concepts of matter and force are necessary. Instead, he adopted the view that matter and force are the result of inference from the phenomena. They became postulates of our images of the world that we form by means of inductive inferences from our limited experimental knowledge of the phenomena.

Helmholtz's philosophical intuitions and views often alternate between a rough form of Kantian abstractionism and a strong commitment to empiricism. Helmholtz seems to have started off broadly as a sort of neo-Kantian, a philosophy that continued to influence his understanding of the use of mathematics and the most formal parts of science and scientific knowledge. Yet, starting with his work on the physiology of perception in the 1860s, he moved progressively toward a general form of empiricism. (Thus, in a way, Helmholtz himself made the journey between neo-Kantianism and empiricism that in his later development of *Bildtheorie* Boltzmann was to trace.) The transition very much coincides with the start of Helmholtz's use of the term *Bild* to refer to the discovered laws of science (Schiemann 1998, 25). By the time the *Bildtheorie* emerges in the writings of Hertz, Helmholtz has adopted a thorough empiricism. Thus, in his preface to Hertz's *Principles of Mechanics*, we find Helmholtz stating: "He [Hertz] is obliged to make the further hypothesis that there are a large number of imperceptible masses with invisible motions. . . . Unfortunately he has not given examples illustrating the manner in which he supposed such hypothetical mechanism to act" (Hertz 1894/1956, xvi). Helmholtz the empiricist is here turning not only against his neo-Kantian young protégé. He is also in a way turning against his own younger self.

It seems then that Helmholtz combined a belief in certain rules of formal inference and abstract principles, such as the principle of causality, with a robust empiricism regarding the sources of knowledge. Certainly, throughout his life he was committed to the experimental method, and he seemed often able and willing to revise his beliefs about fundamental matters of principle in view of new experimental results. In different phases of his career, and in attending to different issues, he altered his degree of commitment to the abstract and a priori or the empirical and a posteriori aspects of knowledge. It is thus not surprising that it remains a matter of debate how committed he was in fact to the apodictic necessity of the rules of inference generated by different *Bilder* or images (Patton 2009, 282).

Helmholtz's work on perception provides some important insight into his philosophical views on models as well. His initial training was in medicine, under Johannes Müller (1801–58), a committed Kantian with respect to the physiology of the visual. Müller defended the idea that the eye could perceive itself, that is, "perceive a priori the manifold, or grid, of points that can be projected onto the retina" (quoted in Patton 2010). He thought that there is a correlation between points in the retina and points on a theoretical projection surface that we can identify with Kant's manifold of pure intuition. On this view, perceptions are copies of—or very much like—those objects that they are perceptions of. However, Helmholtz quickly distanced himself from his mentor and began to develop a "sign theory" of perception instead. On this view, perceptions are not copies of the objects perceived. Rather, they are signs for those objects, standing for them in the same relation that a name stands for its referent. Although his views on perception evolved throughout the years, Helmholtz never really abandoned this basic commitment to sign theory. Thus, in the dispute between nativism and empiricism that he identified in the physiology of perception, he placed himself without any doubt on the empiricist side.[43] Importantly for our purposes here, the sign theory is a predecessor of the concept of *Bild* in Helmholtz's work. Once perceptions are identified with representational signs, there are obvious questions to ask about the rules of inference that operate among them as well as the relation between different sign systems.[44] These questions are all integral to the notion of *Bild*, and the particular emphasis on inference rules that characterizes the German-speaking school of modeling can thus be seen to have roots in Helmholtz's sign theory of perception.

Finally, it is instructive to consider Helmholtz's own work on electrodynamics briefly. Helmholtz liberally employed an analogy with fluid dynamics and understood the electrodynamic fluid to have highly idealized properties. For a start, viscous forces are not conservative because of friction, so Helmholtz decided to model the fluid as endowed with a vorticity vector at each point, a function of its average rotation in the point's surrounding region. He then went on to introduce a few more idealizations, including the notion of "vortex line," or tangent to the vorticity vector at any point, and "vortex filaments," or vortices that are the limiting cases of vortex lines at a region indefinitely small (Patton 2010). He was able to derive many conclusions from these idealizations, including three important theorems in fluid dynamics that are of use today. The interesting point for our purposes is that his practice at this point reveals that he accepted that *Bilder* are justified not by their accord with the systems they stand for but, instead, like any other signs, by their heuristic fertility in generating new theoretical and empirical results. This is of course in line with the emphasis on modeling in the English-speaking school, but Helmholtz's prior commitment to the sign theory adds a couple of important twists. First, given their origin in perception, *Bilder* go deep into our very phenomenological knowledge of the world, in a way that the English-speaking school never really entertained for its mechanical models and analogies. The latter make use of the conventional representations of mechanical systems and are adopted because of their heuristic fertility. *Bilder* also make use of conventions, but for the German-speaking school these conventions are rooted in our sensorial or phenomenological experience of the world. Second, given the formal nature of the sign systems on which they build, *Bilder* provide us with highly articulate formal rules of inference akin to those that operate among sign systems, for example, among the terms in any language. This opens the door to the building of formal systems to characterize such rules or inferences precisely. Thus, while the rules of inference of the English-speaking modelers are typically implicitly built into the mechanical models and analogies they use, the defenders of *Bildtheorie* would often attempt to make those rules explicit in axiomatic formal systems.

Helmholtzianism therefore provided the most fertile background for the development of *Bildtheorie*. In a sense, all the main basic elements are already present in Helmholtz's use of models, or *Bilder*. Hertz and Boltzmann both added to these basic elements to formulate a genuine theory of models, or *Bildtheorie*. Hertz did so from a strictly Kantian viewpoint, one that in many ways parallels Helmholtz's youthful philosophical commitments. Boltzmann did so from a decidedly empiricist framework, one that is closer to Helmholtz's views on perception as well as his more mature epistemological views. Both developed further the fundamental Helmholtzian insight to treat models as sign systems endowed with their own rules of inference, and therefore this became the distinguishing hallmark of the German-speaking school of modeling.

In Hertz's *Principles of Mechanics*, we find the following famous formulation of *Bildtheorie* in terms of a requirement governing permissible scientific images: "We form for ourselves images or symbols of external objects; and the form which we give them is such that the necessary consequents of the images in thought are always the images of the necessary consequents in nature of the things pictured" (1894/1956, 3). The two main elements of this prescription are Helmholtzian in origin. There is first the idea that there is necessity in the relation between images or *Bilder*. The necessity invoked here must be of a logical kind and requires that certain rules of inference be in place, so *Bilder* have an inferential structure very much akin to sign systems indeed. Second, there is the thought that the proper role of *Bilder* is to track the relations of necessity in processes or things in nature, not to copy the things themselves. Whatever necessity is involved in such processes, it is not of the same sort as the necessity involved in *Bilder*. The rules of inference that operate among the *Bilder* need not correspond to any set of laws of nature, but the consequences of rules and laws must be correlated. Thus, Hertz writes: "The images that we here speak of are our conceptions of things. With the things themselves they are in conformity in one important respect, namely, in satisfying the above-mentioned requirement." Later in the book, I shall argue that this requirement properly corresponds to a relation of logical conformity at best and not one of material conformity as it has sometimes been claimed.[45]

In Hertz's hands, *Bildtheorie* achieves a much greater degree of sophistication and explicitness than can be found in any of Helmholtz's writings on the topic. Hertz particularly defends the view that more than one image can correspond to any given target, even provided that the requirement of conformity is met fully. As he writes (Hertz 1894/1956, 2): "The images which we may form of things are not determined without ambiguity by the requirement that the consequents of the images must be the images of the consequents. Various images of the same objects are possible, and these images may differ in various respects." It follows that Hertz denies for images the analogue of Helmholtz's "sign constancy," or one-to-one correspondence. This has a particular consequence, namely, to show that scientific representations are typically underdetermined by those things they represent, whether they are objects, systems, processes, effects, or phenomena.

Another important feature of Hertz's *Bildtheorie* is the introduction of several important additional properties of *Bilder*, which can lead to selecting one among the many *Bilder* available for a particular target. Hertz lists the following four: permissibility, correctness, distinctness, and appropriateness. They are of interest because of what they add to the basic conformity requirement governing *Bilder* outlined above, but, in the context of this book, they are even more interesting because what they reveal is inessential for scientific representation.[46] Thus, *permissibility* is coherence with "the laws of our thought" (Hertz 1894/1956, 2). We can understand it as the requirement that *Bilder* do not contradict themselves, that is, that they do not include internal inconsistencies. Hence, a scientific representation can satisfy the basic requirement of conformity while containing internal logical contradictions. *Correctness* is the requirement of coherence with the properties of the targets. It is, according to Hertz (1894/1956, 40), "determined by the things themselves and does not depend on our arbitrary choice." An incorrect image is thus one whose essential relations contradict the relations of external things. But an incorrect *Bild* is still supposedly a *Bild* since it may exhibit the required conformity. Similarly, distinctness amounts to what we now would call *completeness*: the representation provides an accurate rendition of every aspect of the target. But it is obvious that not all *Bilder*, however conforming, will have this property. All *Bilder* may be correct to some degree, but none will be wholly and entirely so. Finally, *appropriateness* refers to simplicity. The *Bild* does not contain superfluous claims; it does not have properties that have no role at all in inducing appropriate inferences regarding salient properties of the target. Hertz most certainly does not think that every *Bild* is appropriate, and *Principles of Mechanics* is a forceful argument to the effect that the standard representation of mechanics is inappropriate vis-à-vis Hertz's own representation. In other words, the conformity of a representation in no way guarantees its appropriateness, which must be argued for separately.

I will assume here that *conformity* is the only essential requirement for a scientific representation. All other requirements may be desirable, but they cannot be said to be essential. Not only that—the other requirements may and in practice always do need to be traded against each other. As virtues that cannot be satisfied simultaneously, permissibility, correctness, distinctness, and appropriateness can typically not be maximized together, and hard choices will have to be made as to which ones to prioritize. Thus, an incorrect model may be accepted by virtue of its appropriateness or an inappropriate one by virtue of its distinctness, and even an impermissible model, containing flagrant contradiction, may be temporarily accepted if it possesses other virtues. Hence, the *Bildtheorie* in Hertz's hands gains richness and depth. The thesis of the underdetermination of *Bilder* exhibits remarkable contemporary relevance, and the requirement of conformity expresses a substantial element (arguably the only substantial element) in scientific or, more generally, cognitive representation.

Hertz's deep philosophical contributions to representation were developed mainly at Bonn, after he was awarded a professorship there and made the director of its physics institute in 1889. Meanwhile, in Vienna, and in roughly in the same period, Ludwig Boltzmann—an early supporter and lifelong admirer of Hertz's work—had been quick to understand the methodological impact of Hertz's discoveries. Again, under the influence of both Helmholtz and Maxwell, Boltzmann developed his own modeling attitude and his own version of *Bildtheorie*. He undeniably had a particular axe to grind. He was quick to appreciate that the *Bildtheorie* may be used in defense of his atomic hypothesis (Blackmore 1995, introduction). And, indeed, at the time the reality of atoms was just a hypothesis, indeed, one about which skepticism reigned. The energentists were forcefully arguing at that time that our knowledge of the physical world should be grounded on the sole concept of energy, and this was perfectly agreeable with the experimental results known at the time as well as Helmholtzian practice. How then could room be made for the atomic hypothesis?

Boltzmann seems to have realized that the *Bildtheorie*—and more generally the modeling attitude—could come to the rescue of the atomic hypothesis as a model of nature that provides us with a particular form of understanding (de Regt 1999). For it is an obvious consequence of the modeling attitude—as pointed out repeatedly by Maxwell from an early time and made wonderfully precise by Hertz—that two or more different models, endowed with their own distinct ontologies, can equally well represent the same phenomenon. Hence, Boltzmann argued that the atomic hypothesis constituted just one more model, one as legitimate as any other in probing and inspecting physical phenomena. It was, for him, a perfectly legitimate representation of nature if it continued to yield the appropriate empirical results. In his most inspired and confident moments, Boltzmann went further and argued that the atomic model had some of the additional virtues described by Hertz, particularly appropriateness. But that seems to be as far as his modest realism would reach.[47] The fundamental commitment was to the pluralism of representation enacted by his methodological *Bildtheorie*.

Boltzmann's aims for modeling are also less lofty than Hertz's and have better endured the passage of time and the demise of the mechanical age. For Boltzmann, modeling is an essentially pragmatic affair, and the modeling attitude is what results from the application of principles of economy of thought to scientific theorizing. As he writes: "Yet, as the facts of science increased in number, the greatest economy of effort had to be observed in comprehending them [models] and in conveying them to others" (Boltzmann 1902/1974, 215). For Boltzmann, the aim of modeling is not the full description of the fundamental laws of nature. Rather, it is more modest—the provision of local and contextual understanding. This is compatible with some robust ontological commitments but only as part of or relative to a particular modeling activity as practiced in its proper context. The main aim of modeling for Boltzmann is, rather, the provision of information. In the tradition of *Bildtheorie*, Boltzmann assumed that the connection between a *Bild* and its target is akin to the one that obtains between a sign and its target in a sign system and that the two need not bear any resemblance.[48] But he often added to the basic requirement of conformity an additional requirement of information gain. Thus, in discussing thermodynamic models—which he was so instrumental in establishing—he emphasized the essential informational advantage as follows: "If for one of the elements [in the model] a quantity which occurs in calorimetry be chosen—for example entropy—information is also gained about the behavior of the body when heat is taken in or abstracted." A *Bild* must show conformity with its target, and, moreover, it must show conformity of the very sort that provides us with new information regarding the target.

The combination of modern analogues of Hertz's conformity requirement and Boltzmann's information-gain requirement will turn out to be central to the very notion of representation explored in this book. In later parts of the book, in defending what I call the *inferential conception* of representation, I shall develop Hertz's and Boltzmann's requirements and argue that they constitute the only two substantial components of scientific representation. In the present chapter, I hope to have made it clear that the grounds for the inferential conception are already to be found, in essence, in the nineteenth-century modeling attitude. There is thus a sense in which I shall be standing on the shoulders of these two giants at the origin of *Bildtheorie*. That these two requirements help bring into greater relief the pronouncements of James Clerk Maxwell about analogy while emphasizing his modeling practice is further indirect evidence for the historical and conceptual acumen of the notion of representation defended in this book.

The *Bildtheorie*, as developed by Hertz and Boltzmann out of Helmholtz's insights, was a main contributing factor in the development and coming of age of the modeling attitude. The clarification of the role of rules of inference within *Bilder*, the establishment of *Bilder* as formal frameworks with internal structure in terms of principles and consequences, the loosening of the criteria of fit of models to targets to a mere conformity, and, finally, Hertz's and Boltzmann's emphasis on the plurality of representations and their properties brought the modeling attitude to age and propelled it into the twentieth century. The modeling attitude is passed on in this final form as the main methodological legacy of the nineteenth century, and it is embodied in the most advanced and technically sophisticated version of axiomatic mathematical physics that we employ to the present day. In the hands of the German-speaking school, the modeling attitude develops into a formidably precise tool for quantitative prediction and understanding. Maxwell's original insights regarding reasoning by analogy, reviewed in the previous section, are still very much built into it from the start, and historically the development of *Bildtheorie* went hand in hand with the success of Maxwellian field theoretical electrodynamics everywhere. This development takes the modeling attitude out of its insular British origins, eventually promoting it as the main methodological feature of twentieth-century science everywhere.

Chapter Summary

This historical chapter introduces the emergence in the nineteenth century of what I call the *modeling attitude*. This is a stance toward scientific work and discovery, and it continues to this day. It incorporated into scientific activity a new set of techniques, assumptions, practices, and undertakings that became prevalent throughout the second half of the century across Europe but particularly in and around the key centers of Edinburgh and Glasgow, Cambridge, Berlin, and Vienna. The net effect of this innovation was to put model building at the forefront of any scientific inquiry. The chapter traces the roots of this movement in empiricism, Scottish commonsense philosophy, and neo-Kantianism and gives an account of its critical elements, originating in Maxwell's astonishing innovations, that eventually crystallize in Hertz's version of the *Bildtheorie*, as outlined in his magisterial *Principles of Mechanics*.

3

Models and Their Uses

The previous chapter reviewed the historical origin and emergence of what I call the *modeling attitude*: not so much a methodological recipe or even a set of recipes, but rather an attitude toward doing science. An attitude or stance is a rather loose set of normative commitments underpinning some practice, one that bounds and informs the practice within recognized parameters.[1] Thus, an attitude or stance does not by itself prescribe any solution to any given problem. Hence the modeling attitude does not inform us regarding the actual choices of models made in a practice; it can only inform us much more generally about a particular way of doing science, a style of scientific reasoning, a very general framework or template for applying and carrying out scientific work. In other words, the modeling attitude sets the parameters for the practice of model building, but there is a great amount of leeway in that practice that is still consistent with a modeling attitude.

The previous chapter also outlined an argument in defense of and a very general template for an inferential account of representation, one ultimately inspired by Hertz's insightful work. My aim in this chapter is to focus much more closely on the contemporary practices of model building as they emerge in nineteenth-century science. While contemporary practices have changed a fair amount since the nineteenth century, I argue that we can still discern some central elements running through the most varied instances of model building. This is an additional argument for a minimal inferential account of representation deriving from practice directly and irrespective of any history. While the examples that I discuss are historically located—and presented roughly in chronological order—they are not meant to make a *historical* case. They are rather meant to illustrate some general features of contemporary modeling practice and thus to advance the argument in favor of a minimal inferentialism from the practice itself. Thus, while the previous chapter did involve making some claims from history as grounds for an inferential conception of representation, the present chapter aims to put together a case from the practice of model building regardless of its history.

The arguments are thus meant to be complementary. In the previous chapter, I argued that there are elements at the historical emergence of the modeling attitude that militate in favor of a roughly minimalist and inferential conception of representation. Ever since the end of the nineteenth century, a natural stance toward scientific knowledge already conceptually presupposes a bare or minimal account of representation in science. The view is historically embodied in an antecedent notion of what a model is and our broad attitudes toward models' use. In this chapter, I make no such claim from history. Instead, I argue that, independently of its historical origin and/or development, the contemporary model-building practice already suggests that *most, if not all, models function representationally* in just this precise minimal sense.

To establish this claim, we must first establish the reach of the quantifier across a variety of different types of models. The arguments in this chapter have consequently somewhat of an inductive character. An inductive inference is stronger the larger and more varied its inductive base: the cases adduced must be both multitudinous and distinct. The larger and the more varied the inductive base, the stronger the inductive inference. The conclusion at which I aim is that models have representational functions. More tentatively, I shall argue that this function is best understood in a deflationary spirit. Hence, I shall need to describe some cases in a fair amount of detail in order to show more accurately how models work in practice and what kind of conclusions are warranted as regards their representational functions. For this reason, I shall emphasize the variety and diversity of the cases, rather than their large number, by describing in detail four very different cases illustrating different types of modeling.

A Classification of Models in the Sciences

It is by now an established truism in the philosophical literature that there are many types of models, serving many different aims and purposes, and answering to many different techniques and skills. Gelfert (2017) employs a classification from Black (1962), according to which all models can be divided into four classes: (i) scale models, (ii) analogue models, (iii) mathematical models, and (iv) theoretical models.[2] There are now grounds to expand on this classification, to include, for instance, model organisms (Ankeny 2001, 2009), fictional models (see the essays in Suárez 2009a), computational models or simulations (Humphreys 2002, 2004; Weisberg 2013), or models of experiments and data (Leonelli 2016; Morgan 2012). But, ultimately, it seems possible to classify all these in terms of the four types invoked by both Black and Gelfert. For example, model organisms are analogue or theoretical or both. Fictional models, such as the nineteenth-century models of the ether, can be thought of as being in part each of the four types outlined above, and statistical models of data are evidently mathematical—and theoretical if we assume data analysis to import a sort of low-level theory. The Black-Gelfert classification seems exhaustive enough (although not exclusive since many models can be more than one of these types at once).

Weisberg (2013, chap. 2) defends a different tripartite classification of models: concrete, mathematical, and computational. His class of concrete models is at least extensionally a subclass of my joint classes of scale and analogue models, while—as Weisberg (2013) himself admits—a computational model can be regarded as a mathematical model, just one where algorithms rather than structures are primary. I shall instead opt for a more general distinction between static and dynamic mathematical models since I would argue that algorithmic computational models are dynamic mathematical models. In addition, I agree with Gelfert and Black that, even if so generally understood, scale and mathematical models do not cover the whole field and that analogue (not just physical but also conceptually analogue) models are required as well. Finally, I think Black is essentially right that there is a difference between explanatory theoretical models and merely descriptive mathematical models. It is true that theoretical models are often mathematical (although not always since not all theories are expressed, or even perhaps expressible, in the language of mathematics), but the relevant difference is that a theoretical model conveys an explanation—often a causal one—of the phenomenon with which it deals while a mathematical model often simply describes the phenomenon with no explanation either forthcoming or required.

Now, while I shall be adopting Black's and Gelfert's classification, my aim is rather different than theirs. I am not particularly interested in mere classification (Black) or figuring out the ontology of models, that is, what they materially have in common (Gelfert). Rather, my aim is an argument for the general claim that models function *representationally*, in a minimal sense of the term. Admittedly, this representational function need not be models' only function. They can also, for instance, be employed heuristically to develop theory, or to explore a space of possibilities, or to guide experimental intervention, or as a framework in which to classify data. However, I shall argue that *qua models* they all have some representational function, understood minimally in the sense of providing templates for surrogative inferences regarding their targets.

Throughout the book, I shall define the terms *source* and *target* as follows. For any pair {*X*, *Y*}: if *X* represents *Y*, then *X* is the *source* and *Y* the *target* of the representation. Or, in other words, *X* is the *representational source* and *Y* its *representational target*. This definition is very minimal. The sources of models can be concrete objects, abstract or fictional entities, mathematical equations or structures, conceptual archetypes, or sets of sentences in some natural or artificial language. The targets can be concrete, abstract, or fictional systems and their states, objects, and/or properties, processes, or experimental data. There is nothing in the properties of these objects or their relations that I suppose is required for them to fulfill their roles; the only requirement on any source-target pair {*X*, *Y*} is that "*X* represents *Y*" is true. Finally, I make no definitional assumptions as to what makes such a statement true either since its truth makers are likely to be many and varied.

SCALE MODELS

The sources of scale models are often concrete objects built at a scale ($x{:}y$) where x is the size of the source and y is the size of the target. Thus, a scale model reproduces and illustrates some but not all of its targets' properties, and the most obvious property that scale models tend to fail to respect is the *size* of the target systems. The scale is often diminishing. Hence $x \ll y$, and thus targets are often much larger than the sources. This is the case for scale models of bridges, buildings, towns, planetary or galactic systems, the universe. Yet sources can also be much smaller than their targets, with $y \ll x$ (as in models of atoms, molecules, or DNA chains). Very rarely are scale models at the same 1:1 scale, although this is certainly possible within the definition of a scale model and can be helpful too, for instance, in geological or earth sciences, where certain objects or parts of terrains need to be modeled in an artificial environment at the same scale.[3]

Scale models are sometimes frowned on as artificial contrivances, employed merely for illustration or even for fun, and mainly as having pedagogical use. This is certainly the case with, for example, the typical planetarium models of the planetary system or the didactic models of the atom used in the classroom. More recent scholarship has done much to raise the status of scale models both in appreciating its educational purposes and in coming to terms with its many diverse scientific uses beyond the classroom (Sánchez-Dorado 2019; Sterrett 2017; Weisberg 2013). The simplest and most paradigmatic cases of scale models are often wholesale replicas at different scaled sizes of the target (hence the name *scale*). Toy cars or toy buildings are replicas at a smaller size in different materials capturing nothing much more interesting than the geometric proportions of the parts in the target system. Yet scale models have much larger and more extensive uses in science and in engineering, where they serve as a solid base for inference to various relevant properties other than relative sizes. Thus, Sterrett (2020, 398) defines scale models generally as "physical objects or situations, usually specially constructed for the purpose, that are employed experimentally to learn about another imagined or existing physical object or situation."

The more complex models in engineering and applied science are not replicas but carefully contrived reproductions of properties of interest at different, more manageable scales. There need not be one unique scale that applies throughout to all properties, but different properties can be scaled differently. Geometric proportion and relative lengths certainly need not be the one feature of the target that is respected in the source. Sometimes, relative areas or volumes are preserved, and often no scale in any spatial dimension is applied. Instead, the scaled property may be velocity or momentum, when modeling dynamic behaviors, viscosity or fluidity, when modeling hydrodynamic systems, heat or dissipation, when modeling thermodynamic systems, and so on.

Weisberg (2013) describes the case of the San Francisco Bay/Delta model, which serves as an excellent example of a concrete physical scale model of a much larger entity. The model was created by the US Army Corps of Engineers in 1956 in order to settle a controversy about whether to dam the San Francisco Bay. The Army Corps opposed the plan to build the dam because of fears it would be damaging to the ecology, produce strong currents in the bay that would affect sailing and marine traffic, increase salinity to undesirable levels, and create pools of stagnating water. The only way to test for such consequences in real life would of course amount to building the very dam that the corps objected to, so instead it proposed to generate a model to test for the counterfactual claim on a real-life scaled-down model. The model was thus built in a huge warehouse of about an acre in area (less than half a hectare) where replicas of the different interlinked parts of the target area were built, including the San Francisco Bay itself and the adjacent San Pablo and Suisun Bays, the confluence of the Sacramento and San Joaquín Rivers, and a good seventeen-mile-long stretch of the Pacific Ocean beyond the Golden Gate Bridge.

This is overall a large surface area, mostly covered in water, but not particularly deep at any point. The model was thus scaled differently for length (1:1,000) than for depth (1:100). Yet the current average velocities were scaled at merely 1:10, and the tides (two for each lunar day of 24 hours, 50 minutes each) were spaced out at a mere 14.9 minutes (Weisberg 2013, 36). This illustrates the ways in which scaling even a concrete toy model of a larger object is far from a trivial exercise in merely cutting down in size. The best way to model currents or tides in the sized-down model does not involve a general reduction of all quantities by the same margin. It involves instead judiciously scaling down different qualities differently to recover the main features of the phenomena. This will often call for dramatically different ratios in the scales employed to preserve the features of interest (velocity, salinity, viscosity, stagnation, etc.). Preserving the phenomena in a model is thus not a simple matter of producing a physical replica. As I shall argue in later chapters, this reasoning can be built up to generate a general argument to the effect that no wholesale mapping from a representational source to a representational target can in general preserve the dynamic aspects of the phenomena of interest. Scale models turn out to be exemplary in more ways than perhaps anticipated.

ANALOGUE MODELS

Black (1962, 222ff.) points out that analogue models involve a change in medium. Both scale and analogue models enable informative inference from features of the source to relevant features of the target. However, unlike scale models (which, according to Black, attempt to reproduce the phenomena in identical media, however differently scaled), analogue models are meant to capture relations of interest by means of building up completely different systems in different media. Black's distinction is not crystal clear and may be thought to be defective. For, as we have seen above, even in a scale model the materials themselves are bound to be different. For instance, the San Francisco Bay model employs twenty-five thousand small ductile copper strips to model the surfaces under the water and their effects (Weisberg 2013, 8). Obviously, the actual seabed is not made of copper. It follows from Black's definition that most scale models are in fact analogue.[4]

However, I maintain that something roughly like Black's distinction can be defended (Suárez 2012a). The key is that, in an analogue model, we are not merely employing materials in the construction of the model different than those found in the target system. We are employing a completely different type of system, one governed by its own laws. The San Francisco Bay model operates and exemplifies the same laws of hydrodynamics as the actual bay. In an analogue model, this is rarely the case and certainly need not be the case. The laws that operate in the representational source (the model) are domain bound and relative to the discipline involved, and they differ from those that operate in the target. Black (1962, 222) also attempted to understand the relation between an analogue model and its target in terms of structural isomorphism. It will become obvious in later chapters that this cannot possibly be my view, and indeed I shall criticize structural renditions of the inferential powers of analogue models. Instead, I believe Black's own notion of *conceptual archetype*—without any structural trappings—is much fitter for the purpose of characterizing analogue models.

Black (1962, 250) defines a conceptual archetype as "a systematic repertoire of ideas by means of which a given thinker describes, by analogical extension, some domain to which those ideas do not immediately and literally apply." The critical aspect in such descriptions—that which justifies the employment of the conceptual repertoire—is the success in drawing apposite inferences regarding the new domain. Hence, a conceptual archetype is a structured set of claims and statements, endowed with an internal logic, that allows for inference, under some interpretation. An analogical model is then essentially a vehicle for the application of the inferential rule book of a conceptual archetype to a new domain under a significant reinterpretation of the central terms. We can see conceptual archetypes doing their analogical work whenever there is a novel application of the models and theories developed in one discipline to a different area of inquiry, an area where they were not intended to apply in the first place. As Black himself suggests, this requires a form of metaphoric displacement of the original claims, which are literally applicable in one domain but also become applicable, in a secondary meaning or reinterpretation, in another domain. Eventually, what is an analogical model for the secondary domain can become established as theory, and the metaphoric displacement is used up, with the secondary domain becoming part of the intended domain of application of the model (a phenomenon akin to literary metaphors eventually becoming dead).

It is true that this understanding of analogical models as akin to metaphors presupposes an "interactive" account of metaphor (see, e.g., Black 1962, chap. 3; Black 1979; and Richards 1936) or a similar one. On this view, a metaphor entails a transfer of claims between a primary and a secondary domain. A metaphor (and by extension, on this view, an analogue model) involves simultaneous literal and metaphoric meanings, as in "man is wolf to men," which on the face of it requires a transfer from the primary domain (the solitary and predatory behavior of wolves) to the secondary domain (men behaving likewise toward other men). This bypasses other accounts, including Davidson's (1978) influential dismissal of the double-meaning view. On Davidson's view, there is only one literal meaning for any metaphoric expression, one that involves a comparison and a statement of similarity. The view that there could be a transfer of meaning from one domain to another is thereby compromised; there is a sense in which there are not even properly two distinct domains to begin with. Yet, in the case of analogical models, there certainly is a primary domain of application of the model, from which it takes its intended original interpretation and derives its laws, and a secondary domain, to which the laws are applied only derivatively by analogical extension. I adopt the Black-Richards account of metaphor because it makes this phenomenon easier to comprehend as a metaphoric transfer of meaning.

A thoroughly studied example of an analogue model is the so-called Phillips-Newlyn machine, a hydraulic contraption devised to model the British economy in the late 1940s (Morgan 2012). The machine was designed in London in 1949 and then built as the prototype Mark I in 1950 for the University of Leeds, where Walter Newlyn was employed at the time, and where Mark I still sits (Morgan 2012, 183). The definitive Mark II version was manufactured at the LSE, where Bill Phillips worked on it under the patronage of James Meade. (It has been exhibited since 1995 at the London Science Museum.) Thereafter, a UK engineering firm produced several exemplars for Cambridge, Harvard, Melbourne, and Rotterdam Universities (Morgan 2012, 178).

The Phillips-Newlyn machine is an actual hydraulic contraption that shows the money flow in the economy literally as the flow of some viscous red liquid within a vertically structured system of pipes, reservoirs, and pumps (fig. 3.1). It provides a representation of the economy by other means, thus embodying the many metaphors in economic theory that associate money with liquid: *money flow*, *liquid assets*, *liquidity preference ratio*, etc. (Morgan and Boumans 2004). Its workings illustrate the many connections between income and outcome in a national economy, including savings, investments, consumption, taxes, imports, exports, and diverse institutional setups such as the government, the central bank, foreign interventions, and trade. An intended use of the machine was certainly educational, that is, expounding both Newlyn's theory of money flows and the subtle interrelations between institutional setups, actions, and money stocks and flows. Yet the machine prompted and suggested other scientific uses and inquiries. For instance, it helped settle a dispute between economists regarding whether interest rates were fixed by liquidity preferences (Keynes) or the flow of savings in the form of loans rather than investment (Robertson). The machine demonstrated that interest rates are a result of both the public preference for liquid stocks and the rate of flow of savings, thus showing that there was no real contradiction to be solved (Morgan 2012, 209). Both Phillips and Newlyn went on to use insights derived from the working of their machine in their later theoretical work.

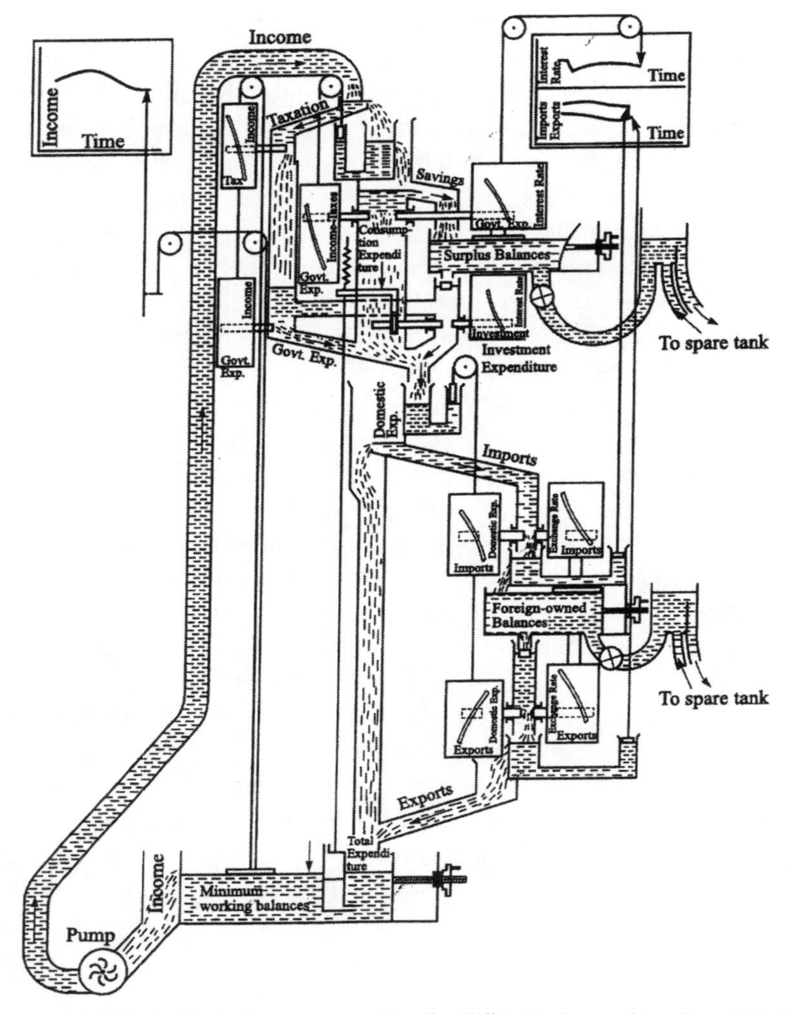

FIGURE 3.1. Nicholas Barr's diagram representing the Phillips-Newlyn machine (Barr 2000, 103 [fig. 11.2]).

The Phillips-Newlyn machine is evidently not merely a sized-down version of the economy. Nothing prevents us from employing units that far surpass those in the real national economy, and there is no articulate sense in which we can think of it as an actual economy at a scale. The machine is not an economic entity but a mechanical system of parts in an oriented gravitational field with liquid running through it obeying the laws of hydrodynamics. In other words, the source system in this representation has an internal dynamic life of its own, which turns out to be essential to the representation. The great virtue of the Phillips-Newlyn machine is precisely its capacity to represent economic processes in time: the flows of liquid assets and their interaction with other aspects of the economy. In reasoning about the dynamic qualities of the primary domain, we can acquire surrogate knowledge about the secondary domain. In this respect, the whole machine is akin to an interactive metaphor for the British economy, illuminating some of its less well-understood dynamic aspects. The entire *conceptual archetype* of engineering hydrodynamics that goes into its construction is thus used in the creation of a model source that appropriately captures aspects of a very different target system, an entire national economy.

MATHEMATICAL MODELS

It is now natural to think of a mathematical model as a description of a natural or social system, phenomenon, experiment, or set of experimental data in the medium provided by the language of mathematics. This extension of analogical modeling to mathematics is a good first approximation to mathematical modeling, but it does not turn out to be a complete rendition since, in the case of mathematical modeling, the interactive account of metaphor is not fully applicable. As we shall see, it is often better to think of mathematical modeling as a kind of literal modeling of some *abstract* aspects of the target system. And it is rarely the case that the functioning of the target system illuminates the workings of the source system in the way that the interactive account of metaphor takes it to be typical of analogical modeling. Instead, in the models, only the primary mathematical domain serves a greater understanding of the secondary domain, not conversely.[5] For all these reasons—and more, to be inspected later in the book—it is best to reserve a separate category for mathematical models.

The historical development of the modeling attitude recounted in chapter 2 of this book already shows that mathematical modeling emerges out of analogical modeling more generally, by a process of abstraction. This fits in with the examples of mathematical modeling that I employ in this book. I nevertheless make no claims to generality and exclude no mathematical models that go beyond my description. For instance, in most experimental sciences, there are mathematical models that can be taken to represent directly the concrete aspects of experience that are recorded as experimental data (such as, e.g., in polling data, psychological learning data, astrophysical observational data, and so on). If so, I would also willingly include them in the present category. However, even such low-level mathematical models of data can be understood to fit my description fully. Thus, it has been argued that, more generally, models of data represent not the raw data but some of its abstract features, namely, those features that are particularly salient for whatever uses we go on to put the models to. Perhaps the best such argument remains one of the earliest and most influential philosophical accounts of data models proposed by Patrick Suppes (1962). Contemporary critiques of Suppes, such as Leonelli (2019) and Bokulich and Parker (2021), if anything make my claim stronger. A data model is also best thought of as a kind of mathematical model.

Suppes argued that data are routinely statistically analyzed in order to correct for systematic and experimental errors and identify and nullify random outliers. This statistical analysis requires data to fit in with what Suppes calls a *model of the experiment*, formally described as an ordered couple: $\zeta = \langle Y, P \rangle$, where Y is the set of all possible experimental outcome sequences of length y, and P is a probability distribution function defined over these sequences. In other words, the model of the experiment describes all possible outcome sequences in an experiment and their theoretical likelihood (roughly, the chance each sequence has of obtaining if the experiment is run as planned in accordance with the theoretical distributions). The model of the experiment is not a model of any actual data but one of what can emerge from a well-run experiment in accordance with some theory. To have a model of the actual data, we obviously need to run the experiment.

Suppose we run N trials on this experiment, each one of length y (i.e., suppose we toss N number of coins y times or toss one coin repeatedly y times in each run and do so for N runs). Then, according to Suppes (1962, 257 n. 3): "Z is an N-fold model of the data for experiment ζ if and only if there is a set Y and a probability measure P on subsets of Y such that $\zeta = \langle Y, P \rangle$ is a model of . . . the experiment, Z is an N-tuple of elements of Y, and Z satisfies the statistical tests of homogeneity, stationarity, and order." These tests are features of conditional probabilities. Roughly, homogeneity requires that each single outcome is equally likely given the conditions, stationarity requires that every sequence be as likely as any other, and order requires that the probability of each outcome is independent of previous or posterior outcomes (hence configuring a set of independent identical distributions for outcomes, i.e., independent tosses of an unbiased coin). In other words, a model of data is for Suppes an abstract rendition of the data made to fit a prearranged model of the experiment that satisfies several formal conditions of probabilistic conditional independence. As Suppes (1962, 256) puts it: "We have taken yet another step of abstraction and simplification away from the bewildering complex complete experimental phenomena."

Suppes's account of models of data is not without its critics, but more recent scholarship has, if anything, moved further away from any notion of raw data underpinning these constructions and further toward viewing these models as abstractions. Considering examples from outside Suppes's preferred cases in experimental physics and psychology also reinforces the view that data can serve evidential purposes only if they already come represented in suitable ways (Harris 2003; Leonelli 2016). More recently still, Leonelli (2019) argues that data in the life sciences now often come in the way of imaginings and reproductions of concrete entities at different stages—often regarded as part of some time series—that are then further computationally streamlined, collated, and composed into even more abstract ideal specimens. Leonelli describes six different phases in the treatment of data: preparing the specimens; preparing and performing the imaging; storing and disseminating the data; computationally coding for image analysis; filtering the images; and conducting a final analysis of the re-created images. In every one of these steps, there are judgments made as to what appropriate simplifications are permissible and what sort of abstract features are salient. Thus, data models already come tailored for specific evidential purposes, whether they are entirely mathematical or classified in part under some of the other three categories that I describe.

A notorious example of a mathematical model of a phenomenon (which may or not be used to model sets of data generated by the phenomenon in question) is the Lotka-Volterra model (Knuuttila and Loettgers 2017; Weisberg 2013). The model represents an obviously abstract and streamlined ecosystem where only two species are represented, a predator species and a prey species. Yet Volterra intended the model to apply to the populations of predator and prey fish in the Adriatic Sea in order to resolve the conundrum of the scarcity of prey fish at the end of World War I. This was puzzling because of the reduced fishing activity during the war, which many anticipated would have led to increased stocks of fish. Volterra (1926) and Lotka (1934/1998, xxi) independently showed that it is possible to model the dynamics of prey and predator populations by means of just two interlocking nonlinear differential equations, namely, $dX_1/dt = X_1(\varepsilon_1 - \gamma_1 X_2)$, and $dX_2/dt = X_2(\gamma_2 X_1 - \varepsilon_2)$, where $X_1(t)$ and $X_2(t)$ are the numbers of individuals in the prey (X_1) and predator (X_2) populations, respectively, at any given time t, and (ε_1, γ_1) and (ε_2, γ_2) represent the (birth and death) rates of the prey and predator populations, respectively. These dynamic equations characteristically have no stationary equilibrium. Their only equilibrium is two indefinite interspersed oscillatory waves (Weisberg 2012, 12 [fig. 2.1]; see also fig. 4.2 below). Now, as it turns out, the model provides an accurate representation of prey and predator populations in the Adriatic, and in addition it solves the quandary that prompted Volterra to propose it in the first instance. The model predicts quantitatively that any general biocide (any agent that reduces equally the number of prey and predator) would in fact relatively favor the prey population. This is known as the *Volterra property* of the model, and it accounts for the relative increase in prey when fishing resumed: the inactivity during the war had given the predator population the upper hand at the expense of the prey population. Yet there is no need to know anything about the evolutionary factors involved in deriving the result. It follows strictly from the mathematics alone.

THEORETICAL MODELS

The *Volterra property* makes sense from an intuitive theoretical point of view as well, of course. More generally, the Lotka-Volterra model itself (i.e., the very equations and the interspersed dynamic behavior of prey and predator populations that the equations describe) becomes intelligible in the light of evolutionary theory, in terms of reproductive pressures in a competitive environment. The reasoning is simple. As the prey population grows, we expect the predator population to increase in its wake, given the increased abundance of food, until a certain threshold is reached at which the predator population number overcomes the prey reproductive rate. At that point, we expect a sudden turn downward in the prey population as it becomes dramatically depleted. But then another inflection point ensues when the resulting competition for scarce food leads to a subsequent collapse in the predator population. This in turn enables the prey stock to recover, and the cycle begins again. Note that this explanation already imparts a theoretical interpretation on the quantities represented in the model in terms of evolutionary competition between the species. The dynamics then comes to represent the interlocking evolutionary forces at play. We are now considering not just a mathematical model but a full theoretical explanation of the phenomena via a theoretically interpreted model.

Max Black thought that it makes sense to distinguish these two sorts of models (mathematical vs. theoretical) in terms of their very different explanatory powers, and I shall follow suit here. A mathematical model per se does not explain the phenomenon it models.[6] A theoretical model, by contrast, does attempt an explanation, often in terms of acting and efficient causes, and at any rate as a result of the application of the concepts and categories of some background explanatory theory. Sometimes, as in the case of the evolutionary explanation of the phenomenon of the depletion of the fish stock in the Adriatic, by means of the Volterra principle, as given above, the theoretical model merely interprets a preexistent mathematical model. But this need not always be so. Theoretical models do not always ride on mathematical models. They do not even need to be expressed in the language of mathematics. A theoretical explanation can be given in some natural or artificial language other than mathematics, and it can have explanatory power that is based on the causes it cites.

A case in point is evolutionary explanations prior to the emergence of the modern synthesis and its overly mathematical models of population genetics. The remarkable debate over the evolutionary origin of whales, for example, illustrates well the nature of theoretical explanation in the absence of any mathematical model. As is notorious, Darwin first intimated that the evolutionary origins of whales might lie in North American black bears and explained the idea by means of the following model: "If the supply of insects were constant, and if better adapted competitors did not already exist in the country, I can see no difficulty in a race of bears being rendered, by natural selection, more and more aquatic in their structure and habits . . . till a creature as monstrous as a whale" (Darwin 1859, 184). Darwin was ridiculed for proposing this explanation, which led him to withdraw it in later editions of *On the Origin of Species*. It can be argued that the ridicule was undeserved since it is clear in Darwin's writing that he was merely proposing a possible model. And, at any rate, the template for the explanation is now considered correct, even if controversy continues to rage as to what extinct terrestrial mammals are in fact the ancestors of whales. The template appeals to individual organisms' reproductive success as a causal effect of their degree of fitness with their environment; and this allows for successive adaptations by natural selection to explain the emergence of new aquatic species as the evolutionary descendants of earlier terrestrial species. (Gingerich [2012] provides a contemporary explanation in nonmathematical terms akin to those employed by Darwin.) One can quarrel with the details of each model explanation, but the template seems reasonably sound, ultimately appealing to the three critical theoretical principles of heritability, variation, and natural selection. These are all good illustrations of apposite and successful theoretical models that are not mathematical.

Throughout the history of science, there are many examples of theoretical models that are not prima facie mathematical, yet they are explanatory in their own terms. Thought experiments often seem to require background theory and in fact often involve a contrast between theoretical models, often with the aim of contradicting one of them. Galileo's leaning tower of Pisa refutation of Aristotelian physics may be an outstanding example. (See, e.g., Stuart, Fehige, and Brown [2018, chap. 5].) Nevertheless, I do not appeal to an exclusionary dichotomy between mathematical and theoretical models. Rather, most of the examples that I describe involve late nineteenth-century highly mathematized models. However, as we shall see, it makes a relevant difference whether the model in question is merely a mathematical description of phenomena, without any further explanatory purpose—and devoid of any causal interpretation—or, by contrast, appeals to theoretical explanatory factors—possibly causal in nature—lying behind the phenomena. So, while not a dichotomy, Black's distinction between mathematical and theoretical models is, with the appropriate caveats, helpful for our purposes and retained in the description of the three main case studies that follows.

A Paradigm Case in Engineering: The Forth Rail Bridge

My first case study is a graphic scale model in engineering, a remarkable feat of Victorian engineering and, moreover, a case thoroughly studied in different disciplines, the rail bridge over the Firth of Forth in Scotland.[7] The Forth Rail Bridge was built between 1882 and 1890, opening officially on March 4, 1890. However, the history of the bridge, its purpose, and the techniques employed in its construction all go back to the mid-nineteenth century, and it is worth recounting in some detail.[8]

The rapid industrialization of Great Britain throughout the Victorian years was notably led by the discovery of the steam engine, and it is often symbolized by the railways. It was government policy to encourage competition among a handful of rail companies (a practice that continues to this day). Cargo traffic from the industrial cities in the Midlands and North of England toward Scotland markedly increased during those years, as did the felt need for passenger carriers from Edinburgh to the towns and cities north of the Forth estuary, particularly Dundee and Aberdeen. The Caledonian Railway Company operated a slow railway line west of the main estuaries, through the town of Perth. The North British Railway company operated an even slower and more cumbersome service on the east coast. This required passengers to embark and disembark for ferry transits through both the Forth estuary (between Lothian and Fife) and the Tay estuary (between Fife and Angus) just to travel from Edinburgh to Dundee and beyond. Allowing the North British Railway to bridge the estuaries became necessary to maintain competition, and government permission and support was therefore granted.

The Firth of Tay is considerably shallower in depth than is the Firth of Forth, and, besides, it leads straight into the city of Dundee, so it was decided that it should be bridged first. The lead engineer in charge was Sir Thomas Bouch, who was then renowned for laying down more than three hundred miles of properly functioning railway in England and Scotland (Petroski 1995, 76). Royal assent was given in July 1870 and the cornerstone laid a year later. The estuary is a mile long at the point where the bridge was being built, and, given the need for an oblique approach to Dundee, a two-mile-long bridge was required. The plan was for a conventional truss bridge with spaced-up girders, and detailed calculations were made for the stresses and strains on the sandy foundation. Fatally, no such detailed calculations were made for wind resistance, even in an area known for strong gales. It was merely assumed that the wind strengths would be typical, and the very rough resistances that were estimated indicated that no reinforcements would be required. The Tay Bridge was completed in September 1877 after six years of labor and officially opened in July 1878. A year later, Queen Victoria crossed the bridge and awarded Bouch a peerage. Just six months later, the famous "Tay collapse" took place, the bridge coming down in an unusually strong gale, taking with it an entire train bound out of Edinburgh and its seventy-five passengers, none of whom survived (Petroski 1995, 74).

The story is worth recounting because it explains the strong public concerns, technical specifications, and safety requirements that led to the construction of the Forth Rail Bridge a decade on. The Tay disaster ended in a public inquiry that in June 1880 reprimanded Bouch even though his calculations for the strength of the winds and the ensuing pressure on the bridge were typical at the time. Yet the conclusion of the inquiry was clear: "The fall of the bridge was occasioned by the insufficiency of the cross bracing and its fastenings to sustain the force of the gale" (quoted in Petroski 1995, 75). When in 1881 the railway board invited Sir John Fowler, its consulting engineer, to put forward proposals for the Forth Rail Bridge, the issue of sideways wind pressure was foremost in everyone's mind. Fowler was an established senior engineer, and he was mainly responsible for the complex approaches to the bridge. The design of the bridge itself was mainly the work of Benjamin Baker, the much younger main assistant engineer. Baker innovated by introducing a different cantilever design for which he claimed to gain inspiration from ancient Oriental bridges (Baxandall 1985, 19).

The principle of the lever is based on that of a simple balance scale, where a weight on the short arm exquisitely balances a lighter weight at the end of the long arm. A "canted lever" is then a balance with a bent short arm that can be anchored in a wall or a stronger fixed structure, thus providing the force that would maintain the longer arm upright. This would allow the upper arm to support any weight consistent with any shearing or compression forces within the material (if not heavy enough actually to break the longer arm). A cantilever bridge is an arrangement of two such levers, with the shorter arms anchored in each shore and the longer arms linking up in the middle up in air, which Baker illustrated by means of a military sketch of an Asian cantilever bridge (fig. 3.2).

The design for Forth Rail Bridge was based on the cantilever principle for several reasons, the main one being that it allows reduced cross sections to be exposed to the sideways wind. Baker presented it as an innovation for the case in hand, but in fact cantilever bridges had been built before in the West, both by James Buchanan Eads in the United States and by Heinrich Herber in Germany. It was partly Baker's extensive knowledge of cultural history (Baxandall 1985, 17), partly the need to convince the public that the new design was radically different than that of the Tay Bridge, and perhaps partly some sense of wounded British pride that prompted Baker to acknowledge no forerunners among his close contemporaries.

The monumental bridge was built according to plans drafted by the diverse engineers; multiple technical decisions were involved in the choice of materials and particularly the shapes and dimensions of the different elements. In a cantilever bridge, the upper spans in the long arms and the supports for the short arms at each end are in tension while the lower spans and the middle piers are in compression, as can be seen in the celebrated anthropomorphic model that Baker employed to illustrate the principle (fig. 3.3).

FIGURE 3.2. Sketch by Lieutenant Davis, RN, of a Tibetan bridge (Westhofen 1890, 218 [fig. 9]).

FIGURE 3.3. Benjamin Baker's anthropomorphic model of the cantilever principle (Westhofen 1890, 8 [fig. 5a]; see also Baxandall 1985, 21 [fig. 4]).

This led to specific choices for the upper spans and the lower spans, with an eye on material resistance and the ever-present consideration of side-wind stresses. Thus, the design avoids large flat surfaces presented to the side winds. The upper spans are L-shaped girders with maximal tension strength and minimal resistance to wind, while the lower tube-shaped spans are designed to resist huge compression strengths. This in turn led to the novel choice of steel as the material for the construction of the girders, steel being particularly ductile in comparison to the wrought iron typically employed and in fact used for the Tay Bridge.

The engineers took advantage of an islet (Inch-Garvie) in the estuary halfway between the shores to build a larger middle pier in addition to piers on each side of the firth. Hence, the bridge spanned essentially six levers placed on three piers firmly installed on land. In contrast with the Tay Bridge, however, the design of the piers is extremely broad-based; each pier is about 120 feet wide at its base but only 33 feet wide at the top (Baxandall 1985, 19). All the specifications for the different parts were detailed in advance in highly precise graphic models in an array of different scales. Many of the key ones are summarized in the central plate figure in Westhofen (1890, 74–75) (see fig. 3.4).

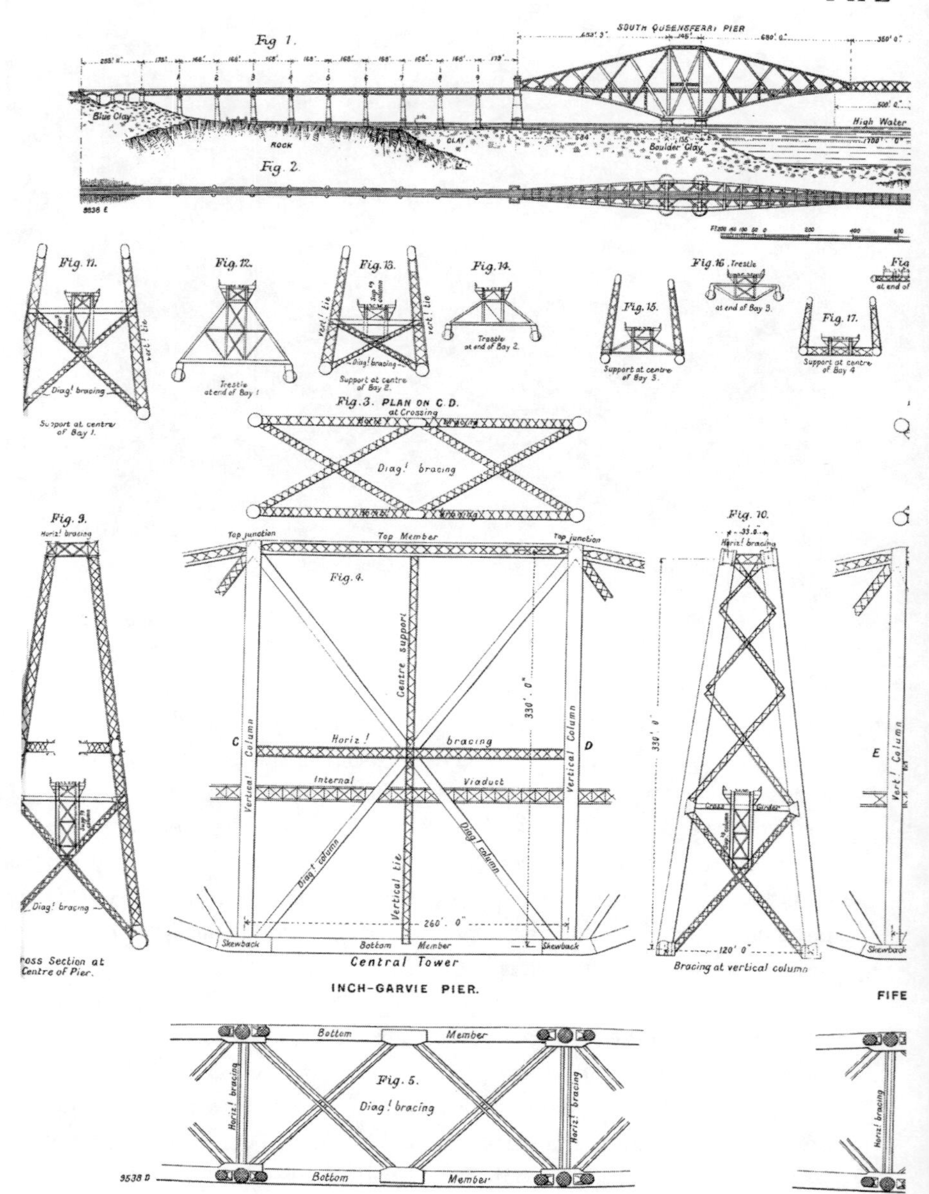

FIGURE 3.4. Plans for the Forth Rail Bridge, cantilevers, piers, and girders (Westhofen 1890, 74–75).

FORTH BRIDGE.

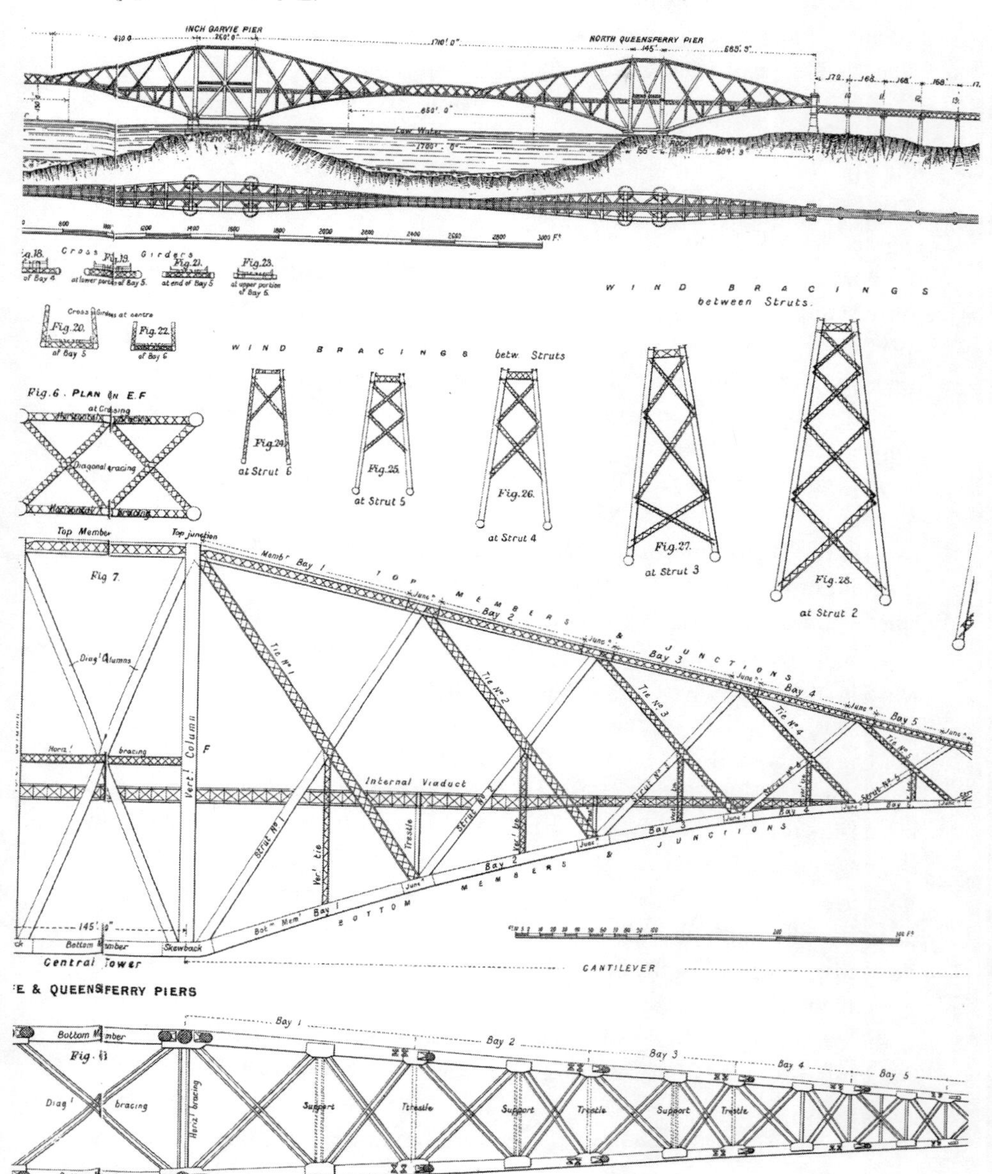

We shall come in due course to the question of how model sources can possibly predate their targets. The engineering model plans for the bridge were the blueprint for the construction of the bridge. Hence, at the time the plans were drawn, the model represented a not-yet-existent entity. This will lead us to considerations regarding the various uses of the inferences that scientific representations license. For now, what matters for my purposes is the complex nature of the simplest of possible scientific models, a scale engineering model for an artifact. The explanation of the model as a representation requires a great deal of background knowledge. To analyze and understand precisely and properly how such a model operates and what it specifically represents in all its detailed nuance, a good deal of understanding is required. This means not just understanding the model's source and its diverse parts and how they relate but also understanding the model's history: who produced it, the purposes for which it was produced, and how it came to be generally employed. Models rarely wear their representational content on their sleeve, as it were. Rather, as Baxandall (1985) magisterially taught us with the Forth Rail Bridge—as with many other artistic objects—gaining the kind of full and deep comprehension of the representational work that a simple source such as a scale model does requires a phenomenal amount of historical and social context regarding its genesis as well as its appropriate and intended use. This is as true for an engineering blueprint or a scale model as it is for any artistic representation in Renaissance Italy.

The Forth Rail Bridge models also bring to the fore another key ingredient in a scale model, namely, its intended use. The model graphs are intended as blueprints for building up certain properties, qualities, parts, or relations in the emerging target system. Afterward, they remain in their intended, common use as guides to those properties, qualities, parts, and relations. Thereafter, the photographs of the actual built-up bridge are similarly intended to be used as portraits of the bridge itself, and they are naturally at a scale. Like Baxandall (1985), I always emphasize intended use, not intentions per se. It is the common use that people make of a model source in the actual world, rather than anyone's state of mind or intention, that sustains the representational role of models.[9]

The Forth Rail Bridge models represent because their legitimate socially sanctioned use is representational. Representational uses are heavily constrained and thick with all kinds of normative assumptions. It is a hugely constrained use, and Westhofen (1890) is a wonderful compendium of all the different normative keys that are essential for the appropriate application of the different graphs and models created with reference to the Forth Rail Bridge. Scale models represent in this sense: whatever other features they may have, it is always the case that a scale model is a sized-down, sized-up, or otherwise convenient version of something else, namely, its target. The model's intended use is to portray or make salient certain features of its target, and it is therefore always employed with an eye to the sort of information that it carries about such targets (whether the information is prospective, as in a blueprint, retrospective, as in architectural plans for buildings no longer in existence, or merely standing, as in a guide or a map of some contemporaneously existing object). We can conclude that scale models are therefore always representations, understood in this minimal sense that appeals only to their intended use as tools for informative inference regarding their targets.

Analogy and Abstraction: The Kinetic Theory of Gases

In the last section, I argued that analogue concrete models can develop their sources ahead of their targets. This is typical in engineering and architecture, in which disciplines maps or plans for buildings etc. act as blueprints in the construction of the target objects. Previously, a few models were reviewed that we now regard as fictions, such as Maxwell's (1861–62/1890) vortex model of the ether. Our next case study also has roots in the nineteenth century. As in the case of Maxwell's model of the ether, the so-called billiard ball model of gases begins its life as an imaginary analogue model, and it slowly transforms itself into a larger theoretical model, the fully fledged kinetic theory of gases. And, as in the previous case in nineteenth-century physics, this one also illustrates the transition from an imaginary conceptual model with concrete instances or realizations to a highly theoretical one. But, in contrast with Maxwell's vortex model, the fully fledged kinetic theoretical model of a gas is both qualitative and explanatory in addition to being cast in the language of mathematics. And what it represents is certainly no fiction, however abstract a gas might be. However, the general lessons apply. Representation may have no real target, and often in the sciences the target systems are largely unknown, or ambiguous, or under construction, or generally controversial and the object of much discussion. That there is much uncertainty about the system modeled is precisely one of the main reasons why it is being modeled. The very building of a model is in such cases a critical part of research into the nature of the target, and the process of model building itself often throws out certain aspects and features of the target and unearths new ones.

We have also already noted how models streamline and simplify their descriptions of their targets. This is necessarily the case for scale models since scales necessarily involve lesser or greater detail. But, as we saw, they simplify in all kinds of other different ways since they are not merely replicas but smartly tailored sources standing for other systems. Analogue models do not just simplify. They metaphorically extend the laws and the entities that are entrenched in some domain to new domains, where they appear prima facie doubtful. In so doing, they do not merely simplify but often caricaturize, even tergiversate over fundamental aspects. Some are concrete, such as many of the actual realizations of the mechanical ether and the Phillips-Newlyn machine. The billiard ball model, a more conceptual kind, is well-known among historians of science and particularly well-known among philosophers. This is a model that eventually develops into a fully fledged mathematical theory of the domain in question, not unlike electromagnetism develops out of the many analogue mechanical models of the ether at the end of the nineteenth century. The key difference, of course, lies in the existence of the target system. Mechanical models of the ether were created in the course of the search for some nonexistent entity and its properties. What the billiard ball model was in search of is the all-too-real molecules confined in gases and their macroscopic properties.

There are other aspects of the case study worth emphasizing, relative to the nature of the idealizations and abstractions introduced into the description. In all the cases of modeling so far discussed, there have been several idealizations and simplifications introduced. We have noted prominent simplifications in scale models, remarkable idealizations and even caricatures in analogue models, and all kinds of formal simplifications and abstractions in mathematical and theoretical models. Some of the inevitable trade-offs involved in the billiard ball model will serve us as a benchmark and as an argument against some accounts of representation.[10] The model has been generally thought to be a central analogy driving through the development of the kinetic theory of gases in the second half of the nineteenth century (Brush 2003). Its most celebrated appearance in the contemporary modeling philosophical literature occurs in Mary Hesse's works (e.g., Hesse 1963/1966). Hesse uses it to illustrate her well-known distinctions between the positive, the negative, and the neutral analogies, notoriously the set of properties shared between analogue and model, those that are openly contradicted, and those that are open and fuel scientific research, respectively. Hesse (1963/1966, 8ff.) traces the model back to Campbell (1920/1957), another milestone in the emergence of the philosophical modeling literature. It is worth quoting Hesse in full:

> When we take a collection of billiard balls in random motion as a model for a gas, we are not asserting that billiard balls are in all respects like gas particles, for billiard balls are red or white, and hard and shiny, and we are not

> intending to suggest that gas molecules have these properties. We are in fact saying that gas molecules are analogous to billiard balls, and the relation of analogy means that there are some properties of billiard balls which are not found in molecules. Let us call those properties we know belong to billiard balls and not to molecules the negative analogy of the model. Motion and impact, on the other hand, are just the properties of billiard balls that we do want to ascribe to molecules in our model, and these we can call the positive analogy. . . . There will generally be some properties of the model about which we do not yet know whether they are positive or negative analogies. . . . Let us call this third set of properties the neutral analogy. (1963/1966, 8)

The distinction has been influential, and it is now rightly regarded as a key element in any analysis of the role of analogy (Bailer-Jones 2009; Cartwright 1999b; da Costa and French 2003; Gelfert 2016; Harré 2004; Hughes 2010; Morgan 2012; Morgan and Morrison 1999; Suárez 1997, 1999a). Hesse famously introduced her distinction in an imaginary dialogue between a Campbellian defender of models and a Duhemian critic. The Campbellian classifies the properties of billiard balls in three groups, depending on their status in the gas molecule analogy. The negative analogy contains the properties that pertain to billiard balls but not gas molecules, such as their color, hardness, brightness. In the positive analogy, there are all the properties that gas molecules and billiard balls share, mainly motion and impact. Then there is a third group of properties that constitute what Hesse calls the *neutral analogy*. In the words of Hesse's Campbellian character, these are "the properties of the model about which we do not yet know whether they are positive or negative analogies": "These are the interesting properties, because . . . they allow us to make new predictions" (Hesse 1963/1966, 8).

The distinction between the positive, the negative, and the neutral analogies stands well today, regardless of the merits of Hesse's own original analysis of the billiard ball model, and has been applied to a range of cases. I shall therefore discuss the model in Hesse's terms, adhering to the distinction between the positive, the negative, and the neutral analogies, but I shall add further detail. I shall argue that, both as regards the historical role of the billiard ball model and as regards the intricate nature of the analogies, particularly the so-called neutral analogy, we can get a genuine grip on the nature of the model as a representational tool only if we move somewhat past the simplifications in Hesse's dialogue.

As regards the history of the model, Hesse presents it as the key driving motive behind the development of the kinetic theory of gases. Her review places the billiard ball model squarely in a British tradition beginning with James Clerk Maxwell. And she obviously locates the origins of later philosophical discussions of the model with Campbell (1920/1957). Now, undoubtedly, Campbell's *The Elements* was an important source for critical approaches to scientific theories and empirical knowledge and remained influential into the 1970s, particularly in Britain. Many Campbellian terms remain in philosophical use today, such as *the dictionary*: the set of rules of coordination that give meaning to the ideas expressed in hypotheses by linking them to the operational concepts that occur in the laws to be explained (Buchdahl 1964, 156). Campbell's dictionary is in some ways akin to logical positivistic rules of coordination, and it resembles both Carnap's and Reichenbach's attempts at coordinative definitions.[11] There is a similar sense in which they link the analytic statements in the theory to the empirical protocol statements.

The critical difference of course concerns Campbell's much more benign attitude to the analogies involved in such rules. For the logical positivists, the theoretical statements are defined, at least in part, through the formal mixed postulates, or bridge principles. By contrast, Campbell thinks that the link between a hypothesis and the empirical laws that give it meaning is not at all formal but instead mediated by "a sort of analogy between concepts and ideas" (Buchdahl 1964, 157). This is undoubtedly a more appropriate approach to the sorts of analogical models employed in the sciences and the different ways in which they can serve the ulterior purpose of generating more precise theoretical models. So Hesse is no doubt right in anchoring the philosophy of science that vindicates the serious use of analogical conceptual models, such as the billiard ball model, in Campbell (1920/1957). Yet, as we shall see, some historical anachronisms are imputed that turn out to be unhelpful. And there are important technical differences between their accounts.

For a start, curiously enough, the term *billiard ball model* does not appear in *The Elements* once. Nor does the more generic *billiard ball*. Instead, Campbell refers to "elastic balls of finite diameter," and obviously billiard balls are a close approximation (although not exactly so—more about this later). A fully developed appearance of the billiard ball analogy in English occurs in Sir James Jeans's influential textbook (Jeans 1940, 12ff.), and Hesse may have picked it up there. This matters, of course, because, if it turns out that the billiard ball model is a retrospective invention (i.e., if it is part of a later twentieth-century reconstruction of the historical development of the emergence of the kinetic theory), it would seriously compromise Hesse's analysis. For her analysis critically relies on the neutral analogies in the billiard ball model being the actual driving factor in the heuristics of the research program that culminates in the kinetic theory of gases. And this could hardly have been the case if the model failed to play such a role in the actual history.

Fortunately, however, a model of perfectly elastic spheres in collision did play a key role in the development of the kinetic theory of gases. It appears prominently in Maxwell's early work on the topic, such as his celebrated "Illustrations of the Dynamical Theory of Gases" (Maxwell 1860/1890). Maxwell's dynamic treatment of gases as analogous to sets of randomly moving, freely colliding, and perfectly elastic spheres predates any similar effort by Boltzmann by about six years (Balázs 2017). However, his spheres are ideal objects, precisely because they are perfectly elastic and the collisions are entirely conservative. This means that all kinetic energy is conserved and no energy is spent, dissipated, or in any way wasted or transformed. This is hardly a realistic assumption and in many ways deeply unlike billiard balls, which do vibrate and of course suffer some energy loss owing to friction and so on.

The billiard ball model as we know it seems to have been introduced by Boltzmann several years later, in the form of what he called quasi-elastic *ivory balls* (see Boltzmann 1877). Notably, in that period—since the seventeenth century and way into the early twentieth century—billiard balls were made of ivory, traditionally elephant tusk. Maxwell goes on to acknowledge the provenance of the analogy in Boltzmann only a year later when, in his review of Watson (1876), he writes: "But Boltzmann's molecules are not absolutely rigid. He admits that they vibrate after collisions, and that their vibrations are of several different types, as the spectroscope tells us. But still he tries to make us believe that these vibrations are of small importance as regards the principal part of the motion of the molecules. He compares them to billiard balls, which, when they strike each other, vibrate for a short time, but soon give up the energy of their vibration to the air, which carries far and wide the sound of the click of the balls" (Maxwell 1877, 246).

So here we finally find the billiard ball model, as we know it, in its full glory. It is a complex result of a combined Boltzmann-Maxwell-Watson effort over the years. Remarkably, at this point in the history, the kinetic theory of gases is by and large established. It is striking then that the billiard ball model in fact appears at a late stage and mainly to introduce a disanalogy between gas molecules and ideal colliding perfectly elastic spheres. Gas molecules are really unlike those ideal spheres, but, naturally, they resemble real and existing billiard balls. This puts the model in an unsuspected light, quite different than the way in which Hesse and the ensuing tradition present it. The billiard ball model is less a generator of a research heuristics and more a correction of an antecedent and highly idealized model. It is moreover a correction perfectly in line with the ongoing heuristics of an already-established research program.

My account of the history vindicates Campbell's (1920/1957) own original emphasis on Maxwell's early ideal spheres model.[12] Maxwell's ideal spheres model provides the dictionary that translates the terms in the kinetic theory into an antecedent and previously understood vocabulary. We shall see later how the translation in effect works and what it entails for the nature of the representation that the model enacts. And, of course, the relation is fundamentally analogical in that characteristically Maxwellian sense that we have already probed: there are positive, negative, and neutral properties in the analogue model, and what they mainly do in practice is to invite positive, negative, and open-ended ways of reasoning into conclusions about the nature of gas molecules. The positive analogy provides the conductor between the model source and its target, a conductor that allows modelers to *reason from one to the other*, in Maxwell's felicitous phrase.

Fictions and Idealizations: Astrophysics Models of Star Formation

The final case study I invoke is a contemporary one that exemplifies both mathematical and theoretical models, but in an unusual order. In stellar astrophysics, the key "observable quantities" are modeled by means of four different theoretical equations. Each of these equations can be interpreted within a particular physical theory, and all are then brought together to generate what is essentially a mathematical model. In the overall model thus constructed, it is possible to adjust the parameters rather freely, very much as we saw was possible in the Lotka-Volterra model. The ensuing model enables a multitude of different simulations of stellar luminosity, brightness, and spectral class in accordance with the fundamental empirical law in stellar astrophysics. The model is illustrative in different ways and, I shall argue, exemplifies well some of the fundamental features of scientific representations more generally.[13]

The standard definition of a star takes it to be any gas cluster uniformly constituted by a mixture of hydrogen and helium, bound together by self-gravity, and radiating energy from some internal source (Prialnik 2000, 1). The application of the technical terms *uniform gas*, *bound*, *self-gravity*, *internal source*, and *radiation*, however precise they are, is rather vague in practice. Self-gravity dominates—otherwise there would be no cohesive body to speak of—but it need not be and rarely is the only active force. Neither is radiation the only means of stellar energy output; there may be other internal sources of energy. The boundaries of a star are typically imprecise since stellar coronas can stretch for thousands of miles and are in constant interaction with the environment. Thus, a star is of course not a closed system. Finally, as regards composition, hydrogen dominates, particularly in young stars, but, as the star burns its fuel, it generates elements with higher atomic numbers, such as oxygen, carbon, and nitrogen, which it ejects into the interstellar medium. This is why stars are said to be the "kitchens" of the universe, the places where the heavy elements that make life possible are "cooked up."

In other words, the main assumptions are highly simplifying of the actual complexity in real stars. They are introduced into stellar structure models mainly on account of their ability to generate the appropriate empirical predictions swiftly. Astrophysics is an observational science, one in which experimentation is at best a metaphor. Astrophysicists occasionally will speak of *experimenting* when allowing distant source radiation to be filtered through nearer gravitational fields (gravitational lensing) or when observing extragalactic bodies in collision or the birth of new stars (supernovas). All this talk is figurative; there is no experiment per se, and certainly no controlled experimentation is going on. The actual astrophysical data is ultimately always observational. However, there are some solid empirical laws that provide plenty of information and a very good base for inferences from data to theory. In particular, the main regularity in observational astrophysics is the so-called Hertzsprung-Russell (or HR) law, which establishes a correlation between the luminosity (L) and the effective surface temperature (T_{eff}) of a main sequence or "ordinary" star.[14] A recent diagram displaying the characteristic correlation curve expressed by the law is reproduced in figure 3.5.

It is important to note that, although the HR law is often referred to as the *empirical* law of stellar astrophysics, the quantities plotted in a typical HR diagram are not directly observed, registered, or detected but rather inferred by means of some simple extrapolations. There are only two astrophysical quantities that are in fact directly observable—in the sense of being genuinely detectable. One is the incident radiation flux from a stellar source into a telescope or detection apparatus, also known as its apparent brightness (I_{obs}). The other is the characteristic set of the spectral lines of radiation or distribution of radiation intensity per wavelength (l), also known as the (electromagnetic, not merely optical) spectrum (I_λ) of its source. The typical distances (d) of the sources can be estimated, occasionally by a simple procedure of measuring their parallax (the relative position of a nearby star against the fixed background of more distant stars at two given times of the year, i.e., at two different points of the earth's elliptical orbit), but more often because of a complex statistical treatment of parallax to known star clusters and other galactic objects. So, strictly speaking, distance is not really an observable quantity, or at any rate not one subject to direct measurement or detection.[15]

Once data have been collected for the values of these three quantities (I_{obs}, I_λ, d) for any given star source, it becomes possible to derive the star's

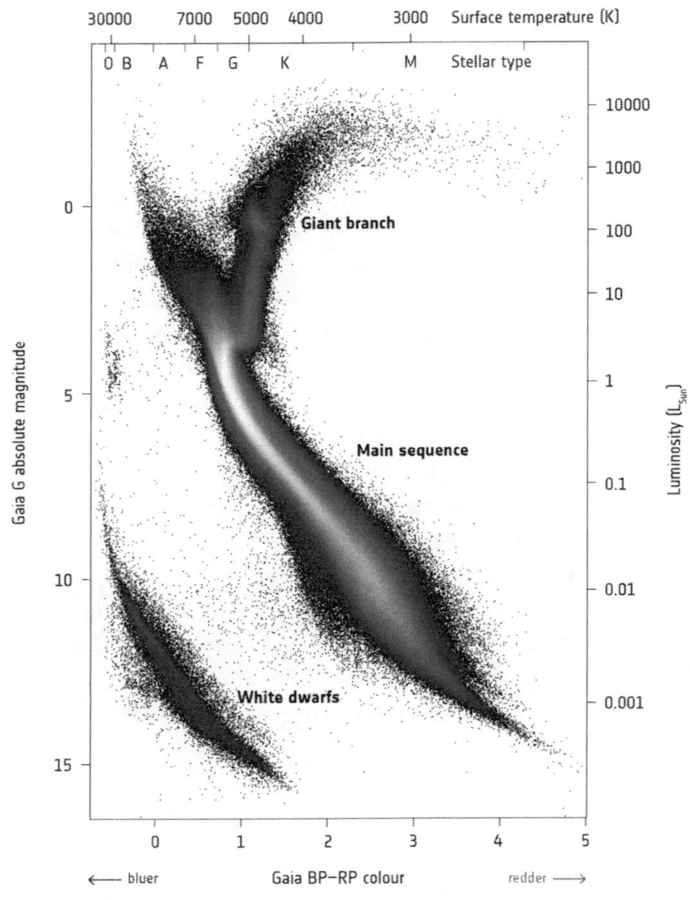

FIGURE 3.5. A Hertzsprung-Russell diagram plotting data from the European Space Agency's Gaia space observatory. CC BY-SA 3.0 IGO.

luminosity (L), its effective surface temperature (T_{eff}), and the chemical composition of its photosphere (the outer layer of the star). The luminosity of a star is its *energy power*: the amount of energy radiated per unit time, a simple function of distance and observed incident radiation flux, that is, $L = 4\pi d^2 I_{obs}$. If we further assume that the star is a blackbody spectrum, then it becomes possible to derive the effective surface temperature of a star from its luminosity, as follows: $T_{eff}^4 = L/4\pi R^2\sigma$, where σ is Stefan-Boltzmann's constant.

Another, similar blackbody assumption allows us to derive the chemical composition of the photosphere from a spectral analysis of the emission and absorption lines in its spectrum (I_λ). Luminosity (L), effective surface temperature (T_{eff}), and chemical composition are then regarded as the basic observational or empirical quantities of stellar astrophysics, and the former two are mapped in a typical HR diagram. The HR diagram is taken to represent the evolution or life cycle of stars, from their birth as objects of high temperature and luminosity, to their maturity as objects of lower temperature and luminosity, and in some cases to their eventual final stage as red giants and white dwarfs away from the main sequence.

In figure 3.5, red giants lie to the upper-right-hand corner of the diagram, indicating a high-luminosity but low-temperature object. By contrast, white dwarfs are low-luminosity and high-temperature objects, so they lie in the lower-left-hand corner. The life cycle of stars begins in supernova explosions, as high-temperature and high-luminosity objects on the upper-left-hand corner of the main sequence. As stars gradually burn their fuel, they become cooler and dimmer, thus moving down in the diagonal in the main sequence. Eventually, at least in the case of particularly massive stars, the outer layers become lighter and cooler as heavier elements are ejected into the interstellar medium. However, even in the last stages, the core of a star continues to produce blackbody radiation. These radiation forces greatly outweigh the bounds of self-gravity, by then much looser, and the star expands hugely. As the outer layers move still farther away from the central radiative core, they cool down even further, and the spectrum lines of the photosphere of the star correspondingly shift dramatically to the red as its surface temperature goes down. That is why in their final stages stars become red giants: a comparatively huge mass of lighter and cooler gas with a still fully functional radiative core. A white dwarf, by contrast, is effectively dead, a star in its terminal stage as the fuel burns out completely, radiation forces cease abruptly, and the star collapses under its own gravity into a tiny and very dense cloud of compact gas made up of comparatively heavier elements. Eventually, the radiation source dies out altogether as its internal power engine ceases to emit any radiation, and the star is dead.

In later chapters, I shall be addressing the nature of the assumptions that are built into the models. They turn out to be critical to an assessment of the practice and theory of representation in science. Stellar structure modeling in astrophysics provides particularly clear instances worth running through in some detail. The assumption that a star is a blackbody (sometimes called the *thermal equilibrium assumption*, or EA) entails that the energy transfer within a star is entirely radiative (Prialnik 2000, 16ff.). That is, energy will transfer across the body of the star independently of the motion of matter within the gas (convection) and independently of its conductive properties (conduction). The EA effectively rules out any transfer by conduction from, for example, hotter to colder parts of the gas. It also makes the model independent of rotational vortices or convection forces within the star. Yet the physics indicates that the gas in a star is unlikely to be fully radiative.

In effect, the EA is a considerable idealization. The radiation and gas temperatures will very rarely be identical throughout the lifetime of the star (radiation temperatures will typically be higher). Only at its very inner core, if anywhere, can the radiation and gas temperatures be approximately identical. Hence, equilibrium and the EA at best hold locally and probably not at all during the most active and spectacular periods of a star's life both within and outside the main sequence. What justifies the EA from a theoretical point of view is the fact that neglecting convective and conductive transfers carries no implications for the temperature distribution within the star. In other words, the assumption that a star is in thermal equilibrium and a blackbody (EA) is not introduced because it is a little bit away from the truth at the edges, or approximately true. Rather, it is introduced in order to facilitate the appropriate inferences with respect to the temperature distribution in a typical star across a wide variety of cases.

Three other assumptions play a key role in stellar structure models (Prialnik 2000, 6–8), namely, isolation, uniform composition, and spherical symmetry. First, a star is assumed to be a noninteracting body isolated in empty space (provisos are made for binary stars and other interlinked clusters). Let us refer to this as the *isolation assumption* (IA). It implies that the boundaries of the stellar gas are sharp and distinct from the background interstellar medium and that no significant interaction across takes place. However, stellar gas differs at best locally in density from the interstellar medium surrounding it. The sharpness of the boundary is debatable and in fact debated by many astrophysicists working in the field of coronal stellar astrophysics. It is, for instance, remarkable that a star's corona (the huge halo surrounding it) is typically much hotter than is its photosphere (the outer layer). The fact that the corona is subject to huge disturbances caused by the star's magnetic flux also suggests that the boundary is not sharp in any sense other than as the source of detectable radiation, mostly in the visible range. (Coronas are plasmas and hence too light to be significant sources of emitted radiation.) If we understand the star to be in fact the whole complex self-sustaining interacting system at all levels, then the IA is not even an approximation since a star is in permanent electromagnetic dynamic interaction with its corona, and vice versa. The IA is again far from realistic and mainly introduced to simplify calculations.

Next, consider the assumption that the star has a uniform chemical composition throughout, the uniform composition assumption (UCA). A star is indeed taken to be composed uniformly of primarily hydrogen and only secondarily helium. Yet, as we have already seen, most old stars always contain large amounts of heavy elements since they have already performed vast amounts of their "cooking." There are standards at play here, and indeed it is easy to see that the UCA takes as its prototype a star in the main sequence, such as our sun. The sun's internal composition is roughly up to 91 percent hydrogen atoms (accounting for roughly 70 percent of its mass fraction) and 9 percent helium atoms (accounting for roughly 27 percent of its mass fraction). The remaining 3 percent of its mass fraction is made up of oxygen, carbon, nitrogen, etc. In older stars, this proportion of the mass ratio will be even bigger. Yet such heavy elements are often disregarded in at least the simplest stellar models. Instead, the sun's composition is taken as the benchmark, with the mass fractions for hydrogen and helium at 70 percent and 30 percent, respectively, assumed as the rule. Once again, stellar models make this assumption—which is not even close to the truth—to derive a very simple set of mathematical relationships between the dynamic properties relevant to the evolution of the stars—the stellar structure equations.

Finally, we reach our fourth assumption—that a star is bound entirely by self-gravity and consequently that the shape of the star is spherically symmetrical. This is the spherical symmetry assumption (SSA). The SSA is again clearly an idealization, both in terms of the forces that it takes to act and with respect to the ensuing shape of the star. For instance, rotational forces within a star can be huge and thus greatly distort its spherical symmetry along the axis of rotation. And there are known to be large internal magnetic forces within the star as well that can also distort its shape. Hence, the SSA is again a misrepresentation of real stars. In most cases of regular main sequence stars, the assumption can be treated as an approximate idealization; but, in some cases (binary and distorted stars mainly), spherical symmetry is only a fiction brought into the models to facilitate calculations. More precisely, the justification for our four assumptions is their convenience and the economy in inference they generate when working together. For jointly these assumptions entail that the dynamic evolution of a star depends only on its initial mass, which becomes the only free parameter. This enables modelers to cut through all kinds of constraints imposed by the physics of stellar interiors and hugely expedites inference regarding the star's life cycle (Prialnik 2000, 7, 81–86).

When all these assumptions (the EA, the IA, the UCA, and the SSA) are combined, models can be built that are effective in deriving both internal and observable properties of stars such as their surface temperature (T_{eff}) and luminosity (L), the sorts of quantities that get plotted in an HR diagram such as that in figure 3.5. The assumptions yield the four mathematical equations of stellar structure (Prialnik 2000, chap. 5): hydrostatic equilibrium, continuity, radiative transfer, and thermal equilibrium:

(1)	$\frac{dP}{dr} = -\rho\frac{Gm}{r^2}$	Hydrostatic equilibrium
(2)	$\frac{dm}{dr} = 4\pi r^2 \rho$	Continuity
(3)	$\frac{dT}{dr} = -\frac{3}{4ac}\frac{\kappa\rho}{T^3}\frac{F}{4\pi r^2}$	Radiative transfer
(4)	$\frac{dF}{dr} = 4\pi r^2 \rho q$	Thermal equilibrium

The variables represent some of the key properties of stellar interiors, such as pressure (P), heat flow (F), rate of nuclear energy release per unit mass (q), temperature (T), etc. These quantities are typically ascribed not to points or regions of the star body but rather to each layer of the stellar interior. Our four assumptions (the EA, the IA, the UCA, and the SSA) have the convenient consequence of allowing us to parameterize the whole star with respect to mass at any layer since the EA, the UCA, and the SSA together imply that the mass increases monotonically with radial distance and that no fluctuations are due to convection forces. The models allow via the equations an estimate for the values of these quantities at the outmost layer—the photosphere—of any star. This then yields values for the observable properties: effective temperature (T_{eff}), luminosity (L), and mass fraction at the photosphere (I_λ). We can in this way model any main sequence star and check that indeed its observable properties fit in with the empirical law of stellar astrophysics and the HR diagram. In other words, the four assumptions (the EA, the IA, the UCA, and the SSA) are brought into the model because they jointly generate the equations that facilitate inferences to the observable properties of the star.

Let me stress that the assumptions are not entirely inane. There are ways to assess the accuracy of at least some of them independently, for some of the stars that we might wish to model.[16] Thus, chemical composition at the photosphere of most stars is not uniform, and this belies the UCA to a certain degree, one that can be measured. And there are established means to account for the presence of heavy elements and slightly diverse ratios of hydrogen and helium. In fact, bringing such considerations to bear is essential in modeling both older stars and the evolution of any given star throughout its life cycle. Thus, stellar models of white dwarfs assume that their cores are composed of a carbon-oxygen mixture with all remaining hydrogen or helium confined to the outer layers (Collins 1989, chap. 5; Hansen, Kawaler, and Tremble 2004, chap. 9). Hence, the UCA can be regarded as an idealization that can in different ways be modulated and adjusted to the case in hand, and similar procedures may be available for the other three assumptions, particularly the SSA and the IA. Regardless of their status as theoretical assumptions, their great advantage is to combine jointly in generating powerful mathematical models that enable appropriate inferences to the sorts of observational quantities of stars that can be mapped onto the HR diagram.

The Argument from Practice

The different models that have been described in this chapter can be roughly classified into four nonexclusive types: scale models, analogue models, mathematical models, and theoretical models. I first employed paradigmatic cases to illustrate each of the classes. Engineering case studies such as the San Francisco Bay/Delta model illustrate the nature of scale models, which are not simply replicas but intelligent contrivances. The hydraulic Phillips-Newlyn machine illustrates analogue models, in which the types of systems and the laws employed as sources can differ dramatically from those that apply to the targets. The Lotka-Volterra model is a powerful abstract mathematical model, and many models of scientific data are mathematical too. Finally, early evolutionary explanations and thought experiments throughout the history of science serve to exemplify the nature of theoretical explanatory models.

I then moved on to a more detailed and technically involved description of three different models. These models first served as benchmarks and illustrations for different theses and arguments regarding the nature of scientific representational practice to be explored in the following chapters. Second, as announced, together with the models discussed earlier in this chapter and in chapter 2, they provided a base for a certain induction to the representational purpose of scientific modeling. The diagrams and plans for the Forth Rail Bridge are instances of scale models, and Benjamin Baker's anthropomorphic model of the lever principle is simultaneously an analogue model and a scale model. Meanwhile, in their development of the kinetic theory of gases, James Clerk Maxwell and Ludwig Boltzmann applied an analogy to perfectly elastic ideal spheres that eventually gave rise to the notorious billiard ball model. The philosophical debate that ensued is intriguing and complex; it gave rise to certain distinctions between forms of analogy that will serve their own purposes later in the book. Finally, an example from contemporary astrophysics modeling illustrated the ways in which highly idealized theoretical assumptions can give rise to powerful systems of mathematical equations. These equations in turn serve to simulate and proficiently guide inferences to observable aspects of phenomena whose ultimate causes are inaccessible to us.

Mine is a diverse set of models with different characteristics, and in developing the case studies I have already emphasized several philosophical distinctions and notions that will serve us well in the discussions to come, including the ways in which models can be blueprints in the construction of their real targets, the use of mechanical analogies in the representation of fictional entities, the different forms that analogy can take, the high degree of idealization of some of the theoretical assumptions that are conjoined in the construction of mathematical models, and the inevitable trade-offs between inferential reach and depth, on the one hand, and any literal realism in the descriptions, on the other hand. These are issues to which I shall return in due course. For now, the main lesson to be derived from these cases is that, in all these instances, models play a representational role. A model is built to inform us about a particular target system or phenomenon. The model sources are selected judiciously with this aim in mind and are thereafter accordingly contrived, designed, and manipulated. There may be other uses to which those models are put, but often the main use is representational; and it is the one use that in most cases will explain the nature, construction, and development of the model as it is or becomes eventually.

That models have intricate, complex, and overriding representational uses is a lesson that derives not from any metaphysics of science or its development, or from any grand scientific methodology, or from an even grander philosophy of reference or metaphysics of relations. It is very much a lesson that emerges from the practice of model building itself, and, while we may not have explored a very large number of models throughout all the sciences to the effect of this conclusion, we do have enough in the way of the diversity of the representational sources that can play a role in modeling. We also have enough in the way of descriptions of how these sources are constructed, the materials employed in their construction, and the questions and the purposes out of which they emerge in their appropriate historical context and attending to their historical development and evolution. Finally, we have enough now in the way of the diverse set of philosophical and methodological distinctions and notions that are put to work in building such effective models, enough at least for the purposes of the inquiry into the nature of scientific representational activity that will take our attention over the rest of this book.

Chapter Summary

This chapter advances a taxonomy of models and provides an additional set of arguments, rooted in contemporary modeling practice, in favor of a broadly deflationary account of representation. It first defends a classification of models into four kinds that originates in Max Black: scale, analogue, mathematical, and theoretical. It then argues that this classification is sufficiently inclusive for present-day purposes while endorsing both Black's conceptual archetypes and his interactive account of metaphor (considered jointly with that of I. A. Richards). Three detailed case studies are then presented that will contribute the benchmarks or exemplars for the rest of the book, namely, the engineering model of the 1890 Forth Rail Bridge, the billiard ball model of gases, and stellar structure models in astrophysics. It is in addition argued that all these case studies provide some minimal inductive base for the conclusion that models in science are *representations* in a deflationary sense of this term.

PART II

Representation

4

Theories of Representation

The first part of the book has dealt with the history and practice of what has been called the *modeling attitude*, a stance toward inquiry that takes model building to be at the heart of the natural and social sciences' attempts to understand and manage our interactions with the world. It was argued in chapter 2 that the history of nineteenth-century physics informs and vindicates this emergent attitude. In chapter 3, I argued that the practice of model building itself supports the view that a main function of this new modeling attitude is representational. This was the result of a broadly inductive argument with a diverse inductive base that, I claimed, generalizes. Regardless of their diverse uses, the various models share a common purpose to represent their targets, and the generalization suggests that this is a general feature of models. Of course, models may have other uses (pedagogical, illustrative, educational, exemplifying, even materially acting in the world), but the representational use is at the heart of what makes them models, in their respective contexts.

The second part of the book turns to this representational purpose of model building. How are we to understand this representational function that all models in practice share? It has been mentioned that representation is surprisingly a rather new topic in the philosophy of science, having been largely ignored in the literature for a long time. However, I have also been at pains to show that there are nonetheless important antecedents in nineteenth-century science, particularly in the nascent mathematical modeling inaugurated by James Clerk Maxwell, Ludwig Boltzmann, Heinrich Hertz, and others in the second half of the century. In addition, other philosophical antecedents were pointed out. Although sparsely practiced during the years of dominance of logical positivism, a philosophy of modeling science can be found in the works of Norman Campbell, Max Black, Rom Harré, and Mary

Hesse. Finally, for the last twenty years or so, philosophers have been paying increasing attention to the topic, partly in the wake of work by cognitive scientists and historians and sociologists of science (Giere 1988; Lynch and Woolgar 1990; Nersessian 2008), and there are now several different philosophical accounts and theories of scientific representation available. They all attempt to illuminate the notion and to accommodate the representational role of most models while doing justice to their diversity. My main aims in this part of the book are to classify the different approaches, to assess them, and, ultimately, to defend my own approach to representation, which I call the *inferential conception*.[1]

In pursuit of all these aims, I shall be relying on the material in the earlier part of the book, in two different ways. First, I shall employ the case studies in chapter 3 as my benchmark. More specifically, the overall requirement I shall impose on a philosophical theory or account of representation is that it should give a proper rendition—even, if possible, an illuminating account—of the sorts of modeling practices exhibited in the various case studies presented in chapter 3. Second, I shall be seeking inspiration for my own inferential account of representation in the historical recounting of the birth of the modeling attitude in chapter 2. I shall argue that a proper account of scientific representation, understood as the practice of model building in the sciences, can and must start with insights already developed in the nineteenth century, particularly in the writings of Maxwell, Boltzmann, and Hertz.

The Analytic and the Practical Inquiries

The relative neglect of the topic of representation has been followed by some very intense catching up in the last two decades. Earlier I pointed out the parting shot that was R. I. G. Hughes's (1997) seminal paper.[2] Since then, representation has undoubtedly grown into a booming and central topic in the philosophy of science, at least judging by the number of conferences and workshops held and books and articles produced in the last two decades. The topic is at the crossroads of attempts in analytic philosophy to come to terms with the relation between scientific theory and the world and attempts in the philosophy and history of science to develop a proper understanding of the practice of modeling in the sciences. I have been emphasizing the latter, which I believe to be central to the topic, and which is part and parcel of my argument in favor of the type of practical inquiry into representation that I favor in this book. This chapter and next, however, are devoted to the prospects of a more analytic approach to the topic.[3]

The interest from analytic philosophy is related to the notion of reference and the metaphysics of relations; the interest from the philosophy of science is related to an attempt to understand modeling practices. These two distinct forms of inquiry into the nature of representation can be distinguished as the *analytic inquiry* and the *practical inquiry*. Although they are not exclusive, they impose different demands and point in different directions. The analytic inquiry seems to have historically preceded the practical one, but the relative importance of the latter has grown to the extent that in recent years it has become dominant. This mainly reflects the intense attention that philosophers have paid to scientific models and modeling practice in the last decades.

The analytic inquiry tends to presuppose that representation is a relation, and it focuses on providing an analysis of this relation. It is useful to recall our terminology as follows. We are referring to the vehicle of the representation as the *source* and the object of the representation as its *target*. (Thus, in a portrait, the canvas is the source, and the person portrayed is the target.) Anything can in principle play the role of the source or the target, so these terms are mere placeholders. Our only assumption is that *X* is the source and *Y* the target when and only when "*X* represents *Y*" is true. The analytic inquiry then, in its most basic form, takes it that representation is a relation *R* such that the assertion that "*X* represents *Y*" is equivalent to the assertion that "*R* holds between *X* and *Y*." The most basic form of such an inquiry requires a uniquely specified dyadic relation that holds between all sources and targets. But the analytic inquiry can take more complex forms, and, as we shall see, there are different ways in which this requirement can be relaxed.

The contemporary literature on the analytic inquiry into representation goes at least as far back as Charles Sanders Peirce, who provided one of the earliest and most influential theories of signs, even though he used the term *sign* ambiguously to refer to different sorts of representations and representational sources alike. In more contemporary terms, his theory lays down a triadic relation between the *source*, a sign in some symbol system (for Peirce, the *representamen*), its *target* (or the *object*), and an interpretation or understanding of this sign (or the *interpretans*). Moreover, in his terms, a representation is a nondegenerate—for example, essentially triadic—relation. It cannot be further decomposed or analyzed into a complex function of more basic dyadic relations between *representamen*, *object*, and *interpretans* (see esp. Peirce 1931, chap. 3).[4]

Peirce famously went on to divide all representations into three types: iconic, symbolic, and indexical. Roughly, the sources of iconic representations bear similarities to their targets, those of indexes bear causal relations, and symbols denote their targets conventionally. Thus, a portrait is typically an icon of the person portrayed, smoke is often an index of fire, and the word *cat* is a symbol for the feline animal. In all cases, the triadic relation is distinct and cannot be reduced to a function of the dyadic relations (similarity, causation, denotation) that hold between source and target.[5]

By contrast, the practical inquiry has avoided questions regarding the nature of the representational relation, focusing instead on the very diverse range of models and modeling techniques employed in the sciences. The presupposition behind this type of inquiry is that these modeling techniques must be properly understood in their context of application. The literature on modeling in science is by now immense, and I have already reviewed some of the historically key texts, including Campbell (1920/1957) and Hesse (1963/1966). In the last two decades, the turn toward what I have referred to as the *practical inquiry* has intensified. This movement takes model building to be the primary form of representational activity, so a very large amount of work has been devoted to studying cases of models and modeling in painstaking detail to figure out in what specific ways models are helpful to the modelers in their diverse pursuits.[6]

One theme shared by all these attempts to understand modeling is the emphasis on *use*. The assumption that runs through all of them is that, without an appreciation of the particular use of a model in its context of application, it is impossible to appreciate its role fully. We need to pull our gaze away from the relation between the entities (equations and so on) that carry out the representational work and their targets in order also to consider the purposes of those who use and develop the representations. Many different issues become visible when this wider vision is adopted, including the phenomenon known as *transference*—of both knowledge and representational techniques—from one field to another. The same representational source may be used to represent several targets in different fields.[7]

Given these two forms of inquiry into scientific representation, it is not surprising that the questions typically also come in two varieties. First, when we say of a model *X* (a graph, an equation, a diagram, etc.) that it represents a system *Y* (a physical object, a phenomenon, a population, etc.), we may ask, in an analytic spirit, what exactly is the relation *R* presupposed between *X* and *Y*? In other words, what is the relation *R* that *constitutes* representation? I refer to this as the *constitutional question*. Second, we may ask of any specific use of a model what kind of features and properties obtain in the wider context of application that allow models to perform their job. In other words, what are the effective *means* that scientists employ to get representations to deliver the required goods? I refer to this as the *practical question*.

Means and Constituent of Representation

The analytic inquiry pursues definition and conceptual analysis, and it emphasizes the constitutional question. It is interested in the relation *R* that must conceptually hold between source and target for the source to represent the target. Thus, theories of the constituents will typically implicitly answer the question, What is scientific representation? Different theories will answer this differently, but they all have in common that they regard the question as legitimate and well posed. According to any of these views, if there is no relation that answers the constitutional question, then there is no way to explain what all models have in common. The shared representational aim that, as we saw in chapter 3, all models have in common would then remain elusive, a mystery without rational explanation or justification.

The practical inquiry by contrast focuses on what I call *means*. It studies those context-dependent properties and features of a particular representing situation that make the source useful for scientists as a representation of the target. It is interested in pragmatic questions regarding the actual workings of models, including (but not limited to) judgments of accuracy or faithfulness. Accounts of the means of representation will typically provide case-by-case analyses of the types of properties, relational or not, of sources, targets, users, purposes, and context—both the context of inquiry and the wider social context—for any given representation.[8] To put it in a slogan, the practical inquiry cares for the means, whereas the analytic inquiry pursues only the constituent. How should these terms be defined? Elsewhere, I have provided tentative definitions as follows (Suárez 2003, 229–30; Suárez 2010b, 93):[9]

> *Constituent*: *R* is the constituent of representation if and only if for any source-target pair (*S*, *T*) *S* represents *T* if and only if *R* (*S*, *T*, *x*), where *x* sums up whatever additional elements come into the relation of representation.[10]
>
> *Means*: For any source-target pair (*S*, *T*) at a given time and in a given context, *R′* is the means of the representation of *T* by *S* if some user of the model employs *R′* (at that time and in that context) in order to draw inferences about *T* from *S*.

Concerning these definitions, the following remarks are in order. First, the constituent is implicitly defined by a necessary and sufficient condition, but the means are simply characterized via a sufficient condition. Second, note the inverted order of quantifiers: *R* is a unique and universal relation for any source-target pair, while *R′* is a context-dependent, time-indexed relation that is not unique even for a particular source-target pair, never mind

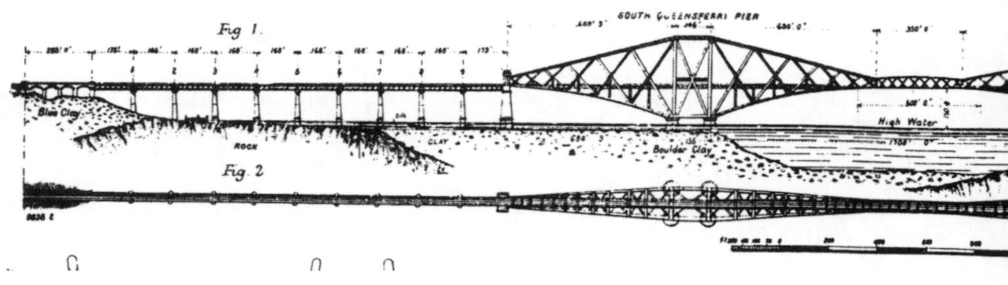

FIGURE 4.1. Sideways plan for the Forth Rail Bridge, including the Forth basin and the surrounding terrain (Westhofen 1890, pl. 3).

universally. Finally, note that the means are pragmatic and context dependent, while the constituent is analytic and does not depend on anything outside the definition of the representational relation itself.

The distinction leads naturally to a further and important distinction. Philosophical discussions concerning scientific representation have in the past often focused on the *accuracy* of scientific models. Fortunately, this is no longer the case, and now representation is carefully distinguished from truth, accuracy, or faithfulness (Contessa 2007; Frigg 2006; Suárez 2003, 226; van Fraassen 2008, 13–15). The distinction is essential to make sense of the phenomenon of scientific misrepresentation, which, as we shall see, is critical if we are to assess the different accounts. Models are often, if not always, inaccurate, and they misrepresent in definite ways. This does not, however, take away any of their representational power. An imperfect and inaccurate model M of some target system T is still a model of T (just as an inaccurate portrait or a caricature is still a representation of the person portrayed). The puzzles regarding the notion of representation are prior to and independent of issues of accuracy.

It may help to focus on a simple and graphic example from the engineering sciences, such as the Forth Rail Bridge in Edinburgh, which we studied in the previous chapter. Recall that this is one of the first steel-built cantilever bridges in Europe, designed by Benjamin Baker on the assumption of a cantilever principle (see fig. 3.3 above). Figure 4.1 is a graphic sideways representation (one of the many appearing in Westhofen [1890]).

The point is that there are two questions one can ask about this graph. First, what object or system in the world does it represent? In our terminology, it is the source of the representation, and the actual bridge is its target (and as we saw this can be the prospective target, as in a blueprint; or the retrospective target, as in a photographic plate). Only once this is established can a further question be meaningfully posed. How accurate or faithful is this graph as a representation of the bridge? If the graph represented something else—for

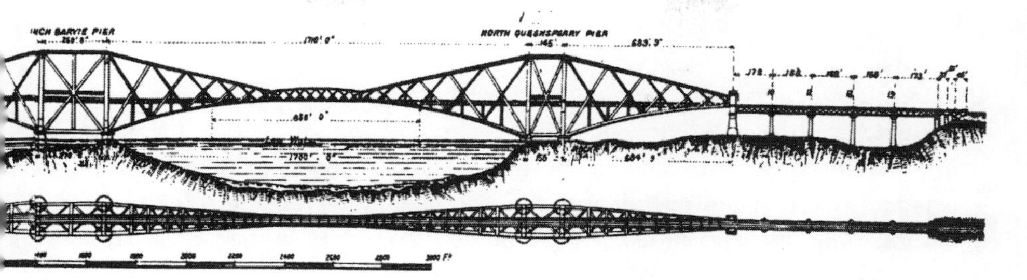

example, the complicated pattern of tensile forces acting on a telegraph line or a child's Meccano construction—the assessment of its accuracy might turn out to be very different. So questions of representation must be settled before judgments of accuracy or faithfulness are formulated. Otherwise, there is no standard against which to draw such judgments. And this shows that the two questions are distinct and moreover conceptually ordered.[11]

The distinction can be accommodated in terms of the definitions of *constituent* and *means* as follows. There are many properties of the graph in which a user can take interest when applying it to find out useful information about the bridge. Even among the geometric properties, we might be interested in the geometric relation between the lines and in using them as a guide regarding the relation between the steel cantilevers, for which the graph presented above is quite accurate. Or we might investigate average distance from the railway to the water surface, for which the graph is less accurate for any particular time and day of the year. Finally, we could gain an interest in the shape of the Forth's basin, for which the graph is even more grossly inaccurate. Then there are nongeometric properties. The shades of gray of the different parts of the graph are by and large inaccurate indicators of the color of the different parts of the bridge, the texture of the paper on which the graph is printed is a very unreliable indicator of the strength of the bridge, etc. All these comparisons are means, each appropriate in its context of inquiry. Some are clearly more effective than others in generating accurate representations, but they all provide some benchmark for accuracy judgments. Moreover, they are all in principle consistent with the claim that the graph printed on the paper represents the bridge. Whatever constituent R is the truth maker of this claim, at best one of the means R' can be identical to it. Hence, means and constituent differ from each other, and the means suffice for accuracy judgments in general. What is the point, then, in practice, of postulating any constituent at all?

Philosophical Accounts of Scientific Representation

Let us now focus on some accounts of the constituent of representation. It is useful to begin by dividing such accounts into different kinds. Two dichotomies will be particularly useful. First, an account of representation in science is reductive if it analytically defines the representational relation *R* in terms of something else. By contrast, it is primitivist if it claims that the representational relation, if there is any, cannot be further analyzed. On such accounts, representation is rather a primitive that can be invoked elsewhere—for example, for explanatory purposes—but its explanatory power cannot be further analyzed into a more elementary constituent.[12]

Another dichotomy divides substantive from deflationary accounts. A substantive account takes it that representation is a robust property or relation of sources and targets. In the next chapter, a few proposals will be presented to this effect. A deflationary view, by contrast, takes it instead that representation is not a robust property or a relation of sources and targets. On this view, the term *representation* just picks out some functional dependencies within a particular context of inquiry. This is the sort of view I defend in this book, particularly in chapters 6 and 7. But, to understand it well, we need to figure out more exactly what we mean by the term *deflationary*.

At the greatest level of generality, a deflationary approach to a concept is defined negatively, as the denial that any substantive analysis of that concept is possible. I will take a substantive approach to some concept *X* to consist in an analysis of *X* in terms of some property *P* or relation *R* that accounts for and explains the standard use of *X*. This is still consistent with many ways in which a deflationary account can deny that such an analysis is possible, and many different justifications and reasons to want to rule out such an analysis. It will help at this stage to set up an illuminating analogy with theories of truth, among which the debate between substantive and deflationary approaches to the nature of truth has been raging for nearly a century. It turns out that the analogy allows us to build up deflationary positions alongside three different dimensions, all of which find application to the domain of scientific representation (Suárez 2015).

As a preliminary caveat, while a substantive account of any concept will tend to be reductionist in the sense of reducing the concept to another set of more fundamental concepts that inform and explain, it need not always be so. Conversely, a deflationary account will not tend to engage in reduction, but it is not logically impossible that it should. However closely aligned in practice, the reducible/primitive distinction is in fact orthogonal to the substantive/deflationary distinction. Thus, a substantive account of *X* will be primitivist if it claims that *X* is a fundamental explanatory primitive that cannot be further analyzed. None of the accounts of representation canvassed here are so, and there are to my knowledge no accounts of truth that are both substantive and primitivist in this sense either. Most substantive accounts are, by contrast, committed to analyses of the concepts that break them down into the component notions, where robust explanatory power is to be found.[13] I shall refer to such accounts as *substantive reductionist* or *reductive naturalist* since they do away with all the pragmatic elements beyond the reductive basis of the relation of representation.

Conversely, most deflationary accounts are nonreductionist in the sense that they tend to claim that the concept in question cannot be analyzed or broken down into any further components. However, this is not out of any logical necessity. It would be possible to argue consistently of some concept *X* that it fails to be explanatory or robust yet is analyzable away in terms of other similarly deflationary concepts. Some approaches to truth in fact adopt this deflationary line, and, as we shall see, there is a similar deflationary avenue to approaches to scientific representation. However, the payoff of any reduction for the deflationist is slim since there is no particularly explanatory advantage to carrying it out. Hence, most deflationary approaches are nonreductive.

Representation and Truth: A Pertinent Analogy

The literature on philosophical theories of truth best exhibits all these options and distinctions. It thus provides an excellent template and benchmark for the distinction between deflationary and substantive accounts more generally. Substantive approaches in this area include correspondence theories of truth, coherence theories, and "utility" theories of truth. Conversely, there are at least three distinct deflationary approaches that one can take: the redundancy theory, abstract minimalism, and use-based accounts. A brief and cursory exploration gives us a good sense of how substantive and deflationary accounts of representation can be similar and how they can differ. What deflationary accounts of truth have in common is a minimal commitment to the disquotational schema with respect to some property *T* (truth or a surrogate) as follows:[14]

The disquotational schema: "*P*" is *T* if and only if *P*.

The disquotational schema is no more than a platitude at first sight. It simply asserts of any sentence or proposition "*P*" that it has the property *T* if and only if the content *P* that "*P*" expresses holds.[15] A substantive approach would add some further conditions, which can include properties of the proposition "*P*" or relations between "*P*" and other entities, such as states of affairs, further propositions, and so on. Thus, the correspondence theory adds the idea that truth is a correspondence or matching of the facts and supposes this to be a substantive metaphysical relation between two relata, namely, the proposition "*P*" and some state of affairs or fact. The coherence theory adds the idea that truth is coherence with the rest of some agent's beliefs (hence, truth is relative to some perhaps ideal or normative agent) and thus understands coherence as a logical relation between "*P*" and other propositions (so the relata need not include facts or states of affairs). Finally, the utility theory—sometimes called the *pragmatist* theory—takes truth to consist in the property of cognitive utility in accordance with the maxim often ascribed to William James: "To be true is to be useful to believe."[16] This entails that *P* has the *T* property by virtue of some relation to a utility function for an agent or agents (again, perhaps a normatively ideal type of agent).

In each instance of a substantive theory of truth, the further conditions stipulated in addition to the disquotational schema set the norms that regulate the proper use of the *T* predicate, and they explain the various uses of truth in practice. Thus, disputes about whether a predicate is true or false are, on the correspondence theory, disputes about the match between propositions and states of affairs; and they can be ultimately resolved in practice only by attending to such matches. For a coherence theory, such disputes concern rather the consistency with background beliefs or further sets of propositions, and those beliefs and propositions must be made explicit and the consistency checked for proper resolution.[17] Finally, a defender of the utility theory of truth would take it that disputes regarding the truth or falsity of any predicate must be resolved by attending to the relation those propositions have to utility calculations on the part of agents—in other words, how useful it is for them to believe in the truth or falsity of "*P*."

Despite their differences, these substantive accounts have in common the ambition to lay down the conditions that must obtain in the world for any proposition "*P*" to be true (whether these further conditions relate "*P*" to states of affairs, further propositions, or utility calculations). In other words, these theories provide—each in its own terms—accounts of the properties *true* and *false* that can be used in explaining why certain propositions can be said to have those properties and others to lack them. Each theory lays down conditions that define the truth concept and moreover explain its use—or, at any rate, the appropriate norms for its use. Deflationary accounts, by contrast, give up on such lofty ambitions, which is tantamount to accepting that the *T* predicate (truth) has no substantive or essential nature, at least not one that can be defined and thus explain the use of the predicate in practice. There are two different ways to deny a substantive nature to truth and reject an explanatory definition of the concept. In the first instance, one can simply deny that an analysis is possible. In the second, one can accept that an analysis is possible while denying that it is the kind of analysis that will shed explanatory light on our use of the concept (Suárez 2015, 39; Suárez and Solé 2006).[18]

The demand for a definition of the truth predicate is rejected altogether in so-called redundancy (Ramsey and Moore 1927, 153–70 [Ramsey's contribution]) or no-theory (Ayer 1936, 117) theories of truth. On these accounts, the predicate *true* is a tool—for disquotation, generalization, etc.—that does not deserve any analysis beyond whatever concerns its use in practice as a "vehicle of semantic descent." There is no point looking for necessary and sufficient conditions for *T* since the *T* predicate is not a genuine concept and possesses no such conditions. This view generalizes to any putative concept *X* that lacks an analysis in terms of necessary and sufficient conditions but that we might find useful in practice as a tool in carrying out discursive or inferential activities. For such concepts, the most that we can aim for is an account of the norms that govern their use in practice. But this account cannot constitute or provide an explanation of those concepts' uses. It merely summarizes the patterns of their use.

The alternative is an account of truth that accepts that the predicate nominally captures some property and can therefore be defined but denies that this is in any way a substantive, explanatory property. In other words, the property that the predicate *T* captures does not go beyond general platitudes concerning its use in practice, and it therefore cannot explain that use (in the sense that a platitude about some practice does not explain why the practice takes place as it does but at best summarizes that use). In Horwich's (1980) Wittgensteinian account, such explanations for the use of the *T* predicate are missing, and to demand them is to impose unfair requirements on the concept that cannot be met. In Wright's (1992) pluralistic account, the explanations will be given by additional constraints that govern the use of the *T* predicate in each domain of discourse but differ too greatly to be unified in any way. The additional norms that regulate the use of the *T* predicate in satire, for instance, are different from those that regulate it in reporting, those that regulate it in ordinary conversational contexts differ from those that operate in scientific contexts, and so on. There is a Shakespearean "multitudinous sea" of such concrete norms of application, and they cannot be subsumed under the abstract nominal definition of the *T* predicate provided by the platitudes. In other words, the platitudes are necessary but not sufficient conditions for the use of the *T* predicate in practice in any context. They amount to the most general but nominal and explanatorily empty description that we can offer for the use of the truth concept.

Let me summarize the application of the contents of this section to the object of our discussion, namely, theories or accounts of scientific representation. These accounts can be classified along the lines of two different dichotomies: reductive or primitivist and substantive or deflationary. These two dichotomies are in principle orthogonal to each other. That is, primitivism is logically compatible with a substantive view of representation, but it is also compatible with a deflationary view. The situation is similar for reductionism. There can be reductionist substantive views, which take it that representation reduces to substantive notions, but there can also be reductive deflationary views, which take representation to reduce to nonsubstantive notions. However, logical coherence on its own is rarely a persuasive argument of plausibility. It does stand to reason that substantive theories will be more naturally reductive and that deflationary views will tend toward primitivism. After all, the reduction of one notion to another is typically motivated by an attempt to figure out the real robust properties underlying that notion.[19] A deflationary approach will, by contrast, normally eschew reduction. Although a reduction of some notion *X* to a further nonsubstantive notion *Y* is in principle a possibility, the cognitive gain seems small.[20]

Substantive Conceptions Exposed

When it comes to representation, the two main reductionist accounts available are indeed also substantive. One claims to reduce representation to the relation of isomorphism, the other to that of similarity. I shall refer to them as the *[sim]* and *[iso]* theories of representation and define them below.[21] In each case, the reduction is carried out by means of necessary and sufficient conditions, and the notions to which they are reduced are substantive. So these theories are also *reductive naturalist* for reasons already discussed. The term *theory* is apt in both cases since they are both ultimately attempts at reducing the concept of representation to what is supposed to be a more fundamental and substantive notion. This section briefly explains their advantages. The next section criticizes and, ultimately, rejects them.

SIMILARITY

The first of the substantive accounts that we will consider attempts to analyze representation in terms of similarity relations, roughly as follows:

> *The similarity conception of representation [sim]*: *A* represents *B* if and only if *A* and *B* are similar.

The motivation behind [sim] is the thought that representation is a vague notion with imprecise boundaries but that similarity is not. The identity theory of similarity is often invoked: *A* is similar to *B* if and only if *A* and *B* share some properties.[22] When combined with this theory, the resulting version of [sim] simply asserts that "*A* represents *B*" is true if and only if *A* and *B* share properties. That may seem too facile, but [sim] is backed up by some strong intuitions. First, it must be stressed that similarity does not boil down simply to physical resemblance. Two objects resemble each other if they are similar in their visual appearance. But there can be similarities in respects other than physical appearance. From a logical point of view, similarity is rather a generalization of resemblance: every resemblance involves a kind of similarity, but not vice versa. So [sim] does not assert that resemblance is necessary and sufficient for representation, and this is a good thing since it would otherwise be easy to refute [sim]. For instance, the graph of the Forth Rail Bridge is similar in some relevant respects to the bridge itself. The relative distances and ratios between the cantilevers and the girders are preserved, as is the average distance between the bridge's three land stands relative to the distances between the beams in each stand. In other words, there are geometric similarities. But the resemblance between the graph and the actual bridge is limited, particularly given the huge differences in scale.

Thus, judgments of similarity require only a listing of properties and a set of identities across them. In most cases, this turns out to be a trivial exercise (objects can be similar in their color, texture, geometry, etc.). Hence, the reduction fulfills the requirements of a substantive account. It provides precise conditions for any source to stand in a representational relation with any target, in any context. It hence illuminates and clarifies the notion, providing exact conditions of application. And, since this is a definition in terms of necessary and sufficient conditions, it exhausts the extensive set of situations that can be said to qualify for representation in terms of the properties of the objects, systems, or processes involved.

This becomes even clearer for many analogue models, and some of the uses of the models that we studied in chapter 3 seem at first to fit the description provided by [sim]. Consider, for example, the Phillips-Newlyn machine. This can hardly be said to *resemble* its target, the British economy. But, while there are striking differences in appearance between the machine and its target, there are similarities in the rates of flow of water through the machine and the flow of money in the economy. Such similarities are in fact the very point of the model. It is by inspecting them, it can be argued, that we get a grip on the relevant features of the British economy. These similarities license any inferences to properties or qualities of the economy on the grounds of observing the effects of pressing levers and operating the different parts of the machine.

This also holds for the billiard ball model. Clearly, billiard balls do not resemble gas molecules. The differences in size and behavior are striking. Rather, as Hesse (1963/1966, 8) points out, there are undeniable similarities and dissimilarities between both systems: "When we take a collection of billiard balls in random motion as a model for a gas, we are not asserting that billiard balls are in all respects like gas particles, for billiard balls are red or white, and hard and shiny, and we are not intending to suggest that gas molecules have these properties. We are in fact saying that gas molecules are *analogous* to billiard balls, and the relation of analogy means that there are some properties of billiard balls which are not found in molecules." If we take the system of billiard balls as the *source* and the system of gas molecules as the *target*, then we can apply [sim] to stipulate that the system of billiard balls represents the system of gas molecules if and only if billiard balls and gas molecules are similar.[23]

This statement then invites the following analysis. We can list the properties of the source system as $\{P_1^s, P_2^s, \ldots, P_i^s, \ldots, P_j^s, \ldots, P_n^s\}$ and those of the target system as $\{P_1^t, P_2^t, \ldots, P_i^t, \ldots, P_j^t, \ldots, P_n^t\}$. The claim then is that some of these properties are identical: $P_1^s = P_1^t, P_2^s = P_2^t, \ldots, P_i^s = P_i^t$. Hence, $\{P_1^t, P_2^t, \ldots, P_i^t\}$ constitute the positive analogy. What about the remaining properties? On Hesse's account, some properties of the source system of billiard balls are not at all in the target system of gas molecules and constitute the negative analogy: $\{P_{i+1}^s, \ldots, P_j^s\}$ are not identical to any of the properties of gas molecules $\{P_1^t, P_2^t, \ldots, P_i^t, \ldots, P_j^t, \ldots, P_n^t\}$. The remaining properties $\{P_{j+1}^s, \ldots, P_n^s\}$ constitute the neutral analogy (since we do not know whether they are shared). Since this is an epistemic criterion, we can nonetheless suppose that all properties of billiard balls are as a matter of objective fact in either the positive or the negative analogy. They are either really shared by both balls and molecules, or they are not, we just do not know at this point. The methodology of inquiry is then driven by the felt need to figure out which among the neutral properties are really in the positive analogy and which really in the negative one. In other words, the similarities constitute the positive analogy and underpin the propriety of the model, that is, the reason why the source is appropriate for the target; while the dissimilarities configure the neutral analogy and provide the hints and suggestions for further inquiry, development, and application.[24]

Furthermore, it appears that similarities do underpin some of the sorts of inferences that it is helpful to draw from concrete physical model sources to their concrete physical model targets. Consider, in this respect, all kinds of toy models, the simplest forms of scale models that one can think of, such as springs for DNA chains, balls-and-strings systems for the planetary system, matches and sticks arrangements for chemical bonding, and so on. In all these cases, it is arguably some similarity of form—physical resemblance—that guides the inferences. In our terminology, this means that the similarities in form are the means of the representations with which we are working. And this seems undeniable in all those cases in which we have firsthand evidence of the grounds for the relevant inferences (regardless of whether they are explicitly acknowledged by the agents themselves). Thus, when it comes to inferences from the Phillips-Newlyn machine, we have reports to the effect that the similarities between the machine's stocks and flow rates and the analogous quantities for the stock and flow of money in the British economy led to strikingly novel predictions: "They all sat around gazing in some wonder at this thing [the machine] in the middle of the room. . . . Then he [Phillips] switched it on. And it worked! 'There was income dividing itself into saving and consumption. . . .' He really had created a machine which simplified the problems and arguments economists had been having for years" (Lionel Robbins quoted in Morgan 2012, 209).

Nevertheless, what seems apparent for concrete physical sources or their graphic depiction need not be the case for more complex models. In the cases of the mathematical and theoretical models that we have explored (Lotka-Volterra, the kinetic theory of gases, the equations for stellar astrophysics), there seem to be no apparent similarities. A mathematical equation is as dissimilar to any physical ecosystem, gas, or star as it could be. When scientists reason from the model sources to the model targets in these cases, they do not seem to be exploiting any given similarities. The representation is rather more conventionally mediated by the symbols and mathematical relations. And these are not on the face of it more similar to any physical system than they are to further mathematical relations. If there is any extended sense of similarity to be applied, we still need to figure it out. The next proposal that we shall consider can be understood as an attempt to do just that.

ISOMORPHISM

The second proposal (more a family of proposals) for an analysis of representation links it essentially to structural morphisms of distinct kinds. An initial rough approximation would state the following:

> *The isomorphism conception of representation [iso]*: *A* represents *B* if and only if *A* and *B* instantiate isomorphic structures.

The sources of the more mathematical models do not appear to hold similarity relations to their targets, and similarity does not seem to be the means of such representations. At least, certainly, no one reasons from the formal features of the Lotka-Volterra equations or the fundamental equations of stellar astrophysics to similar features in their targets. In these cases, the sources and the targets of the representation are too obviously dissimilar to be the relevant grounds for the sorts of inferences that the models enable. However, there may be a way of understanding the content of a mathematical model that enables some sort of assessment of similarities at the level of structure.

As a first approximation (somewhat of an oversimplification, as we shall soon see), two structures $S = \langle D, R_i \rangle$ and $S' = \langle D', R'_i \rangle$ are isomorphic if there is a one-to-one mapping from the domain D of S into the domain D' of S' that preserves all the relations $\{R_i\}$, $\{R'_i\}$ defined on either. For example, two distinct and different objects can nonetheless share their structure (i.e., they are "structurally identical" despite differing in their materials, outlook, semblance, construction, and a bunch of other properties) if their parts are "structured" in the same way. Identity is a limiting case of similarity (in fact a case of maximal similarity since no object can share more properties with any other object than itself). There is then a sense in which these two objects appear to be "structurally similar." So perhaps, at least in the case of mathematical models, we merely need to consider this limiting sense of similarity, structural identity, and we can go on to assert [iso] as the correct theory of scientific representation. Would this work? Does [iso] provide a suitable understanding of the representational relation between mathematical models and their targets? And, if it does, can we perhaps find a way to extend this theory to cover all the other cases of similarity?

Unfortunately, attempts to provide greater precision show that the name *structural identity* is dubious at best, first because there are weaker relations of homomorphism, partial isomorphism, epimorphism, and so on that are defined through the correspondingly weaker forms of morphism, and second because on closer inspection the full formal definition does not deserve the name except as a misnomer. Let me explain. The starting point in any case is a critical observation regarding the nature of isomorphism. The term is sometimes misemployed by association with biological morphology, where two organisms having identical morphology can be understood to imply a degree of isomorphism. But such a meaning is at best metaphoric or derivative (Thompson 1917/1992). The only precise meaning of isomorphism is mathematical and, as we saw, involves a mathematical mapping or function between extensional structures that preserves the relations defined over each of the domains.

The mathematical definition has several instructive consequences. First, it makes clear that physical concrete objects, processes, abstract or fictional entities, or their intensional properties cannot hold isomorphism relations. For only mathematical structures do, and neither of these objects is a mathematical structure. Any theory of representation that asserted that *A* represents *B* if and only if *A* and *B* are isomorphic would be restricting scientific representation to mathematical structures in the very dramatic sense that only mathematical structures could play the roles of sources and targets in any representation. Yet most models that we have so far studied throughout the history of the emergent modeling attitude involve sources or targets that are patently not simply mathematical structures. So such a theory would be mistaken from the word go, and it would shed no light on scientific modeling (other than very specific applications of formal model theory, such as perhaps in group theory). Such a theory does not seem plausible in general for scientific models, even if some scholars have come close to embracing it, as we shall see in the next few sections.

The formulation of [iso] offered above is chosen carefully so as not to restrict the reach of a structural account of representation unduly. It aims to give an isomorphism conception its best chance as a full theory of scientific representation. According to [iso], any two objects that stand in a representational relation "instantiate" isomorphic structures.[25] The definition of [iso] has thus the virtue of making sense of scientific representation across the board since *A*, *B* can be any two objects playing the role of source and target in a representation. The claim that *A* represents *B* is then shorthand for the more complete and correct statement that *A* and *B* instantiate isomorphic structures. In what follows, *A* will denote the source and S_A the structure that *A* instantiates, and *B* will denote the target and S_B the structure that *B* instantiates. Isomorphism is then the demand that there exists a one-to-one function that maps all the elements in the domain of the structure S_A onto the elements in the domain of the structure S_B, and vice versa, while preserving the relations defined in each structure. This requires for a start that S_A and S_B have the same cardinality (since it is a one-to-one mapping), and it suggests that *A* and *B* in their natural interpretation have the same cardinality of constituent parts.

More precisely, suppose that $A = \langle D, P_j^n \rangle$ and $B = \langle E, T_j^n \rangle$, where *D*, *E* are the domains of objects in each structure and P_j^n and T_j^n are the *n*-place relations defined in the structure. *A* and *B* are isomorphic if and only if there is a one-to-one mapping onto $f\colon D \rightarrow E$ such that for any *n*-tuple $(x_1,\ldots,x_n) \in D$: $P_j^n[x_1,\ldots,x_n]$ only if $T_j^n[f(x_1),\ldots,f(x_n)]$ and for any *n*-tuple $(y_1,\ldots,y_n) \in E$: $T_j^n[y_1,\ldots,y_n]$ only if $P_j^n[f^{-1}(y_1),\ldots,f^{-1}(y_n)]$. In other words, an isomorphism is a relation preserving mapping between the domains of two extensional structures, and its existence proves that the *relational framework* of the structures is the same. We can now see how the claim that isomorphism preserves or amounts to structural identity is misleading. In scientific modeling, *A* and *B* are distinct, as are the structures S_A and S_B that they instantiate, because the elements in the domains *D* and *D′* are parts of *A* and *B*, respectively, and hence also necessarily distinct. (The only case where this is not true is that of genuine identity where $A = B$, which is, as we shall see, an interesting limiting case of isomorphism but very rarely picks out any genuine representational relation at all.) It is rather the "superstructure" of the logical properties of each set of the relations in isomorphic structures that is shared between *S* and *S′*. For that reason, I prefer to use the phrase *identity of relational frameworks* (Suárez 2003, 228).

For suppose the structures S_A and S_B are isomorphic. Then, while they do not share their domains (which are different sets of different elements *D* and *D′*) or their relations (which are nothing but *n*-tuples of elements in their respective, different domains), they do share the overall architecture of their relations. A reflexive relation over some subset is carried over to another reflexive relation over the corresponding subset, a symmetrical relation over a set of elements is carried over to another symmetrical relation over the set of corresponding elements, etc. Hence, two objects that instantiate isomorphic structures (say, object *A* instantiates S_A, and object *B* instantiates S_B, where S_A and S_B are isomorphic, in agreement with the definition) are ipso facto structurally similar because they share a relational framework.

We can now apply [iso] to all the models discussed in this book so far. Here is a brief list with an indication as to how the application would roughly go in each case (no detail is needed at this stage). In the case of simple scale models and toy models, where resizing is considerable, the instantiated structures are the relations between geometric parts of the objects. If the arrangement is appropriate (such as in perfect replicas), the relative proportions should be preserved in spite of the rescaling, so there should be isomorphism between the geometric ratios even though most other properties (such as the materials, resistance, tension, etc.) will not exhibit any degree of relevant isomorphism. For the Forth Rail Bridge, similarly, the point is to focus on the sort of geometric structures that are faithfully reproduced in graphs such as that in figure 3.4 above, ignoring all other properties. In analogue models such as the Phillips-Newlyn machine (fig. 3.1), we must concentrate on the relative properties of flows and stocks, ignoring everything else. And, for billiard balls, the only meaningful isomorphism applies to the properties in the positive analogy

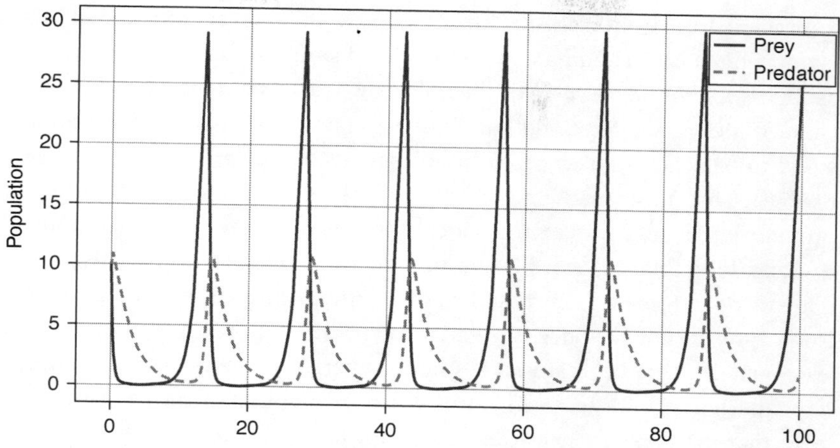

FIGURE 4.2. The Lotka-Volterra equation in a phase space diagram illustrating the oscillatory nature of the populations. CC-BY-SA.

that gas molecules share with billiard balls, namely, presumably, collision impact and collision elasticity.[26]

Finally, a mathematical equation is not per se a mathematical structure (except in the way of the merely syntactic string of symbols that compose it, which, as has already been noted, is irrelevant to its representational function). Nonetheless, it is relatively easy to see how to reformulate a differential equation in time as a dynamic structure in phase space. Every time-differential equation describes a trajectory in time for a particular quantity or quantities of interest. (Think of velocity, which is the time differential of position and describes the trajectory of a point in the very simple phase space of three-dimensional physical space.) So, for example, the dynamics of populations in a Lotka-Volterra model are relatively easy to capture in a phase space diagram, such as the one in figure 4.2. This shows how the numbers in the prey and predator populations oscillate because of their interactions. Similar phase space structures are instantiated by the relevant properties and qualities of stellar interiors in stellar astrophysics models.

This all puts an isomorphism theory of representation in the best possible light, and it gives [iso] a good fighting chance. However, I shall argue in the next section that no substantive theory of representation will ultimately prosper. Moreover, if anything, the prospects for [iso] are more dismal even than those for [sim]. And another thing that follows from the formal definition is that [sim] enjoys a certain advantage over [iso] on account of what we have found here to be the necessary requirements for the application of a mathematical isomorphism. These requirements led us to conclude that isomorphism can be understood as a kind of similarity in what I referred to as the *relational frameworks* of the objects whose instantiated structures are isomorphic. By contrast, neither similarity nor resemblance can be understood to be a case of isomorphism. Judgments of similarity apply to all sets of objects, including mathematical structures. Yet judgments of isomorphism properly apply only to mathematical structures. More generally, similarity (and resemblance) can be unproblematically applied to both response-dependent and intensionally defined properties, while this is not the case for isomorphism. For instance, similarity judgments can be applied (and often are applied) to perceptual and aesthetic experiences (as when we compare our experiences of taste, touch, beauty, elegance), but it is unclear, to say the least, that these experiences can be said to instantiate any structures in any meaningful way.

Chapter Summary

This chapter inaugurates the theoretical part of the book, and it is arguably the most analytic one in character. It begins by introducing a distinction between an analytic inquiry into the nature of representation and a practical inquiry into its practice. While the book unambiguously sides with the latter, attempting to carry out a practical inquiry into the variety of representational practices in science, this chapter takes an analytic approach seriously too as a precondition for meaningful philosophical work in the area. Several important conceptual and terminological distinctions are also introduced between substantive and deflationary accounts, on the one hand, and reductive and primitivist accounts, on the other. It is argued that a substantive account will naturally tend to be reductive and that a deflationary account will tend toward primitivism. Two substantive reductionist conceptions of representation attempting to reduce it to either similarity or isomorphism are explored. It is argued that they prima facie describe well some of the means typical of some scientific representations, including the much-discussed Lotka-Volterra model in population ecology.

5

Against Substance

So far, we have found some reason to accept that similarity relations are sometimes the means (recall: at particular times, in given contexts, for some purposes) of representation in cases of concrete physical scale and analogue models. We have also found good reasons for the view that isomorphism relations are or underwrite the means of representation of at least some mathematical models (including data models and theoretical models in mathematical language). However, it is important to recall that the means of a representation for a purpose, at a given time, in a particular context, need not be the constituent of that representation (a universal relation across contexts, purposes, and times).

In this chapter, I run through five different arguments against [sim] and [iso].[1] The first argument is the simple empirical fact that neither [sim] nor [iso] can be applied to the full variety of uses of representational devices that crop up in the practice of science. Hence, an analysis of the *means* of representation in terms of just one of these conditions would be unduly restrictive and local for a *substantive* theory of representation. However, as I pointed out above, the defenders of [iso] and [sim] have an easy retreat. They can argue that [iso] and [sim] are meant as substantial theories of the *constituent*, not as the *means* of the representational relation, and that they are meant to describe the relation between *A* and *B* that must obtain for *A* to represent *B*, independently of what relations are actually employed by inquirers in drawing inferences about *B* on the basis of *A*. The retreat is perfectly honorable and legitimate, for it is in line with the pretensions of a substantive theory of representation.

However, the other four arguments show that, even in those cases where [iso] and [sim] appear to apply, the analysis they yield is incorrect. In other words, the isomorphism and similarity conceptions cannot on their own *constitute* representation. The second argument is that [iso] and [sim] lack some of the logical properties of representation. The third argument is that they do not allow for misrepresentation or inaccurate representation. The fourth argument is that [sim] and [iso] are not necessary for representation. They fail in some cases of successful representation. The fifth and final argument is that neither [iso] nor [sim] can be sufficient for representation because they leave out the essential directionality of representation.

The Argument from Variety

The first argument is essentially empirical since it is based on the observation that there is a great diversity of representational models in science, each with its own means of representation. We can state it simply as follows:

(a) *The argument from variety*: [sim], [iso] do not apply to all representational devices.

Although similarity and isomorphism are among the most common means of representation in science, neither one on its own covers even nearly the whole range. Let us look at each in turn.

KIND 1 (SCALE MODELS)

Similarity is almost always the means for concrete physical representations of concrete physical objects. An engineer's toy bridge may be similar to the bridge that it represents in the proportions and weights of the different parts, the relative strengths of the materials, and the geometric shape. It is by reasoning based on these similarities that the source does its representational work. There are also important dissimilarities, such as size, that make the representation only a partially successful one, but similarity again seems to be a good guide to determining which parts are representational and which are not. By contrast, isomorphism, which is well defined only as a relation between mathematical structures, does not apply directly to the relation between two physical objects typical of this kind of model. But it does apply to some abstract structures that are instantiated by these two objects, such as their geometric shape.

However, the representational use of the toy bridge is almost always grounded on actual reasoning about its properties, along with those of the real bridge, and not on the properties of the structures instantiated by either bridge. The means of representation are not in this case captured by the [iso] conception because it misidentifies the relata, which are the physical objects themselves and not the structures instantiated. To make this point vivid, suppose, for instance, that two concrete toy bridges exemplify the same geometric structure, isomorphic to that of the larger bridge. We typically treat these two bridges as two different *means* and as distinct representations of the same object, but an isomorphism analysis of the *means* of representation does not allow us to do that. The relationship R that each toy bridge holds to the larger bridge is the same.[2]

Alternatively, consider the graph of the Forth Rail Bridge. The range and depth of the dissimilarities between the source and the target become greater in this case. A piece of paper containing the graph of a bridge is similar to the bridge it represents only with respect to the geometric shape and proportions between the different points; nothing else is interestingly similar. This restricted similarity of structure is perhaps better captured by the alternative conception [iso]. Maps, plans, and graphs are cases in which isomorphism is sometimes the means of scientific representation. So we see that the class of scale models, broadly defined, is very large and that the means vary greatly, alternating between similarities and morphisms, depending very much on each case.

KIND 2 (ANALOGUE MODELS)

Analogue models are preeminently based on the similarities between the model source and the target. We noticed this in the case of the Phillips-Newlyn machine, where it was the similarities between the rates of flow and the evolution of stocks in the machine and their corresponding flows and stocks of money in the economy that were at the heart of the inferences that were drawn and the ensuing discoveries about the economy. The fact that the properties that are shared are dynamic (i.e., the evolution of stocks and flows and their interdependencies) complicates things for the standard accounts of representation as similarity. But it does not take away from the assessment of similarities as the relevant means if what is effectively at stake is a comparison between dynamic processes.

There are many illustrious historical examples of analogue models, both concrete and abstract, symbolic and conceptual. Among the most distinguished are the variety of models of the ether created by the Maxwellians during the second half of the nineteenth century, including Maxwell's own vortex model, reviewed in chapter 2.[3] In all these cases, there are purported similarities (analogies) between the mechanical systems that are the representational sources of the models and the different properties and features of the ether.

These are all peculiar cases in that we now believe that there is no ether. Defenders of substantive notions of representation, such as [sim], may accept the conclusion that seems to follow, namely, that these models are not genuinely representational. However, I argue that the models are representational and evidently so, for nothing at all is different in the way they are used with respect to other clearly representational models. Thus, from the point of view of a practice-based conception of representation, no account of the relation of representation that entails that these models fail to be representations does justice to the way they get used in practice (more on this in later chapters). Now, are these models representational *because* they nonetheless hold similarity relations to the (imagined) ether? In other words, is similarity nonetheless the means of their representation? This is not perhaps as implausible a thing to say as it may seem at first, although note that it commits its defender to an account of similarity as an ostensive yet failing relation since there is no object for the models to be similar to.[4] Whether similarity can be so depleted of content as to be rendered an empty relation is still an open question. Thus, it is correspondingly also an open question whether similarity is indeed the means of the representation in these cases.

The billiard ball model is a conceptual analogue model that leads to a theoretical model for a gas. This case also appears to be harder for both conceptions. A system of billiard balls is not prima facie in any relevant sense similar to any state of nature. Still, we can refer to the relation of similarity or isomorphism that can obtain between two token instances of these things. If we do, we must make sure that we are referring to a similarity between the *dynamic* properties of the systems, collectively taken. A system of billiard balls is similar to a system of gas molecules only in its dynamic properties and in no properties of the entities taken individually at any one time other than their elasticity. Mutatis mutandis for isomorphism: this obtains only between the mathematical structures exemplified by the dynamics of the systems and not between the structures exemplified by the individual entities.

KIND 3 (MATHEMATICAL MODELS)

A mathematical equation, written down on a piece of paper, represents a certain physical phenomenon but is not similar to it in any relevant respect. If the equation is dynamic, one can focus on the phase space structure defined by the equation and on that structure that is best exemplified by the phenomenon: if the equation is an accurate representation of the phenomenon, isomorphism will obtain between the two, and, as noted earlier in this chapter, isomorphism is a case of similarity.

But even [iso] is problematic here. In most cases of mathematical representation, it seems far-fetched to assume that the *means* of the representation is an isomorphism. It trivially is the case that the dynamic phase space structure exemplified by a differential equation must be isomorphic to the dynamic structure exemplified by the phenomenon if the equation accurately represents the phenomenon. But, when scientists reason about a differential equation to inquire into the phenomenon it represents, they rarely include an investigation of the formal properties of these phase space structures. What they actively do is look for solutions to the equation given certain boundary conditions and then check whether some parameters of those solutions correspond to observed features of the phenomena. The isomorphism that obtains is not what they explicitly reason about, so it is not in this case the *means* of the representation.

Consider, for example, the Lotka-Volterra model. Its two equations jointly define a phase space diagram for the size of the respective populations (fig. 4.2). But, when scientists employ the model in order to predict new rates given different initial population sizes and values for the relevant parameters, $(\varepsilon_1, \gamma_1)$, $(\varepsilon_2, \gamma_2)$, they use the phase space diagrams only for illustration (Kot 2001, 107ff.; May 2001, 41ff.). They operate on the equations themselves and go about solving them for the different parameter values: $dX_1/dt = X_1(\varepsilon_1 - \gamma_1 X_2)$, and $dX_2/dt = X_2(\gamma_2 X_1 - \varepsilon_2)$, where, recall, $X_1(t)$ and $X_2(t)$ are the numbers of individuals in the prey (X_1) and predator (X_2) populations, respectively, at any given time t. So, even though an isomorphism can be set up between certain quantities in the phenomena (population sizes at different times) and certain quantities in the dynamic phase space representation, that relation is not the means employed by most scientists when applying the model. As an account of the means of representation [iso] does not fare well even in the case of the most mathematical models.

The same case can also be made for the redundancy of the relevant isomorphism in licensing the relevant model inferences in the sorts of data models explored by Patrick Suppes (1962, 2002). This is particularly clear in the light of the considerations regarding their evidential use raised by Leonelli (2016, 2019) that we reviewed in chapter 3. The fundamental difficulty here is that the targets of data models are not themselves structures but typically objects (photographs, sightings, recordings, inklings, or marks), processes (singular motions, measurement processes), or phenomena (or effects). Van Fraassen (2008) argues that there is no difference in modeling practice between a target and a very low-level structural representation of the target. However, besides involving a controversial circularity (see, e.g., Nguyen 2016b), the assumption also trades on ambiguity regarding the relata of any putative representational relation. Recall that my very pragmatic starting point is that any type of object can in principle play the role of the source or the target of a representation.

KIND 4 (THEORETICAL MODELS)

The sorts of nonmathematical yet theoretical and explanatory models that we reviewed in chapter 3 are hardly instances of means of representation via [iso]. Consider, for example, the early evolutionary explanations of species transitions that we reviewed in the previous chapters, such as the ones on offer in the disputes over the origin of whales and other mammalian aquatic species. The model postulates a certain degree of fitness to the extant environment of some organisms that causally explains reproductive success and the ensuing evolutionary dominance. Prior to the mathematical models of population genetics, these models merely appealed to the individual features of organisms as the repositories of such fitness. The nonquantitative principles of heritability, variation, and natural selection acting on the individual organisms thus did suffice for the putative explanation. In a similar vein, Galileo's thought experiment from the tower of Pisa fails to appeal to any mathematical quantities. Its refutation of the Aristotelian model is, rather, qualitative, appealing merely to the thought-experimental fact that two bodies of distinct weight fall at the same rate and that they could not do otherwise when considering their possible linkages. This conclusion is, thus, independent of any details regarding the nature of free fall and the forces active therein. It was only with Newton that such quantitative mathematical models of free fall came about.

Essentially the same point can also be made, in a different context, for the sorts of theoretical assumptions that underpin the stellar structure models reviewed in chapter 3. We found there that those four theoretical equations combined give rise to a set of predictions for the fundamental values of the observational quantities that appear in the main empirical law of stellar astrophysics, the HR law (see fig. 3.5 above). No similarity or isomorphism relations seem prima facie to be invoked, although one can imagine a description of the HR law in terms of a phase space structure with time as a parameter. In such a dynamic representation of a singular star's life cycle, the main sequence describes the time evolution of the star from early inception, plotted as a dot in the top-left-hand end of the main sequence as described in the diagram, until its old age at the lower-right-hand end of the main sequence, then followed by a sudden move as a red giant toward the upper-right-hand side and perhaps ending in the lower-left-hand side as a white dwarf. There would then be a way to lay out an isomorphism between the motion of the dot in such a dynamic HR diagram and the actual life cycle of the star, measured in terms of the fundamental values of its observational quantities, its effective surface temperature, and its luminosity.

However, this is emphatically not what astrophysicists do in working out the evolution of a given star, or its life cycle. Instead, the model is employed analytically or in computer simulations to derive plausible values for all the quantities of interest at different times in the life cycle of the star (Collins 1989; Prialnik 2000). In other words, the isomorphism that results between the structures is not the effective means of the representation. Whether the model has explanatory power can be a matter of debate (as we saw in chap. 3), but it seems an obvious, if not moot, point that whatever explanatory power it has depends entirely on the resolution of the equations and not on the phase space diagram that they can be taken to represent. One can conclude that the means of effective representation of many theoretical models—even those written in the language of mathematics—rarely are isomorphism or similarity relations.

Note that I am not saying that other mathematical models might not lend themselves to such uses where isomorphism is the apparent means (or *conductor*, in Maxwell's colorful language, reviewed in chap. 2) for the surrogative reasoning that takes modelers from the features of the source to those of the target. Some examples that come to mind are pendular movements or, more generally, dynamic systems in the Hamiltonian formalism, which we can model effectively by means of phase space diagrams. In those cases, the means are ostensibly isomorphism relations, but these cases do not seem to generalize even across the mathematical sciences, never mind to the nonmathematical theoretical models that are more common in some social and life sciences. However, as I already pointed out, the defender of [sim] or [iso] is after an account not of the means but of the constituent of scientific representations. And it is perfectly possible that the effective means of many representations belie its actual constituent, the genuine and only relation by virtue of which any model source represents its target. So there remains a tantalizing possibility that [sim] or [iso], as described, is the correct substantive theory of scientific representation in general, even if each separately fails to account for all the means. The remaining four arguments are directed explicitly against such substantive accounts of scientific representation.

The Logical Argument

The second argument has a logical rather than an empirical character, and it relies on a conceptual analysis of the relation of representation.

(b) *The logical argument*: [sim], [iso] do not possess the logical properties of representation

A substantive theory must make clear that scientific representation is indeed a type of representation, that is, that it shares the properties of ordinary representation. It is widely accepted that representation in general is an essentially nonsymmetrical phenomenon, that a source is not represented by a target merely by virtue of the source representing the target. There may be contexts in which symmetry obtains, but even there it is not automatic. Merely because a painting portrays a person in some context, it does not follow that the person stands for the painting in that context. Merely because an equation represents a phenomenon, the phenomenon cannot be said to stand for the equation. And so on. Representation is also minimally nontransitive and nonreflexive. A theory of scientific representation must do justice to these features.

Nelson Goodman (1968/1976, 3–10) put these logical properties of representation to use against resemblance theories, and his argument carries over against [sim] and [iso]. I shall pursue here an illuminating analogy with painting, a particularly apt analogy in this context as [iso] and [sim] both assume that scientific representation is essentially an object-to-object relation rather than a word-to-object or mental state–to–object relation.[5] That is, both [iso] and [sim] assume that both relata of the relation of representation are concrete entities endowed with properties.[6]

The argument is, however, independent of the analogy and is in no way exhausted by it. The purpose of the analogy is to call attention to the logical properties of object-to-object representation in general, thus suggesting that scientific representation must display these properties too. It could, however, turn out that scientific representation is not a kind of object-to-object representation, or not entirely so. Thus, the analogy fails, but so do the [iso] and [sim] conceptions that I criticize here. To defeat the argument, one would have to show that [iso] and [sim] have the logical properties of representation in general, and this is patently not the case.

Representation is nonreflexive. Diego Velázquez's portraits of Pope Innocent X (fig. 5.1 reproduces one of them) represent the pope as he was posing for Velázquez, but they do not represent themselves. Admittedly, some of the reproductions that came out of the Velázquez studio in Seville were inspired by his original portrait, which hangs in the Doria Pamphilj museum in Rome. What's more, Velázquez astounded the world with the striking built-in reflexivity of *Las meninas*, which represents the act of its being painted (as well as the Spanish royal family). A creative obsession with representing the very elusive act of representation has been part of art since at least the

FIGURE 5.1. After Diego Velázquez's *Pope Innocent X* (1650). Isabella Stewart Gardner Museum, Boston.

Quattrocento. But, even in these cases, the representation typically adds to the object and subtracts from it: the source and the target are not identical.

It would be equally wrong to claim that the pope represents the painting. We can put aside issues about whether a nonexisting object can be said to represent. Even when the pope was sitting down posing for Velázquez, it would not have been right to claim that he represented the painting. Representation is nonsymmetrical (if not asymmetrical) since it is often one-way. Apparent cases of symmetry, such as some of Escher's drawings or two mirrors placed opposite one another and reflecting each other, turn out to be cases in which there is a distinct representational relationship going each way.

The late painter Francis Bacon was obsessed with Velázquez's portrait, and he produced many variations of his own, all of them intending to represent the Velázquez canvas. The Bacon portraits (see, e.g., fig. 5.2) represent

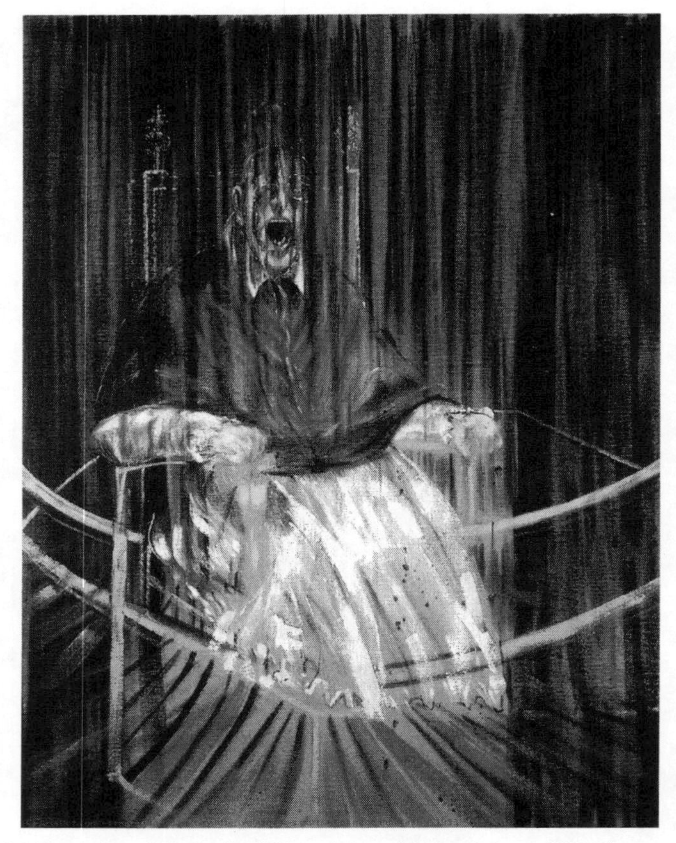

FIGURE 5.2. Francis Bacon's *Study after Velázquez's Portrait of Pope Innocent X* (1953).

not the pope but the Velázquez canvas. His variations allow us to infer much about Velázquez's obsessions and skills, his use of colors, and so on. But nothing can be reliably inferred from the Bacon paintings about the pope. Alternatively, suppose that a tourist takes a photograph of the room where *Las meninas* hangs in the Prado Museum. That photograph represents the canvas, not the Spanish royal family. Representation is nontransitive (if not always intransitive) since apparent cases of transitivity turn out on inspection to involve different representational relations between *A* and *B* and between *B* and *C*.

However, similarity is reflexive and symmetrical, and isomorphism is reflexive, symmetrical, and transitive. A glass of water is similar to itself and similar to any other glass that is similar to it. Mutatis mutandis for isomorphism. A geometric structure (a square) is isomorphic to itself, and always isomorphic to any other structure (another square of perhaps a different size) that is isomorphic to it, and isomorphic to any structure that is isomorphic to a structure that is isomorphic to it (an even larger square). Neither [sim] nor [iso] can account for the constituent of representation.

The Argument from Misrepresentation

The central argument against substantive theories of representation is also the most involved since it appeals to the myriad ways in which models can distort, idealize, abstract, ignore, pretend, fictionalize, or openly lie about their targets as well as equivocate with respect to what those targets might in fact be.

(c) *The argument from misrepresentation*: [sim], [iso] do not make room for the ubiquitous phenomena of mistargeting and/or inaccuracy.

Misrepresentation is a ubiquitous phenomenon in the representation of ordinary life. It comes in two varieties. There is first the phenomenon of mistaking the target of a representation, or as I call it *mistargeting*: often we mistakenly suppose the target of a representation to be something that it is not. Suppose that a friend of mine has disguised himself to look roughly similar to Pope Innocent X in the relevant respects. In seeing the Velázquez canvas for the first time, I am struck by this resemblance, and, ignoring the history and true target of the representation, I go on to suppose that the Velázquez represents my friend. This is a clear case of misrepresentation, but there is no failure of similarity to explain it. Indeed, misrepresentation by accidental similarity would be impossible if [sim] were true precisely because similarity would then always warrant representation. The same argument goes mutatis mutandis for isomorphism, and it is an argument that can easily be transferred to cases of scientific mistargeting. Consider the case of the quantum state diffusion equation:[7]

$$|d\psi\rangle = -i/h H|\psi\rangle dt + \Sigma_j(\langle L_j^*\rangle L_j - ½ L_j^* L_j - ½\langle L_j^*\rangle\langle L_j\rangle)|\psi\rangle dt + \Sigma_j(L_j - \langle L_j\rangle)|\psi\rangle dt.$$

This equation represents the evolution of the quantum vector state of a particle subject to a diffusion process. (The first term is just the usual Hamiltonian in the linear Schrödinger equation. The other two terms account for random diffusion and interaction with a larger environment.) A mathematician who knew nothing about quantum mechanics would be able to solve this equation for some boundary conditions. By accident, the motion described might correspond to a particular classical particle's Brownian motion. This accidental fact does not on its own turn the equation into a representation of the particle's motion, however, because the essential directionality of representation is missing.

It is easy to make the same point for the Lotka-Volterra model. The equations can be taken to represent any phenomenon in which two quantities oscillate roughly as in figure 4.2 above. But this may be an entirely spurious or accidental correlation between two variables that have nothing particularly to do with each other or at least are not causally related in the way that prey and predators are in the Adriatic Sea. It is questionable that any such correlation is thereby represented by the Lotka-Volterra model, certainly not given the intended interpretation and common usage of the equations in the model. Nothing like the Volterra property would emerge from a spurious correlation since there need be no explanatory fact underlying the correlation, and, hence, the model would be predictively inane. The point has been argued persuasively in the general case by Putnam (1981, chap. 1) and in the scientific case by van Fraassen (1994, 170) and need not be rehearsed in detail here. It has long been noted within the philosophy of art too. Thus, Richard Wollheim writes (1987, 54):

> The connection between seeing-in and representation was noted by theorists of representation both in antiquity and in the Renaissance. Yet almost to a person these thinkers got the connection the wrong way around: they treated seeing-in as—logically and historically—posterior to representation. For they held that, whenever we see, say, a horse in a cloud, or in a stained wall, or in a shadow, this is because there is a representation of a horse already there—a representation made, of course, by no human hand. These representations, which would be the work of the gods or the result of chance, wait for persons of exceptional sensitivity to discern them, and then they deliver themselves up.

For Wollheim, like Putnam, van Fraassen, and I myself, the skill and activity required to bring about the experience of seeing-in (the appreciation by an agent of the representational quality of a source) are not consequences of the relation of representation but conditions for it. This is a fundamental point about representation in general, one so often misunderstood that it bears returning to repeatedly. For suppose that Putnam, van Fraassen, and Wollheim were all wrong on this point. A mathematician's discovery of a certain new mathematical structure (defined perhaps by a new equation) not known to be isomorphic to a particular phenomenon would amount to the discovery of a representation of the phenomenon, independently of whether the mathematical structure is ever actually applied by anyone to the phenomenon. The history textbooks would have to be rewritten so that it is Riemann, not Einstein, who gets credit for first providing a mathematical representation of space-time.

The second form of misrepresentation is the even more ubiquitous—perhaps universal—phenomenon of inaccuracy. Most representations are to some degree inaccurate in some respects. And [iso] cannot account for inaccurate representation at all. For, on this conception, either a model is a representation of and thus isomorphic to its target, or it is not a representation at all. But [sim] requires that the target and the source must share some, although not necessarily all, of their properties. Hence, it can account for the type of inaccuracy that arises in an incomplete or idealized representation of a phenomenon, that is, one that leaves out particularly salient features, such as the highly idealized representation of classical motion on a frictionless plane. But this will not always help us understand inaccurate representation in science, where the inaccuracy is much more often quantitative than qualitative. For example, without general relativistic corrections, Newtonian mechanics can at best provide only an approximately correct representation of the solar system. Some motions would not be quite as predicted by the theory even if all the features of the solar system were to be accounted for. The interesting question is not what properties fail to obtain but rather how great the divergence is between the predictions and the observations regarding the values of the properties that do obtain. Unfortunately, [sim] offers no guide on this issue.

The Nonnecessity and Nonsufficiency Arguments

These arguments turn explicitly on the set of conditions that define the constituent of representation. It is worth inspecting them separately.

First:

(d) *The nonnecessity argument*: [sim], [iso] are not necessary for representation. Representation can obtain even if [sim], [iso] fail.

It is trivial that any object is in principle similar to any other object. In fact, the point is often made that, if all logically possible properties are permitted, then any object is similar to any other object in an infinite number of ways; that is, there is an infinite number of properties that we can concoct that will be shared between the objects ("being on this side of the moon," "being neither black nor blue," etc.). If so, similarity would be necessary for representation but in a completely trivial way.[8] For it would be a necessary condition not only on representation but also on nonrepresentation.

The defender of similarity might retort that it is not fair to include those logically possible shared properties that have nothing whatever to do with the representation itself (such as "being on this side of the moon"). We need to restrict ourselves here to only those properties or aspects of the source and the target that are relevant to the representational relation: *A* represents *B* if and only if *A* and *B* are similar *in the relevant respects*. It is not the case that any source is in principle trivially similar in the relevant aspects to what it represents. Suppose that I am interested in representing in a painting the color of the ocean in front of me. I might represent the ocean by painting some blue and green stripes on a piece of paper. Representation obtains in this case if the colors on my paper are similar to those of the ocean, and it fails otherwise: that is the only relevant property. Any other logically possible similarity, such as "being on this side of the moon," is irrelevant to this representation.

There are two important objections to this move. First, what is the criterion of relevance invoked? This criterion must presumably link relevance to the representational relation itself, for otherwise there would be no reason to expect *relevant similarity* to be necessary for representation. The shared properties that are relevant are precisely those that pertain to the representation. So we obtain that *A* represents *B* if and only if *A* and *B* are similar in those respects in which *A* represents *B*. While this might be illuminating about the actual use of similarity in practice, it is circular as an analysis of representation.

The defender of similarity might retort that relevance is a fully intuitive notion of straightforward application in practice, a primitive notion in no need of further analysis. That this is not so is made most vivid in the analogy with art, and, to illustrate this point, I invoke *Guernica*, the well-known painting by Picasso (see fig. 8.2 below). There are similarities between parts of this painting and many objects, such as a bull, a crying mother holding a baby, an enormous eye. They all seem undeniably relevant to the representational content of the painting if any similarities are, yet none are good guides to the actual targets of the representation. And there are at least two well-documented targets. Picasso was interested in representing the first ever carpet-bombing of an entire civilian population: the bombing, under Franco's consent if not direct order (Preston 1993, chap. 9), of the Basque town of Guernica by Hitler's Condor Legion and Mussolini's Aviazione Nazionale in 1937. In addition, *Guernica* represents the threat of rising fascism in Europe, which is the reason why it was hugely effective in bringing world attention to the Spanish Republic's cause. This is all historically well documented.[9] The point is that none of the targets of *Guernica* can be easily placed in the relevant similarity relation with the painting, and mutatis mutandis for isomorphism.[10]

The case in science is not significantly different. An equation—that is, the actual physical signs on paper—is as dissimilar as it could be from the phenomenon that it represents. Mutatis mutandis for isomorphism, as we have already seen in the case of inaccurate representation. We are perfectly happy with the claim that Newtonian mechanics provides a representation of the solar system even if Newtonian mechanics, which has no general relativistic corrections, is empirically inadequate and nonisomorphic to the phenomena of planetary motion. A possible retort on behalf of [iso] and [sim] is that we should concentrate entirely on the subset of properties, or the substructure, that corresponds to those motions that are predicted correctly. But in cases of quantitative inaccuracy this normally will not help. Newtonian mechanics arguably does not describe any actual planetary motion in a quantitatively accurate way.

Second:

(e) *The nonsufficiency argument*: [sim], [iso] are not sufficient for representation. Representation may fail to obtain even if [sim], [iso] hold.

The previous four arguments already point to a feature of representation that is not captured by the [iso] or [sim] analyses: the essential directionality of representation. This was perhaps most apparent in the discussion of the *argument from misrepresentation*. The object that constitutes the source of a model has no directionality per se, but in a genuine representational relation the source *leads* to the target. Neither similarity nor isomorphism can capture this capacity of the representational relation to take an informed and competent inquirer from consideration of the source to the target. But it is this feature that lies at the heart of the phenomenological asymmetry of representation. Consider, for instance, two "identical" glasses. They share all their (monadic) properties, and they are hence as similar as they could be. But neither of them leads to the other unless they are in a representational relation, and then only that which is the source will have the capacity to lead to the target. Or consider the trajectory in phase space described by the state vector of a quantum particle. Unbeknownst to us, this trajectory may well be isomorphic to the motion in physical space of a real classical particle. But, unless the phase space model is intended for the particle's motion, the representational relation will fail to obtain. Hence, neither similarity nor isomorphism is sufficient for representation.

There is an additional reason why isomorphism is not sufficient for scientific representation. It is related to the notions of exemplification. Goodman (1968/1976, 52ff.) provided a useful analysis of the notion of exemplification as a special class of representation. For Goodman, if *x* exemplifies *y*, then *x* denotes *y*, and vice versa; but *x* can denote *y* without exemplifying it: exemplification requires denotation both ways. My sweater exemplifies red if and only if it both denotes red and is denoted by red (i.e., the sweater is used to refer to red and *is* also red).

Now let us suppose that this analysis of exemplification goes through for structural representation. Then, whenever object *x* exemplifies structure *y*, it both represents *y* and is represented by *y*. It follows then that, for an object *A* to represent some object *B* by means of [iso], the structure exemplified by *A* must be isomorphic to the structure exemplified by *B*. But that just means that, if the supposition is right, for *A* to represent *B* there must be a structure that represents *A* isomorphic to a structure that represents *B*. And we will now want to ask how the structures represent the objects in the first place.[11]

Now, recall the discussion of the quantum state diffusion equation in the *argument from variety*. We established then that isomorphism is not the means of the representation in this case. Could it be the *constituent* of the representational relation? Perhaps mathematical representation by means of differential equations is precisely the type of representation for which the means/constituent distinction is appropriate. But [iso] stipulates that to represent a phenomenon by means of an equation we need to describe the structure instantiated by the phenomenon independently. For remember that isomorphism is defined as a relation between mathematical structures. However, as we have seen, for the purpose of establishing a theory of the constituent of representation, this will be circular: *A* represents *B* if and only if the structure that represents *A* is isomorphic to the structure that represents *B*.

For instance, the quantum state diffusion equation for a localizing particle describes a random walk motion in a phase space structure. This structure represents not the particle's motion but a representation of it, namely, the motion of the vector in Hilbert space that corresponds to the state of the particle. But representing a representation of *x* is in no way equivalent to representing *x*, and we are left with the question of how *x* is mathematically represented in the first place. So isomorphism is not in general sufficient for representation. The case of representation of a well-established physical phenomenon by means of a differential mathematical equation is, perhaps paradoxically, the hardest case for [iso] to accommodate.

Amending and Weakening Similarity and Isomorphism

A recurrent theme in these arguments, one that became explicit in the discussion of the nonsufficiency argument, is the appeal to the essential directionality of representation: a necessary condition for *A* to represent *B* is that consideration of *A* leads an informed and competent inquirer to consider *B*.

I will refer to *A*'s capacity to lead a competent and informed inquirer to consider *B* as the *representational force* of *A*. Representational forces are relational properties of sources in particular contexts of inquiry. They are determined at least in part by correct intended uses, which in turn are typically conditioned and maintained by socially enforced conventions and practices: *A* can have no representational force unless it stands in a representing relation to *B*, and it cannot stand in such a relation unless it is intended as a representation of *B* by some suitably competent and informed inquirer in the appropriate normative context of use.

Note that I am careful to refer to the essential *directionality*, not intentionality, of representation. There has been an important debate in the philosophy of mind in recent decades that assumes that a source's representational force is nothing but the intentionality of the mental state of an agent that employs the source to represent some target. This assumption is friendly to my critical analysis since intentionality has the right properties to ground the arguments that I propose against similarity and isomorphism. However, I do not need or wish to make this assumption here. At best, given the present lack of consensus about what intentionality is, this assumption would be akin to trying to explain a child's ability to ride a bicycle by appealing to his sense of balance: even if nobody doubts that there is a connection, it is not very explanatory since we do not have a clear understanding of how children develop their sense of balance.[12] Analogously, it cannot be sound methodology to invoke a difficult and obscure notion (intentionality) to explain a difficult but not particularly obscure human activity (representation). That is why, at this stage at least, I prefer a plain intended-use theory of representational force that leaves room for further specification, which may include intentionality but need not do so.[13]

Can [iso] and [sim] be made to work simply by amending them to account for this directional component? The amended versions would look as follows:

> [sim]′: *A* represents *B* if and only if (i) *A* is similar to *B* and (ii) the representational force of *A* points to *B*.
>
> [iso]′: *A* represents *B* if and only if (i) the structure exemplified by *A* is isomorphic to the structure exemplified by *B* and (ii) the representational force of *A* points to *B*.

The first thing to notice about these amended versions is that they abandon the aim of naturalizing representation. Representation can no longer be established by means of a scientific investigation of the facts of the matter, for there are elements in the relation of representation, namely, the representational forces in part 2, that essentially involve value judgments and are not reducible to facts.[14]

But, in fact, [sim]′ and [iso]′ cannot be correct. Certainly, the additional clause stipulating the correct intended use of the representation turns conditions [iso]′ and [sim]′ into sufficient conditions for representation, and the nonsufficiency argument no longer applies. Depending on how we explicate intended use, the logical argument might also lose its force. But the other arguments still apply. The nonnecessity argument is, if anything, strengthened as the necessary conditions on representation are now stronger. The argument from variety shows that neither [iso]′ nor [sim]′ can describe all the means of representation; while the misrepresentation and nonnecessity arguments show that they do not provide a substantial theory of the constituents of representation. Simply adding further conditions to [iso] or [sim] to make room for the essential directionality of representation will not help.

The problem, therefore, lies not with what [iso] and [sim] lack but with what they have an excess of. Let us then look at some attempts to weaken the conditions on representation imposed by [iso] and [sim]. These programs either continue to be actively pursued or could conceivably be pursued. So my conclusions have a correspondingly tentative and provisional character.

SIMILARITY WITHOUT IDENTITY

One assumption that was built into [sim] is the identity-based theory of similarity. This theory seems natural, gives a high level of precision to the concept, and makes it possible for us to quantify and measure degrees of similarity between objects (as ratios of properties shared). But it may be mistaken.

There is evidence within experimental work in cognitive psychology for a nonidentity-based understanding of similarity, one that emphasizes the essential role of contextual factors and agent-driven purposes in similarity judgments (Tversky and Gati 1978, 1982). In the terminology that I employ here, this turns [sim] into a nonreductive naturalist theory. Let us suppose that the similarities between two objects are not simply a matter of sharing properties but a more complex contextual matter. We do not have a very good understanding of what this relation might be, but it can be argued that on such theories of similarity there is typically no reason to expect similarity judgments to be symmetrical. The fact that *A* is similar to *B* does not ipso facto require *B* to be similar to *A*. If Tversky's account of similarity is right, the logical argument does not cut as strongly against similarity as it seemed. Yet it nonetheless applies for any theory of similarity must concede this: similarity comprises identity; identity is a limiting case of similarity. Thus, however similarity is conceived, it must be reflexive. For, if something is not similar to itself, it is not similar to anything else. Here, representation and similarity depart, certainly, for most representations patently do not represent themselves.

However, the combination of Giere's (1988) emphasis on the essentially pragmatic character of similarity judgments with a nonidentity-based understanding of similarity would undeniably bring similarity and representation closer. And, indeed, in response to earlier versions of the inferential conception, Giere (2004, 743) went on to propose that we understand representation as a four-place *activity*: "The activity of representing, if thought of as a relation at all, should have at least four places . . . with roughly the following form: '*S* uses *M* to represent *W* for purposes *P*.'" This theory would be successful to the extent that it builds the source's representational force, with all its normative import, into the relation of representation itself. Once again, this would turn it into a nonreductive naturalistic theory in the terminology of this book. The nonsufficiency argument would have no force against such a theory, and neither would the mistargeting part of the argument from misrepresentation or the nonsymmetry part of the logical argument. Yet, as we shall see, reflexivity and the nonnecessity argument remain standing blocks for any versions of this theory developed in recent years.

HOMOMORPHISM

Elizabeth Lloyd, a prominent defender of the semantic view, suggests that "in practice the relationship between theoretical and empirical model is typically weaker than isomorphism, usually a homomorphism, or sometimes even a weaker type of morphism" (Lloyd 1988/1994, 14 n. 2). The [iso] condition does indeed get weakened in a variety of ways, which solves some but not all the problems that I have raised. For instance, following the pioneering work of Krantz, Luce, Suppes, and Tversky (1971), Brent Mundy (1986) employs the notion of homomorphism and shows how to apply it to measurement theory, space-time geometry, and classical kinematics. We say that an extensional structure *A* is faithfully homomorphic to an extensional structure *B* if and only if there is a function that maps all the elements in the domain of *A* into the elements in *B*'s domain while preserving the relations defined in *A*'s structure. More precisely, suppose that *A* and *B* uniquely exemplify the structures $<D, P_j^n>$ and $<E, T_j^n>$, where *D*, *E* are the domains of objects in each structure and P_j^n and T_j^n are the *n*-place relations defined in the structure. Then *A* is faithfully homomorphic to *B* (Mundy 1986, 395) if and only if there is a mapping $f: D \rightarrow E$ such that for any *n*-tuple $(x_1, \ldots, x_n) \in D$: $P_j^n[x_1, \ldots, x_n]$ if and only if $T_j^n[f(x_1), \ldots, f(x_n)]$. The correspondingly weakened version of [iso] is:

The homomorphism conception of representation [homo]: A represents B if and only if the structure exemplified by B is homomorphic to the structure exemplified by A.

Homomorphism is, unlike isomorphism, neither one-to-one nor onto, so the cardinality of A and B can differ. This feature was notoriously used by Krantz, Luce, Suppes, and Tversky (1971) to show that [homo] rather than [iso] is appropriate for theories of measurement. The important advantage that [homo] enjoys over [iso] is then the ability to deal with partially accurate models. Parts of a source may not represent any of the aspects of the homomorphic target. So the hope is that [homo] will be able to refute the part of the argument from misrepresentation that refers to inaccurate representation and its consequences for the nonnecessity argument. The solar system may be represented only by the part of the Newtonian model that asserts the number of planets and their average proximity to the sun, without specifying their precise motions. The highly developed structural theory of measurement as homomorphism into the real number continuum allows [homo] to provide precise estimates for these numbers (Krantz, Luce, Suppes, and Tversky 1971; see also Díez 1997). It seems clear that the move to [homo] weakens the nonnecessity argument (although, interestingly, it does not dispel the force of the art analogy in that argument).

However, all the other arguments apply against [homo] too. This includes the argument from variety, the mistargeting part of the argument from misrepresentation, and the nonsufficiency argument. The logical argument is significantly weakened but not avoided. Homomorphism is not symmetrical, but it is reflexive.

PARTIAL ISOMORPHISM

Another proposal to weaken [iso] may be provided by Mikenberg, da Costa, and Chuaqui's (1986) notion of partial structure and the corresponding notion of partial isomorphism introduced by Bueno (1997). A partial structure $\langle D, R_{i1}, R_{i2}, R_{i3}\rangle$ defines for each relation R_i a set of n-tuples that satisfy R_i, a set of n-tuples that do not satisfy R_i, and a set of n-tuples for which it is not defined whether they satisfy R_i. Given two partial structures $A = \langle D, R_{i1}, R_{i2}, R_{i3}\rangle$ and $B = \langle E, R'_{i1}, R'_{i2}, R'_{i3}\rangle$, "the function $f: D \rightarrow E$ is a partial isomorphism if (i) f is bijective, and (ii) for every x and $y \in D$, $R_{i1}(x, y)$ if and only if $R'_{i1}(f(x), f(y))$ and $R_{i2}(x, y)$ if and only if $R'_{i2}(f(x), f(y))$" (Bueno 1997, 596; French and Ladyman 1999, 108). The corresponding theory of representation would then be:

The partial isomorphism conception of representation [partial iso]: *A* represents *B* if and only if the structure exemplified by *A* is partially isomorphic to the structure exemplified by *B*.

The advocates of partial isomorphism argue that the introduction of R_{i3} serves to accommodate the partiality and openness of the activity of model building. That may be so, but as a theory of representation [partial iso] fares even worse than [homo]. Since according to (i) f is bijective, it follows from (ii) that $R_{i3}(x, y)$ if and only if $R'_{i3}(f(x), f(y))$, and hence partial isomorphism reduces to three separate isomorphisms. To this day, it remains to be seen whether [partial iso] can avoid the inaccuracy part of the argument from misrepresentation and correspondingly weaken the nonnecessity argument.[15] Even if this could be done, [partial iso] would be at a disadvantage with respect to [homo] since the logical argument weights even more strongly against [partial iso]: partial isomorphism, unlike homomorphism, is symmetrical.

STRUCTURAL REPRESENTATION WITHOUT ISOMORPHISM

Other writers within the structuralist tradition have been more cautious. It does not follow from the claim that theories (or models) are or contain structures that the relation that constitutes representation is a structural one. The arguments that I have presented so far suggest that we should look elsewhere for the constituent of representation, perhaps even in those cases in which the source and the target of the representation are structures.

Chris Swoyer (1991, 452), for instance, rightly claims: "Structural representation is not a necessary condition for representation in the ordinary sense of the word, since with sufficient perseverance—or perversity—we can use anything to represent virtually anything else, and in many cases the two things won't have any interesting structural similarities at all. And it is not sufficient for ordinary representation, since if you can find one structural representation of something, you can usually find many." Swoyer is also precisely right in characterizing structural representation as having the "potential" to be used in surrogative reasoning about its target.

After having considered six different phenomenological constraints on structural representation, Swoyer proposes the notion of an Δ/Ψ-morphism. (Swoyer's constraints implicitly rule out isomorphism, homomorphism, and partial isomorphism as the relation of structural representation, thus adding grist to the mill of my critique of [iso] and its cousins.) Consider the representation of some structure B by means of another structure A, and consider two subsets of B's domain Δ and Ψ. Then Swoyer's notion is as follows:[16] A structure A *structurally represents* another structure B if and only if there is a (neither necessarily one-to-one nor onto) mapping $c: B \rightarrow A$ that preserves all the relations defined over Δ and counterpreserves all the relations defined over ψ, where ψ is nonempty. Since ψ is nonempty, structural representation serves always to carry out surrogate reasoning about its target. Swoyer's notion does not meet the logical, misrepresentation, and nonsufficiency arguments presented here (in particular, Δ/Ψ-morphisms are reflexive); nor is it meant to do so since it is not meant as a theory of scientific representation in general. Yet his work shows that [iso], [homo], and [partial iso] do not correctly describe even the means of *structural* representation.

Misrepresentation Revisited

The arguments described in the previous section are by now nearly two decades old, and they have met with considerable success.[17] I believe that they played a role in getting philosophers to abandon hope of naturalizing away the notion of scientific representation in terms of dyadic relations between sources and targets. And the field has moved essentially in the way that I hoped, toward practical inquiries, with increasing attention paid to situating each modeling practice within its context. Nonetheless, a few scholars aimed to rebut the arguments, particularly the nonnecessity and logical arguments, often in conjunction with attempts to patch up or build on either of the two substantive views in dispute.[18] The final section reviews some of the proposed amendments and refinements of substantive accounts. The conclusion is skeptical regarding the prospects of any of these attempts to build up a substantive theory of the constituent of representation. Nevertheless, a lot of the work done in response has certainly helped illuminate the diverse means of representation in their contexts.

It is also perhaps easy with the benefit of hindsight to recognize that the attempts to build up [sim] and [iso] have met different fates. The most robust responses have come from defenders of similarity. This was perhaps to be anticipated since I already conceded that similarity was the more general relation (Suárez 2003): different kinds of structural morphisms can be understood as cases of similarity, but not vice versa. The different avenues that could, I pointed out, be pursued in defense of a sophisticated defense of similarity have indeed been pursued with some degree of success. These attempts to build up [sim] intriguingly add to our understanding of similarity as a common means of representation even if they cannot account for the constituent. By contrast, the ensuing debate has shown the limitations of any structuralist conception of representation as well as how uncommon it is as a means of representation, with some rare but interesting exceptions. Still, there are lessons to be learned from the failures of structuralist conceptions of representation as well.

One of the most powerful attempts to put an effective type of similarity to use in understanding modeling is that of Michael Weisberg (2012, 2013). Weisberg employs influential work by Tversky and his collaborators (see, e.g., Tversky 1977) on the psychology of similarity judgments to build up what he calls the *weighted feature matching* account. The Tversky-Weisberg theory seems substantive since it can be formulated as a necessary and sufficient condition on scientific representation. It essentially takes the route that I mentioned above of rejecting the identity theory of similarity to fill in the details of [sim] differently. On this account, a source S represents a target T if and only if the comparative similarity of target to source is large, where comparative similarity is measured by means of the following complex function: Sim $(S, T) = \theta \cdot f(S \cap T) - \alpha \cdot f(S - T) - \beta \cdot f(T - S)$, where S and T are the full set of salient and relevant features of source and target and θ, α, and β are relative weights that typically depend on context.[19] The set-theoretical relations $\{\cup, \cap, -\}$ express set-theoretical union, intersection, and complementation. Thus, $(S \cap T)$ is the intersection of S and T, that is, the set of properties shared by the source and the target, while $(S - T)$ is the set of properties of the source that are absent in the target, that is, the equivalent of Hesse's negative analogy. But there is a third element, namely, $(T - S)$, which is the set of properties in the target that are missing in the source. These are properties still to be discovered by means of further inquiry. The neutral analogy is a guide toward these properties, but there is never a guarantee that it would exhaust it.

Nevertheless, because of the inclusion of this third element, the Tversky-Weisberg account renders similarity nonsymmetrical: Sim (S, T) need not equal Sim (T, S) because S and T can be endowed with a different range and number of salient features or attributes and θ, α, and β can have different values in each case. Obviously, transitivity does not follow either, for similar reasons, given the different values of the parameters for each of the nested relations between, say, S, T, and W. In other words, Sim (S, T), Sim (T, W), and Sim (S, W) have distinct parameters. So, the Tversky-Weisberg account provides an answer to some of the main criticisms raised against [sim] and avoids the awkward conclusions that (i) a model represents a target only to the extent that it is represented by it and (ii) a model that represents another model thereby also represents the latter's target. This is undoubtedly good news for similarity.

Yet it is still possible to challenge the Tversky-Weisberg account on several grounds, and these are all powerful reasons why it cannot provide a substantive theory of scientific representation. First, the logical argument has not been answered completely since, even on this sophisticated understanding, similarity remains a reflexive relation. This is easy to see just by replacing T with S in the formula given above: Sim $(S, S) = \theta \cdot f(S \cap S) - \alpha \cdot f(S - S) - \beta \cdot f(S - S)$, which entails Sim $(S, S) = \theta \cdot f(S)$, a positive value. Second, it is important to notice the pragmatic and contextual elements of Weisberg's approach, which emphasizes that the relevant features can be selected only within the context of use. In other words, the sets of features of S, T to be considered are relative to judgments of relevance in the appropriate context of inquiry. Implicitly this assumes that there are no context-independent or context-transcendent descriptions of the features of sources and targets. This is essentially a deflationist insight and a formidable obstacle in turning Tversky-Weisberg similarity into a substantive account of representation. Third, and relatedly, the idea that every source or target object or system can be described in terms of different salient properties and features in each context presupposes some antecedent notion of representation. It seems to assume, more specifically, that sources and targets are *represented as* having sets of features within a context and that it is these "pre-representations" that are fed into the formula for the similarity metric. If so, rather than explaining and defining representation, the Tversky-Weisberg approach to similarity presupposes that a notion of scientific representation is already in place.[20]

This objection to the Tversky-Weisberg approach is also raised and interestingly enhanced in a recent paper by Wei Fang, who has been a notable late contributor to the debate over scientific representation. Fang argues that the Tversky-Weisberg similarity measure mistakenly presupposes a set-theoretical representation of the features of the source and target systems involved. This means that each feature of the systems of interest (both the source and the target) is taken to be isolated from the other features (i.e., all features are always noninteracting and independent from each other) and can thus be represented extensionally in terms of those elements in the domains of each of the sets that each feature denotes. Fang (2017, 1758–59) proposes an alternative account in terms of holistic fit between models and targets, where models are considered organically, as it were. It is the model as a whole that licenses the right inferences toward their targets, and it is not possible to break down this work into work done by each of the features separately. I am broadly sympathetic to Fang's approach, as will become evident in the next chapter. However, it is important to note that the more narrowly focused objection that I am raising here to the Tversky-Weisberg approach holds on the mere observation that this approach already presupposes a particular form of representation of the systems involved. That observation is enough to generate the objectionable circularity, regardless of what the right account of the constituent of representation is, or indeed regardless of whether such a constituent exists.

A fourth and final objection is that, while the Tversky-Weisberg ingenious measure of similarity goes further than any similarity measure relying only on Hesse's positive and negative analogies, it cannot capture the impact of what I have elsewhere called the *inverse negative analogy*, that is, the properties of the target that are explicitly denied in the source (Suárez 2016, 451). It turns out that Hesse's proposal fails to distinguish those properties of the model source that are absent in the target (in the billiard ball model, these include, arguably, the ball's color and shine) from those other properties of the source that are explicitly denied in the target (limited elasticity, finite escape velocity, and, in general, those properties of a gas system *as a whole* that billiard balls, even regarded *as a system*, patently do not have). The former properties can perhaps be ignored altogether since they do not play a role in the dynamic processes that ensue in either billiard balls or, naturally, gas molecules. The latter properties cannot, however, be so dismissed. Unlimited elasticity and infinite escape velocity are fundamental properties of gases but patently not of billiard balls. It would seem wrong to say that these properties are altogether not part of the representation. They are part of the representation, and, even though they are extremely dissimilar in source and target, they too facilitate inference making from source to target.[21]

In other words, even the Tversky-Weisberg proposal does not account for all the typical ways in which models go wrong or misrepresent their targets. More specifically,[22] the form of misrepresentation that involves "lying," "pretending," or "positively ascribing the wrong properties" does not fall under the account. This is a problem if the account is ever to achieve generality as a substantive theory of representation, given how pervasive simulation is in model building.[23] I conclude from these objections that the Tversky-Weisberg approach fails as a substantive theory of the constituent of scientific representation, in our terminology, even if it provides some nuanced detail for the judgments that allow modelers to set similarity relations as the practical means of inquiry into target systems.

Another approach to similarity develops insights from the literature on creativity. On this view, similarities between representational sources and targets are not so much discovered as they are created by manufacturing the source and tailoring the specific comparative links to the selected aspects of the target as described on purpose (see Ambrosio 2014; Sánchez-Dorado 2019; and Sterrett 2017). It is evident that, as an account of the constituent of representation, this will not do since those similarities that are created are heavily dependent on context. Just as we saw is the case with the Tversky-Weisberg account, the description of both target and source requires antecedent representations of both even as they undergo modification and construction in tailoring to each other. It is these antecedent representations that partition the objects involved in an appropriate manner for the representational job in hand.

Some of the issues at stake in this very recent discussion strike me as strongly reminiscent of the earlier discussions of metaphor reviewed in chapter 3 and particularly of Max Black's (1962) and I. A. Richards's (1936) interactive accounts. Recall that the central element of the Black-Richards approach is the recognition of the mutual interaction in a metaphor that the primary theme and the secondary theme (the *tenor* and the *vehicle* in Richards's [1936, 89ff.] account) exert on each other. This account of metaphor is dynamic too since both tenor and vehicle are altered when they are combined in a metaphor. The creative approach to similarity in modeling is equally suggesting that both representational source and representational target are fundamentally altered when brought under such a representational relation. However, recall that the key insight in the Black-Richards interactive view is that the dissimilarities between tenor and vehicle matter cognitively in a metaphoric expression just as much, if not much more than, the similarities. While this is exciting work still to be developed, my conjecture is that the creative similarity approach is bound to reveal the fundamental role played in the representational relation by the dissimilarities between the model source and the model target—the sets of properties that are manifestly distinct—in our effective use and dynamic development of models. Far from being an endorsement of [sim], this recent work surreptitiously abandons, I would argue, any hope that similarity can ground representation. Nevertheless, the approach is very valuable and relevant as a description of the dynamic aspects of model building whenever similarity relations are involved as means of the representations. It also suitably emphasizes the irreducible elements of creativity involved in model building and the sets of intricate judgments that must be made to establish a representational relation.[24]

As regards the structural representation literature, one exciting development in the last fifteen or so years is Bartels's (2006) sophisticated and nuanced proposal to build up a substantive theory of representation in terms of homomorphism. In what is explicitly a response to the nonsufficiency and misrepresentation arguments, Bartels carefully distinguishes the "representational content" of a model—or generally any scientific representation—from its "representational mechanism." The latter is the actual linking of the source of a representation with its target, and Bartels acknowledges that this is a pragmatic act that must be understood within the context of its use.[25] The representational content of a model, by contrast, is given by what can be called the *structural possibilities* of the model. This is where homomorphism plays its role, establishing the sorts of structures that fall under the representational remit of the model—its domain of possibility. What the model represents is then a match for one of these possibility structures, as determined by the representational mechanism (Bartels 2006, 12).

The ingenious addition of a representational mechanism gets the account off the charge of nonsufficiency. Indeed, Bartels admits that homomorphism is not sufficient for representation but sufficient only for what he calls *potential representation*. It is only the conjunction of homomorphism and the representational mechanism that can be said to be jointly sufficient for representation. Plausibly, the representational mechanism also gets around the mistargeting aspect of the misrepresentation argument as well as the logical argument. How? Well, the representational mechanism implicitly fixes what I called the essential *directionality* of representation: the one-way relation that representational sources have toward their targets. The mechanism is directed in the same way, which plausibly means that actual representation is not generally reflexive (the source's mechanism does not point toward itself), symmetrical (the mechanism is not typically reciprocal), or transitive (nested representations can have different mechanisms at each point in the chain). As for mistargeting, this is obviously taken care of by the essential directionality of the representational mechanism. Thus, the only arguments standing are the variety and the nonnecessity arguments, and Bartels claims that homomorphism will underpin any potential scientific representation: "B represents A only if B is a homomorphic image of A" (Bartels 2006, 8). Hence, homomorphism is necessary for representation, and all the variety of representational means in the last instance reduce to it.

Together with Francesca Pero (Pero and Suárez 2016), I have argued that there is no reduction of scientific representation to homomorphism. Our result merely strengthens the conclusion arrived at by Swoyer (1991) and already reviewed here. Even when considering the variety of structural representations (i.e., those representations that are indeed relations between sources and targets that are uniquely defined as structures), homomorphism is not always or even often the effective means of such representations. The argument is technically involved and requires a careful distinction among all the possible morphisms that can obtain. It is nevertheless possible to leave those technical complexities aside[26] and illustrate the basic difficulty by means of the billiard ball model.

My treatment of the billiard ball model emphasizes the crucial role played by the *dissimilarities* between representational source and representational target. Recall that Hesse (1963/1966) presents the model as consisting of a positive, a negative, and a neutral analogy between macroscopic balls on a billiard table and gas molecules in a container. In the positive analogy (the properties the two systems share), she includes motion and impact; in the negative analogy (the properties of billiard balls that gas molecules lack), she includes color, hardness, brightness; and the remaining properties of billiard balls are in the neutral analogy. While she does not mention them, there is a further group of properties of interest, namely, the properties of the gas—according to kinetic theory—that are not properties of billiard balls. These were previously referred to as the *inverse negative analogy*, for they are the converse of Hesse's negative analogy. For instance, the macroscopic features of a gas considered as a whole, such as volume, density, or pressure, are obviously lacking any correlate in any macroscopic features of billiard balls. The kinetic theory of gases allows us to derive such macroscopic properties of the gas in the container considered as whole from previous macroscopic states and from the microscopic states of the gas molecules, statistically averaged. Nothing like that evidently obtains for billiard balls. Rather, the billiard ball analogy is here positively misleading since nothing follows from average speed of the balls about the pressure exerted outward by the set of balls. The obvious missing ingredient is free expansion, a defining thermodynamic property of gases with no correlate in any dynamic property of any mechanical system of solids and one that evidently billiard balls cannot ever exhibit.[27]

More generally, scientific models do not simply abstract away properties of the systems of interest by ignoring them. As I mentioned earlier in connection with the Tversky-Weisberg similarity measure, models also lie or simulate by pretending that the target system has some properties that in fact it resolutely does not. They impute such inverse negative analogies when they implicitly assume them as presuppositions in the model. In sui generis terminology, models *abstract*, but they also *pretend* or *simulate*. Yet homomorphisms obtain only between those properties shared by the target and the source systems (Pero and Suárez 2016; Suárez 2016) and can at best accommodate abstraction. Nevertheless, while Bartels's proposal does not overcome the objections to substantive theories, his introduction of an essential and directed representational mechanism can be regarded as a friendly addition to a more deflationary or functional account of scientific representation. For the central questions of representation now turn on the nature of this mechanism, which cannot be described as a property of representational sources and targets. On the contrary, the elucidation of Bartels's representational mechanism must now call essentially for some inquiry into the intended uses of the representation. And that is just what a deflationary account of representation recommends.

Chapter Summary

This chapter considers the range of arguments against the substantive accounts of scientific representation presented in the previous chapter in terms of similarity and isomorphism. It identifies five arguments, namely, the argument from variety, the logical argument, the argument from misrepresentation, and the nonnecessity and nonsufficiency arguments. It finds them convincing overall and a good basis on which to reject any substantive accounts of representation generally. It ends by considering the prospects for more sophisticated accounts of representation as both similarity and isomorphism that, while not substantive or even avoiding entirely defining the notion of representation, might be fruitfully employed as part of a practical inquiry into the means of some scientific representations.

6

Scientific Theories and Deflationary Representation

In the remaining two chapters in this part of the book, I defend my own account of representation and modeling in science, the so-called inferential conception. First, in this chapter, I provide some framework—both historical and conceptual—in terms of debates regarding the nature of theories and models and specifically what is known as the *semantic view* of scientific theories. I then expound on deflationary conceptions of representation generally, and I discuss in detail an allied deflationary cousin to the inferential conception, R. I. G. Hughes's denotation-demonstration-interpretation (DDI) account of representation. I argue that there is a simple and natural extension of the DDI account that essentially satisfies all deflationary constraints on representation. My main claim in this chapter is that deflationary proposals regarding representation, such as Hughes's or my own, are not in conflict with any of the extant views on scientific theory, including prominently the semantic view. In the follow-up chapter 7, I develop the inferential conception as a further refinement of the extended DDI account. In this chapter, I already anticipate that both deflationary accounts are perfectly compatible with a particular version of the semantic view of scientific theories.

The Semantic View of Scientific Theories

The argument against reductive-naturalist accounts of representation deployed so far can be confused for an argument against certain views of the nature of scientific theories. In recent, state-of-the-art work, Rasmus Winther (2020a, 2020b) distinguishes three such views: *syntactic*, *semantic*, and *pragmatic*. My main aim in this first section is to argue that none of these views stands in conflict with a deflationary, pragmatic, and inferential conception of scientific representation. I shall focus particularly on the semantic view of scientific theories because it lends itself more readily to the charge of incompatibility and has in fact explicitly been argued to require a very different structuralist account of representation.[1] By contrast, I shall argue that there is a version of the semantic view that provides a perfectly hospitable framework for the conception of representation that I defend in this book.[2] This is appropriate since the semantic view takes inspiration from the modeling tradition reviewed in the first part of the book. It ought to be possible to develop the inferential conception of representation against the background of this account of theories.[3]

The semantic view has a long pedigree by now and has been defended by such influential philosophers as Patrick Suppes (2002), Wolfgang Stegmüller (1979), Ronald Giere (1988), Elizabeth Lloyd (1988/1994), Bas van Fraassen (1987, 1989), and, most recently—in the form that will interest us most here—R. I. G. Hughes (2010). However, it has come under heavy criticism, and many authors now regard it as discredited (see, e.g., Cartwright 1999b; Frigg 2006; Halvorson 2012; Morgan and Morrison 1999; and Worrall 1984). While the critics have powerful and convincing arguments against a particular version of the semantic view, I shall argue that there is another version of the semantic view that remains viable and that it is perfectly amenable to the concept of representation defended in this book.

The semantic view of scientific theories arose in response to the perceived failings of the so-called syntactic or received view of scientific theories, so a quick review here is helpful. The received view has antecedents in logical empiricism, and it often gets associated with the work of Rudolf Carnap (1939, 1956), Carl Hempel (1965), and Ernest Nagel (1961). On such a syntactic conception of scientific theories, a scientific theory is essentially identified with a set of statements in predicate logic, typically standard first-order logic. But it is important to stress that the syntactic conception is not committed to these sentences being mere syntactic markers devoid of interpretation. Rather, it accepts that they come with a semantic interpretation of their own, one provided by the scientific theories in question.[4] Thus, in his influential text, Nagel (1961, 90) summarizes the contents of a theory in three parts: "(1) an abstract calculus that is the logical skeleton of the explanatory system; (2) a set of rules that in effect assign an empirical content to the abstract calculus . . . ; and (3) an interpretation or model for the abstract calculus, which supplies some flesh for the skeletal structure." In the early reception of logical empiricism, Nagel's parts 2 and 3 were sometimes merged or conflated, in the operationalist fashion of Bridgman (1936), with the so-called bridge principles providing for a dual carriageway, so to speak, for both epistemic support and semantic meaningfulness (often in the form of a verificationist criterion of meaning).[5] The most considerate views of Carnap (1956), Nagel (1961), and Hempel (1965) are, however, more faithful to the account of the nature of theories that first emerged in the discussions of the Vienna Circle in the 1920s and stayed unaltered until the emergence of the semantic conception in the 1970s.

As a crude example, take Newtonian mechanics, whose universal law of gravitation can be axiomatized in a simple first-order calculus as the statement that, for any two massive objects x and y,

$$\forall x, y\colon F(x, y) = G\frac{m(x)\cdot m(y)}{r(x,y)^2},$$

where $F(x, y)$ is the force that operates between x and y, $m(x)$ is the mass of x, $m(y)$ is the mass of y, $r(x, y)$ is the distance that separates x and y, and G is the universal constant of gravitation. Thus, as regards Nagel's part 1 above, the symbols operate as syntactic markers and obey the typical logical and mathematical operators (addition, division, multiplication, and so on). But, in addition, in a scientific theory such symbols are suitably interpreted in terms of the entities, properties, and relations that they make implicit, if not explicit, reference to—in our example, mass (m), distance (r), and Newtonian force (F). This provides the model interpretation (Nagel's part 3 above) for the "logical skeleton" that is embodied in the universal law equation. As for the rules of correspondence that furnish the required empirical content, they are given by the links of the symbols listed above as suitably interpreted by the model interpretation to the empirical results of observations in appropriate experiments. Thus, the interpreted abstract calculus accounts for the observations of the relative positions of the planets in the solar system in relation to that of the sun—or, given the universal character of the law, the relative motions of projectiles on the surface of earth—and so on. In other words, the rules of correspondence (Nagel's part 2 above) provide the empirical content of the theory, while the model interpretation (Nagel's part 3 above) provides its semantics, a prerequisite in fact for any application of the theory.[6]

Notably, the logical empiricists defended the view that the model interpretations of scientific expressions—that is, their semantics—come in two different types: those that are directly referential (ostensibly picking out their referents from among observable things, objects, or their properties) and those that are derivative (indirectly picking out their referents through their logical ties to those that are directly referential). In the logical empiricist framework, the distinction between these two separate semantics relies on an antecedent distinction between observable and theoretical terms. Carnap (1956, 40) best articulated the most sophisticated version (which others followed), in which "the total language of science, L, is considered as consisting of two parts, the observation language L_O and the theoretical language L_T." The observational vocabulary V_O is then the set of descriptive terms in L_O, that is, those terms in L_O that are not merely formal or logical constants.[7] And those terms in V_O get their meaning by ostension since they are "predicates designating observable properties of events or things [or their observable relations]" (Carnap 1956, 41). Similarly, the theoretical language L_T is further subdivided into logical and descriptive constants, and the resulting descriptive vocabulary V_T is the set of the theoretical terms of science. A scientific theory T is then identified with a finite set of postulates formulated in L_T, and a set of correspondence rules C is then called forth to connect the postulates that make use of the theoretical terms in V_T to the statements in the theory entirely formulated in the observational language V_O. In terms of the distinctions in Nagel (1961) that we saw, the theory amounts to a logical calculus (part 1) together with a model interpretation (part 3); while the correspondence rules (part 2) are strictly extraneous to the body of the theory but nonetheless necessary for its application.

In the 1970s, this received view of scientific theories came under sustained attack.[8] The main criticisms have three sources: (i) the difficulty in drawing any sharp distinction between the observational and the theoretical vocabularies, (ii) the ensuing difficulties with precisely carving out the empirical content of a theory, and (iii) the language relativity that seems inherent in any view of theories as sets of sentences. These difficulties have by now been explored extensively (some would perhaps say ad nauseam), yet they remain to some extent controversial.[9] I would add a fourth difficulty with the received view, one that has received less attention but gains currency today in the wake of discussions regarding inconsistent models and theories and impossible or inconsistent fictional assumptions in theoretical model representations.[10] These practices come into conflict with the strict requirements of consistency imposed by the model interpretation, which inevitably requires all sentences that compose a theory to be mutually consistent (or at any rate those sentences that are to receive an interpretation, however partial).

The semantic conception arises then out of these perceived difficulties with the received view. The criticisms of the observational/theoretical distinction (i.e., part i above) are well-known since they derive from the notorious thesis of the theory ladenness of observation (Feyerabend 1958; Hanson 1958), according to which there are no purely observational terms because there are no theory-free observations of entities or properties. It is notable, however, that some recent commentators are of the opinion that the received view does not really require such a sharp distinction.[11] However, the source of the objections that seem to have historically greater bite against the received view is really part ii. Thus, Putnam (1962/1975) and van Fraassen (1980) both argue that it is possible to make statements about the observable domain or phenomena by means of the theoretical vocabulary V_T of a theory. And, vice versa, it is possible to make assertions concerning the unobservable part of the world by means of the observational vocabulary V_O, however the distinctions between the two vocabularies are drawn (Putnam 1962/1975, 227). It follows that the observational vocabulary does not per se characterize narrowly only whatever a theory says about the observable phenomena. Conversely, it does not exhaust what a theory says about the observable phenomena. And, since it seems most plausible to suppose that the empirical content of a theory is precisely whatever the theory says about the observable phenomena, there is no way in which we can characterize this content merely by attending to a difference in the terms of the vocabulary of the theory. This conclusion holds generally, regardless of the theory in question and how it is linguistically expressed.

The other great source of criticism of the received view concerns its inherent language relativity (this is critical part iii above). Undoubtedly, since the received view identifies theories with sentences expressed in some logical language—and regardless of whether this is a first-order or a higher-order language—it does in a sense *reify* or *embody* scientific theories in some language or other. The question is then whether relativism follows regarding the identity conditions and perhaps even the content of the theory. The paradigmatic counterexample to the identification is the formulation of quantum mechanics as first matrix mechanics in the hands of Heisenberg in 1925 and soon after as wave mechanics by Schrödinger in 1926. Both formulations are provably equivalent, yet, arguably in the terms of the received view, they cannot be identical since they are expressed as distinct set of postulates in different mathematical languages or formalisms.[12]

The semantic view of theories thus emerges in the 1970s as the candidate conception that can overcome all these objections. The main claim of any semantic view—to which I will refer as the *semantic bare claim*, or claim 1—is the assertion that a scientific theory is a set of models. This is obviously meant to contrast with what can be referred to as the *syntactic bare claim*, namely, that a theory is a set of sentences. When expressed in this way, the semantic and the syntactic conceptions obviously differ since models and sentences are evidently different entities. Yet, as we shall see, there are different versions of the semantic view, and how they fare depends critically on what they identify as the relevant sense of *model*. Thus, the semantic bare claim is often accompanied by an additional claim regarding the nature of models, and much will hinge on how this additional claim is filled in. There are roughly three different senses of the word *model* that have regularly been applied: (i) a set-theoretical structure (Suppes 2002), (ii) a state-space description of dynamic systems (van Fraassen 1989), and (iii) a representational model of any of a diverse array of kinds, including sense ii and in some cases even sense i and sometimes others (Hughes 2010; Suárez 2010a, 2010b). It turns out that the second category (sense ii) can be generally subsumed under either sense i or sense iii, so I shall focus only on the alternatives sense i and sense iii.[13]

Suppose that we were to adopt the *structural claim* (claim 2): "Models are set-theoretical structures." The conjunction of the semantic bare claim (claim 1) and the structural claim (claim 2) renders a particular version of the semantic view that I shall refer to as the *structural semantic view* (SSV). According to this version of the semantic view, "theories are sets of set-theoretical structures," and this is the version of the semantic view of theories that has been dominant in the literature at least since the early 1970s, supposedly on account of its strength in resolving the conundrums and difficulties raised by the old syntactic account.[14] One of the earlier examples of the successful axiomatization of a scientific theory in set-theoretical predicates precisely concerns Newtonian classical particle mechanics (McKinsey, Sugar, and Suppes 1953; Suppes 1957; see also Winther 2020a, sec. 3.2). Suppes and his collaborators write down a complex set-theoretical predicate: $\{P, T, s, m, t\}$, where P and T are sets, m is a unitary (one-place) function, s is a binary (two-place) function, and f is a ternary (three-place) function. The elements p in P are the set of particles, while those t belonging in T are the elapsed times or intervals. For any particle p, the function m represents its mass as $m(p)$. For any particle and given time t, s represents the position of the particle at that time as $s(p, t)$. For any two given particles p, q belonging in P, the function f represents the force that particle q exerts on particle p, at time t, as $f(p, q, t)$. We can then write Newton's second law as an axiom for this set-theoretical predicate as follows:[15] $m(p)\cdot(d^2/dt^2)\cdot s(p,t)=\Sigma_{i=1}^{\infty} f(p,q,t)$, where p and q identify two particles, $m(p)$ is the mass of the first particle, $s(p, t)$ is its position at time t, and $f(p, q, t)$ is the force exerted by q over p at that time, and the equation follows on the assumption that there are no other particles or forces exerted on p. The axiomatization is innocuous and clearly applicable, but the usual interpretation of $\{P, T, s, m, t\}$ as a set-theoretical structure $\langle P, T, s, m, t\rangle$ renders it in the format of a structural representation, as required by the SSV.

Nevertheless, there is a more recent alternative to the SSV, the *representational semantic view* (RSV), which conjoins the semantic bare claim (claim 1) not with the structural claim (claim 2) but with a distinct *representational claim*

(claim 3), namely, that "models are representations." On the RSV, "theories are sets of representations." The structural and representational versions of the semantic view are equivalent only if it is the case that representations are set-theoretical structures, in which case in claim 1 we would be able to substitute *structures* for *models*, thus rendering the conjoint claim that "all theories are sets of set-theoretical structures." That is, we have returned to the SSV again. In other words, the RSV entails the SSV if and only if it turns out that all representations (including $\{P, T, s, m, t\}$ in the Newtonian example given above) are set-theoretical structures, that is, if and only if a version of the structural claim (claim 2)—let us refer to it as *structural claim* 2′—holds, namely, that all representations are structures.[16] The point is that this structural claim must also be false for the RSV to be genuinely distinct from its structural cousin. Otherwise, the RSV would simply boil down to the SSV. Both the RSV and the SSV respect the original semantic bare claim (claim 1) and are thus entitled to be referred to as *the semantic view*, but—as long as structural claim 2′ is false—they have essentially very different consequences for the nature of theory.

One of my aims in earlier chapters in this part of the book has been to mount an argument for the failure of structural claim 2. This chapter offers an alternative, one in which the representational claim (claim 3) is appropriately conjoined with the bare semantic claim (claim 1). In other words, I suggest that the RSV is the more appropriate semantic framework in which to understand the nature of theories and concomitantly the nature of the representations with which scientific theories provide us.[17] There are different reasons why the RSV is a more suitable view of scientific theory, and some of the most practice-oriented ones will emerge throughout the rest of this book.[18] There is, however, one reason that I would like to advance at this point, namely, that the claimed superiority of the semantic over the received or syntactic view accrues only to the RSV. The SSV, by contrast, arguably ultimately boils down to an extension of the received view, so it in fact inherits most of the difficulties associated with the latter.[19]

Nonetheless, some advocates of the semantic view have accepted the essential equivalence between the received view and the SSV, and they have argued that the latter has the advantage only of being close to scientific practice in mode of presentation (see Hughes 1989; Lloyd 1988/1994; or Suppe 1989). However, as far as the SSV goes, this is difficult to maintain since set-theoretical structures are very rare as the typical mode of presentation of any theory outside the most formal branches of mathematics. Besides, the SSV has additional difficulties of its own, independently of those derived from its ultimate reduction to some form of the received view. Just as it is implausible that all theories can be construed as sets of sentences, it is similarly implausible that all theories can be construed as set-theoretical structures. And, certainly, there *are* theories that are ostensibly presented as sets of sentences in some natural language, in the life sciences and the social sciences alike. (Galileo's kinematic theory was first presented notably in Italian and merely illustrated with geometric figures, Darwin's theory of evolution by natural selection was first introduced entirely in English with virtually no mathematical language, and so on.) Thus, some of the additional objections to the SSV derive from the difficulties—already reviewed in previous chapters—involved in the attempt to subsume the extremely large and variegated forms of models into the unique class of mathematical models.[20]

Now, defenders of the semantic conception wedded to the SSV have been particularly attracted to versions of what in the last chapter I defined as [iso]. This is hardly surprising since, if theories are—as stated by the SSV—always set-theoretical structures, then it is natural to think that the relation between such theories and their target systems must also be structural or at any rate readily formalizable in the language of set theory.[21] In other words, the SSV naturally suggests, if it does not entail, a structuralist conception of representation such as [iso]. If so, the semantic view of theories would fall with [iso].[22]

By contrast, I am urging that the only claim that is essential to the semantic view is the bare semantic claim (claim 1), which makes no assertion whatever as to what models are. It is after all a claim about theories, not about models, and it merely asserts that theories are sets of models, as opposed to linguistic entities.[23] Any further claim regarding the nature of those models (such as whether they are structural entities and, if so, whether they represent by virtue of some sort of structural relation) is on this view supererogatory to the essence of any semantic view. Thus, the structural claim (claim 2) is an additional and strictly unnecessary claim for a semantic view of theories, and indeed it is rejected in the RSV that I am proposing. By contrast, the RSV makes no commitment at all to theories being mathematical structures, be they set theoretical or any other kind.[24]

Even if one thought that models are best understood as structures (which obviously I do not), one could still disagree with the notion that representation is a structural relation (Suárez and Pero 2019). For, as it turns out, the nature of the relata does not determine the nature of the relation. At any rate, the inferential conception developed in this chapter and the next is neutral on the issue of the nature of scientific theories, and it can be adopted by defenders and opponents of the semantic view alike. Its opponents can accept it as well as the concomitant claim that scientific models are representations in this deflationary inferential sense without thereby committing to the semantic bare claim (claim 1) regarding the nature of theory. Thus, the inferential conception of representation is, first of all, compatible with a sophisticated syntactic view of scientific theory such as Carnap's (1956). It is also compatible with the view of theory as an unverifiable heterogeneous mixture of models, techniques, activities, and practices such as that defended in Suárez and Cartwright (2008). But, third, defenders of the semantic view can also easily adopt the inferential conception of representation, but only if they avoid any commitment to the SSV and adopt the RSV instead. To sum up, the inferential conception is compatible with any of Winther's (2020a, 2020b) three main views regarding scientific theory: syntactic, semantic, and pragmatic.

Deflationary Accounts of Representation

We have seen that the key to upholding an RSV of scientific theories is the rejection of structural claim 2′, namely, the assertion that all representations are structures. Whether this claim is true depends on what one means by *representation*, and there are different accounts of the notion of representation imparting different meanings to the term. In this section, I follow the preliminary classifications outlined in chapter 4 (in the section "Philosophical Accounts of Scientific Representation") and go on to probe into their qualities as candidates to fill the representation role within a distinct and robust version of the RSV. Ultimately, I shall be defending an inferential conception of representation. However, it is worth noting that there are other options that can fill the role appropriately and without at any point falling into any surreptitious commitment to the problematic structural claim 2′.

Recall from chapter 4 that the different accounts of representation can be divided into two types: substantive and deflationary. The substantive accounts claim that representation is some substantive explanatory relation between representational sources and targets. The deflationary accounts by contrast claim that there is no property or relation at stake—or, at any rate, no property or relation substantive enough to carry out any significant explanatory burden. Substantive accounts have traditionally been the norm in discussions in the philosophy of science, particularly since the advent of the semantic conception in any of its varieties from the 1970s on. Patrick Suppes, Bas van Fraassen, and Ronald Giere have been thought to defend substantive accounts of representation as isomorphism, embedding, and similarity, respectively. Their views turn out to be subtler than has been supposed, and van Fraassen's most recent and considered views are explicitly deflationary (van Fraassen 2008).

(More recent defenders of substantive accounts of representation include Pincock [2012], who defends structural isomorphism, and Weisberg [2013], who defends similarity. Others include Bartels [2006], who defends homomorphism, and Contessa [2007], who defends a substantive version of the inferential conception, which he refers to as the *interpretational conception*.)

The DDI Account

Deflationary approaches to representation can be traced back to R. I. G. Hughes's (1997) DDI account.[25] Other explicitly deflationary accounts are advanced in Boesch (2017), Downes (1992, 2021), Odenbaugh (2019, 2021), Suárez (2004, 2010a, 2015), and van Fraassen (2008). Yet other accounts that can be understood to be deflationary in at least some of their most plausible versions include Elgin (2009), Frigg and Nguyen (2020), and Giere (2004, 2009).[26] But in what sense can such accounts be claimed to be deflationary? We can now return to the discussion of the varieties of deflationism in chapter 4 and apply those distinctions to some specific accounts of representation that might fulfill their conditions. In this section, I describe the main features of Hughes's DDI account, while, in the next, I argue that that account can be extended in a way that fulfills the required conditions for deflationism and, moreover, that the extension makes it more plausible. In the next chapter, I shall employ this extended version of the DDI account as a template, against the background of the liberal RSV of theories, to develop my own deflationary account of representation, the inferential conception.

According to Hughes's DDI account, a scientific model is a three-part "speech act," or a complex activity. In the terminology of the previous chapters, the DDI account asserts that, when a source A represents a target B, then typically (even though not necessarily) the following three conditions are met: (i) A stands for B in the sense that A denotes B, (ii) some demonstration is carried out by an agent on A, and (iii) the results of this demonstration are then interpreted to apply them to B.[27] The three steps of denotation, demonstration, and interpretation can be summarized in a by now often-reproduced diagram (fig. 6.1).

As an illustration, Hughes (1997; see also Hughes 2010) deploys the model that Galileo introduces in the "Third Day" of his *Discourses concerning Two New Sciences* (Galileo 1638/1974; see Palmieri 2018 for discussion). Galileo there describes a kinematic problem in exclusively geometric terms. He then goes on to solve this problem in geometry, only then proceeding to apply the solution to the original kinematic problem. He ultimately thus deduces

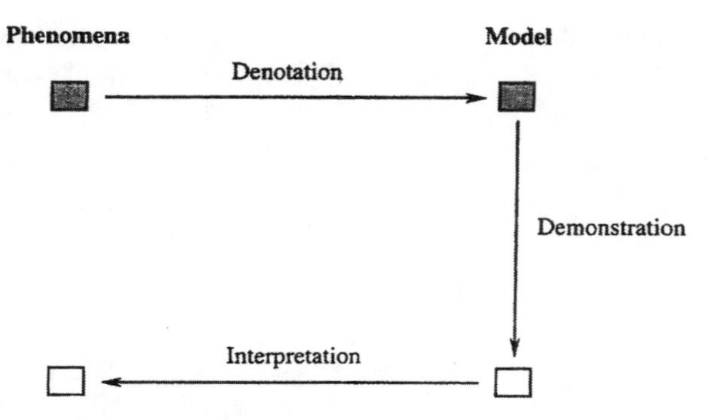

FIGURE 6.1. R. I. G. Hughes's original DDI model.

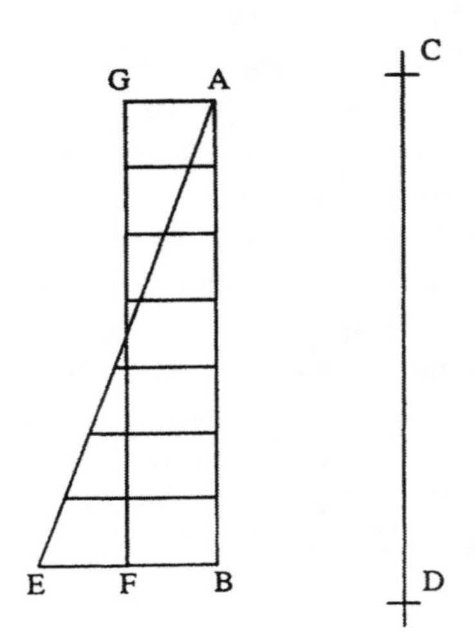

FIGURE 6.2. Galileo's geometric model of kinematics.

that the space *s* traversed by a body in uniform motion with constant velocity in a given interval *t* is equal to that traversed by a uniformly accelerating body initially at rest, provided that the final speed of the accelerating body is twice that of the body in uniform constant motion. Galileo's geometrically modeling the kinematic problem can be illustrated conveniently by the DDI model as follows. First (the first denotational step listed above), the kinematic situation must be described by means of a geometric diagram that therefore denotes it (fig. 6.2).

Thus, Galileo denotes the spaces to be traversed by the separate segment CD, the time *t* that the body takes to traverse this space *s* by means of the segment AB of a parallel line, and the speed of the body at any instant of the interval *t* by another segment of a line perpendicular to the first line (the point A at the start, i.e., null speed, the segment BE at the end). Therefore, the speed linearly increases as in uniformly accelerated motion. Second (step ii above), a demonstration must be carried out on the diagram in its own geometric terms. Galileo uses the diagram to demonstrate the geometric fact that the area of a rectangular shape ABFG is identical to the area of a triangle ABE where E is twice the value of G. Finally (the third step listed above), we need to interpret this result back in the terms of the original kinematic problem. By interpreting the overall area covered as the space traversed by the body in its motion over the *t* interval, Galileo infers that the time *t* that a body in uniform motion takes to traverse *s* is identical to the time taken by a body that was uniformly accelerated from rest to twice the initial speed of the uniform motion throughout. He thus "reaches his answer by changing the question": "A problem in kinematics becomes a problem in geometry" (Hughes 2010, 153). Whatever the merits of the model, it seems clear that the combination of acts of denotation, demonstration, and interpretation constitutes an overall complex act of representation, in this instance, in accordance with the DDI account.

The DDI account is ostensibly deflationary in not providing necessary and sufficient conditions for representation and in claiming that representation is a set of speech acts (Hughes 1997, S329). Thus, although Hughes did not explicitly endorse deflationism, he did explicitly reject any attempt at defining representation, and he demurred at reifying representation in terms of self-standing relations, however they might be abstracted from the corresponding speech acts.[28] Nevertheless, we might be troubled—from a deflationary perspective—by the overt appeal to denotation, which is typically understood as a substantive relation, not an activity. A deflationist certainly would be uncomfortable with any attempt to turn such a substantive relation into an individually necessary condition, jointly sufficient for an act of representation together with the other two. It is important to warn against that reading, which would undoubtedly compromise the strength of the deflationism. In the rest of this chapter, I endeavor to show that there is a reading of the DDI model that is fully in agreement with deflationism and, moreover, best agrees, I believe, with the intended spirit of the original proposal.

In chapter 4, I distinguished at least two broad senses in which an approach to a concept and to the concept of representation can be said to be deflationary: the redundancy no-theory theory sense and the minimalist use-based sense (the latter in turn coming in two varieties, corresponding to Wright's [1992] pluralist abstract minimalism and Horwich's [1980] practice-based minimalism). The DDI model is straightforwardly deflationary in the no-theory theory sense since Hughes is opposed to the definitional project and does not provide an analysis of the concept of representation. But is it deflationary also in the use-based sense? It is not prima facie clear that all three speech acts involved in an act of representation are, as Hughes has it, use based or a mere summary description of regularities in their use in practice. It can be argued in fact that, as originally presented, Hughes's account is deficient from the point of view of use-based deflationism. The appeal to denotation is particularly surprising in this context since denotation is commonly understood to be a substantive referential relation between a denoting sign and a denoted object.[29]

This suggests that Hughes's original proposal is hybrid. Denotation is a relation between source and target, while demonstration and interpretation are best understood as activities on the part of an interpreter or user, performed on the source, in relation to the target. There is, of course, an activity of denoting, but this is commonly understood either to consist in the establishment of a relation or sometimes even to ride on an already metaphysically established relation. In other words, we cannot employ some source A to denote some target B without ipso facto establishing a relation of denotation between A and B or, even worse, without surreptitiously employing a relation of denotation already established. In the terms of the discussion in chapter 4, the denotation is archetypal of what Peirce disparaged as a dyadic relation, while demonstration and interpretation necessarily involve an agent or a user and must be defined in relation to this agent's or user's purposes. So, if demonstration and interpretation are to be understood as relations, they must be conceived as essentially triadic in Peirce's sense. In the example used by Hughes to illustrate his account, it is indeed tempting to think of Galileo's geometric diagram as in itself denoting the kinematic problem, independently of any activity carried out by Galileo himself.[30]

The temptation does not arise for *demonstration*, which cannot be understood as a relation, the establishment of a relation, or the surreptitious use of a preexisting relation, for—recall—the demonstration is performed entirely on the source object without regard for the target. It is, in other words, a piece of reasoning carried out by someone entirely within some space of reasons or inferential framework provided by the model source (which contributes what I shall later refer to as the source's properly licensed *vertical* rules of inference). Hughes's helpful schematic diagram of the DDI account of theoretical representation (Hughes 2010, 154, reproduced as fig. 6.1 above) bears this out nicely. The workings of demonstration are indeed represented there as a vertical displacement line (arguably dynamic, with time running downward), while denotation and interpretation are represented horizontally and instantly at a given time. There is no scope to conceive of demonstration as a relation because, at this stage in the modeling process, the target is simply not considered—in the sense that the demonstration would proceed in the same way regardless of target.

Nevertheless, as I shall point out in due course, the dynamically evolving demonstrations carried out by scientists on any model source often rely on aspects of the equally dynamically changing interpretations. More specifically, in the terms about to be introduced, the *vertical* licensing rules that are applicable in model demonstrations can and often are sensitively dependent on both the evolving inferential capacities and the representational force of the model source with respect to some target or aspects of the target. Much modeling activity in fact responds to this constant pushback that the target exerts on the source, and vice versa. However, this does not affect the fundamental issue at stake at this point, namely, that Hughes's demonstration must be understood as an activity carried out entirely on the source, in the context of some modeling practice, in response to the rules of inference that are effectively licensed in that context. Such demonstrations rely on no other relations than those that operate within the space of reasons that applies to the model source, as interpreted, at any given time.

Finally, *interpretation*, the third and last element in Hughes's DDI account, can also be understood as a relation, but one might question whether this is the most apposite reading in the context. There is an established sense in model theory that takes an interpretation to be a function mapping the elements of the language into a domain of independent entities endowed with their own properties.[31] As we saw in the first section of this chapter, the interpretation or interpretative mapping is what, in this model-theoretical sense, provides any given set of sentences with a semantics under which they can be said to be true or false. However, this is not the kind of interpretation appealed to in the DDI account since, to the extent that the model source contains any sentences at all, they already come fully interpreted in terms of the model itself. More typically, as we saw in chapters 2 and 3, a model source will be some abstract or concrete complex object, diagram, equation, etc. The issue of interpretation thus does not concern the provision of a semantic interpretation in the model-theoretical sense of any given set of sentences in any language. Prima facie, at least, it instead concerns the application of some object or set of objects in the model sources (whether they be concrete physical entities, diagrams, or equations or imagined abstract renditions or versions of such entities, diagrams, or equations) to other objects in the model targets (whether they are real concrete entities in the world or imagined abstract entities). Thus, the third step in the DDI account corresponds not so much to an activity of interpretation as to an application of the model source to the target in order to derive results of interest regarding the target itself. This is entirely in keeping with the intended general condition on modeling, an activity ostensibly driven by inquiring into the properties of target systems by means of surrogate model sources that stand for the targets for such purposes.

The third application step is in fact critical to much modeling activity because it both imposes constraints and provides resources for the demonstration step. It thus pays to study it closely without prejudice, that is, regardless of whether it turns out to be an interpretation, and regardless of the sorts of sources that must be involved.[32] It is important to emphasize at this point that the application step is independent from and does not rely on any putative relation of denotation in the DDI model and thus that it cannot in any way be reduced to the relation of denotation.[33] First, the denotation relation stipulates the target of a given source in a representational relation but not what parts of the target object correspond to what parts of the source object. Thus, in Hughes's Galileo example, the mere statement that the geometric diagram denotes the kinematics of the system does not determine which parts of the diagram represent which parts or aspects of the kinematics. Second, the denotation relation does not stipulate how the source is to be partitioned into significant parts in the first place, and there is quite a bit of leeway for appropriate judgments here.

This is critical since the application of the source to the target requires a partition of the source into relevant parts and properties (a system of interrelated parts) and the relating of such a system to another system of interrelated parts in the target system. Nothing in the mere denotational relation of the target by the source will settle any of these claims. Thus, in the Galileo example, the geometric diagram clearly distinguishes vertical from horizontal lines and the area therein contained. Similarly, the kinematic problem situation clearly identifies time intervals, speed of motion at every instant, and constant or accelerated motion across the interval. These distinctions, features, and properties of source and target are all products of the activities whereby modelers (in this case, Galileo himself) apply representational sources to their targets. They are in no way fixed by the mere statement that the source denotes the target (in Galileo's example by the mere statement that his geometric diagram denotes a kinematic problem).

Hence, as an activity, interpretation is a bit of a mixed bag, including at least two different activities on the part of modelers. First, there is the ascription of some sort of system description to both the source and the target that partitions each object into elements and their relevant features. Second, there is the setting of a sort of correspondence or mapping between sources and targets, as described and partitioned. Both are activities within the modeling practice of applying models to their targets without which the third interpretation stage is rendered impossible. However, the partitioning activity is by contrast entirely within the modeling activity that constitutes the study and description of each of the systems separately, so only mapping can possibly issue in a sort of relation between sources and targets akin to denotation.

Hughes's account is thus prima facie a hybrid account including relations (denotation, mapping) and activities (demonstrating and ascribing or partitioning). It is ostensibly deflationary in only one of the senses explored in chapter 4, namely, the no-theory theory. It explicitly opposes the need for any necessary and sufficient conditions on representation. We can, however, wonder whether the account cannot be extended in a way that would allow a full and consistent application of deflationism in any of the stronger use-based senses explored in chapter 4. What would such an extension look like? Well, the activities laid down by Hughes are already normatively regulated within a modeling practice, one that imposes its own standards of correctness (in the kinds of demonstrations and partitions that are considered acceptable). But what about the putative relations? While both denotation and mapping are prima facie conceptually distinct from any activities, they can also be said to be the end products of some activities, and a deflationist account of representation would then want to focus on a study of the activities in themselves. So, in what follows, I intend to fill in the details of whatever activities in modeling practice eventually issue relations of denotation or mapping between modeling sources and their targets. I shall in the first place propose an extension of the original DDI model, a more complex account that fully addresses the rich tapestry of activities that make up all three steps in modeling practice. The development will eventually lead us to consider a much simpler, two-part inferential conception of representation.

The Denotative Function–Demonstration–Inferential Function Account

A fully deflationary strategy would recommend replacing the relations of denotation and mapping with their corresponding functional activities. I believe that there are credible candidates for such activities (see Suárez 2015, 44ff.) that we can call *denotative function* and *inferential function*, respectively. The resulting denotative function–demonstration–inferential function (DFDIF) account is an extension of the DDI account that is more faithful to modeling practice since it explicates the two standout relations in Hughes's original account fully in terms of the activities that underpin them. Both replacements are, moreover, independently suggested by the recent literature on the nature of fictional assumptions and judgments in scientific practice and the important and central role of scientific models representing fictional or imaginary entities, processes, or phenomena.[34] To give just one illustrious example, Maxwell's (1861–62/1890) vortex model of the ether is a model, a scientific representation, if there is one, regardless of the ontological status of its various components, including both vortices and idle wheels and for that matter the ether itself.[35] No relations are involved, yet undeniable representational work has been done.

The requirement of denotation would rule out such models as nonrepresentations, contrary to the thesis defended in this book that all models are representations. Yet the requirement can and must be weakened in favor of the activities that typically but not infallibly serve to establish denotation successfully. Catherine Elgin (2009, 77–78) points out the way when she writes: "A picture that depicts a unicorn, a map that maps Atlantis, and a graph that charts the increase in phlogiston over time are all representations, although they do not represent anything. To be a representation, a symbol need not itself denote, but it needs to be the sort of symbol that denotes." While this second sentence expresses a residual commitment to the idea that representation requires some notion of denotative function to be in place—and it will become clear in the next section that I too reject it as a strict condition on representation—it nonetheless rightly emphasizes that, for a source to function as a representation, it does not need to denote its target. And, while there are different ways of understanding what denotative function amounts to, I argue that they all coincide in one respect: it is a prerequisite for successful denotation, but not conversely. That is, nothing that denotes can fail to have denotative function, yet a source can have denotative function—and hence appropriately be said to represent—while lacking any denotation.

In replacing *denotation* with *denotative function*, we substitute a functional surrogate for an actual relation. How can this denotative function be understood? There are two possible characterizations, one coming from metaphysics and the other from a study of the logical form of the functional role that it displays. As a deflationist, I would of course urge the logical characterization, but in fact nothing much hinges on the choice since both alternatives yield the required conclusion, namely, that denotative function is a prerequisite for denotation, but not vice versa.

Let us see precisely why. The characterization in terms of logical form—which I shall adopt here—merely stipulates that a source A has denotative function with respect to some target B if and only if, were B real, then A would denote it. The statement must be read as a generic conditional, not a counterfactual, so that all real and existing objects that are denoted by some of our symbols continue to be denoted. There is no need to go further into an analysis of the statement and its truth conditions. We can simply observe that, under such a conditional reading, any source that denotes also has denotative function, but not vice versa. The logical form of the statement makes this clear. The metaphysical characterization, by contrast, interprets this conditional statement in terms of possible world semantics. It stipulates that a source A has denotative function toward a particular target B if and only if A denotes B in all (or most or even just a few) nearby possible worlds. This also allows for objects without denotation such as unicorn pictures to have denotative function and therefore properly represent—under the proposed replacement. For, while in the actual world there are no unicorns for a unicorn picture to denote, we can easily see how a nearby possible world (one identical in all respects to our own world except in the single respect that unicorns inhabit it) is one where the very same unicorn picture denotes. Hence denotative function is rendered into the modality of possible denotation, while denotation per se remains tightly linked to our actual world. In other words, according to such a metaphysical characterization, it is also the case that denotative function is a prerequisite for denotation since possible denotation is a condition for actual denotation, but not vice versa.

The critical difference between denotation and denotative function can also be expressed in terms of different sets of success conditions. In Suárez (2015), I defended the position that denotation is a success term (since denotation requires the existence of what is denoted) but denotative function is not (since it does not require the existence of whatever it is functionally designed to denote). Now I prefer to distinguish two different success conditions, one pertaining to a relation, the other to the function that it performs in practice. It then becomes clear that the success conditions for denotation include those for denotative function, but not vice versa. And such a logical relation between them is made manifest by either the logical form or the metaphysical characterizations just given. Hence, I would now say that a symbol can successfully possess denotative function toward an imagined entity, as in the case of unicorn pictures. (The advantage of this proposal is that it leaves open the possibility that a symbol fails both to denote and to have any denotative function, that is, that it fails to meet either set of success conditions.)

Thus, Maxwell's (1861–62/1890) model of the ether does not successfully denote what it purports to denote. We now take it to have no referent. But it certainly does successfully purport to denote, and it was so understood by everyone who considered it at the time. Like any other nineteenth-century physicist, Maxwell was committed to the existence of the carrier of electromagnetic waves even though his attitude to the vortices and particularly the idle wheels in his own model is highly nuanced. Recall from chapter 2 that he thought of them not as literal descriptions of the mechanisms underlying electromagnetic phenomena but as useful analogies. Both assumptions seem to function representationally nonetheless, and both display the hallmarks of denotative function. One denotes the actual vortices in the ethereal fluid, the other the small counterrotating mechanical devices required for the mechanical consistency of the model. Maxwell's methods for demonstration and application are no different than those of Galileo's in his purportedly nonfictional model of the kinematics of motion. Whatever representational work actual denotation carries out, it is just as effectively carried out by denotative function. And, since the latter is the more general class that encompasses both, it seems within reason to substitute it into the appropriately extended version of the DDI account.

Similarly, the mapping part of the application step in Hughes's model has the critical function within modeling practice of transferring over the results of the demonstrations carried out on the source onto the target. The deflationist instinct is then to focus entirely on that function. For example, Galileo interprets the overall area of the triangle as the space traversed by the body in its motion over the *t* interval. This is a sort of mapping or correspondence since it connects an element in the source (an area in the geometric figure) with an element in the target (the space traversed by the body in motion in the kinematic system). But what is its function? For Galileo, the use of the mapping is to allow him to carry out some inferences to conclusions about the target, namely, that the time *t* that a body in uniform motion takes to traverse *s* is identical to the time taken by a body uniformly accelerated to a greater speed. Hence, the functional role of the mapping is to constrain the set of inferences that are permissible about the target by technically licensing what are the legitimate surrogate inferences.

The deflationary thought is then that such constraints and licenses are in practice identified independently of any underlying relation between the source and the target or their properties, regardless of description. Rather, those relations are *constituted* within the very practice of ascription, partition, demonstration, and licensing of the relevant inferences.[36] In other words, taking area to stand for space traversed licenses a rule of inference to the statement that "equal areas correspond to equal times traveled." The relation between area in the source and time in the target is a consequence of the licensed inferences within the practice, and not vice versa. While there is a sense of correspondence or mapping here, this is not really a mapping between features of the source and features of the target but one between statements or claims about the source and statements or claims about the target. Yet a mapping between statements about A and statements about B need not require a mapping between features of A and features of B under any description of A and B. More specifically, it does not in any way require that either A or B be real, independently existing entities. We routinely make claims about all sorts of imaginary objects.

The deflationary thought is henceforth that the mapping between claims is the product of the inference rules licensed within the modeling practice; it therefore stands alone as feature of the practice itself. Talk about mapping is entirely responsive to those features of the modeling practice that license proper reasoning from the source to the target. It need not answer to any antecedent properties of source or target or any relation thereof.[37] The propriety of the mapping depends on the inference rules that are operative within the practice. No further assessment is available independently or from anywhere beyond that practice. Thus, in the Galileo example, we would be at a loss to find any similarities, any isomorphisms, or any other matching relations between the area of the geometric figure and the space traversed in a certain interval in the kinematic system—until of course the inferential rules are laid down that license us to proceed from statements about the former to statements about the latter. The dual norms of ascription (the partition of both source and target into various elements) and mapping (from claims about the source to claims about the target) make up what we can refer to as the *inferential function* of the source toward the target. The corresponding extension of the DDI account puts this notion in place of Hughes's underdescribed (and potentially misleading) notion of interpretation. The resulting account of representation—the DFDIF account—is consequently deflationary in all the relevant senses, including the use-based sense: it essentially accounts for representation entirely as a set of activities within a normative practice of model building.

The Plurality of Theories and Theory Views

Let us pause and take some stock. I first argued for a version of the semantic view of theories according to which theories are sets of representations. This was to be contrasted with the version of the semantic view according to which theories are structures, which fails for many reasons. I then argued for a fully deflationary account of representation in the form of a suitably extended version of Hughes's DDI account, the DFDIF account. On this account, representations are essentially a triple set of activities within a practice. So, while this account is inconsistent with the structuralist version of the semantic view, it seems perfectly coherent and at home with a representational version of the semantic view (the RSV).

Now, when we conjoin these two views (the RSV and the deflationary account of representation), we obtain the straightforward result that *theories are sets of activities within a practice*. This is an unusual claim to make in the context of a semantic view of theories, although R. I. G. Hughes came close to stating it in his later works (see, e.g., Hughes 2010). Yet it is not such an unusual claim overall these days. In fact, different approaches seem to be coalescing in roughly this direction. Thus, Winther's (2020a) entry on theory structure for the *Stanford Encyclopedia of Philosophy* distinguishes a third approach to theory besides the syntactic and the semantic views, one that he terms the *pragmatic view*. Winther emphasizes five distinct aspects of theorizing that deserve attention and, according to him, are ignored in the standard semantic and syntactic accounts. First, the idealized accounts in the syntactic and the semantic views are *limited* in what they describe as the predictive and explanatory work of theories. Second, theories include not just statements and models but metaphors, analogies, values, and policy views. In other words, they are *plural* in ways the standard accounts cannot accommodate. Third, theories are *informal*, or include elements that are hard, if not impossible, to formalize, such as guesses, heuristics, hunches, rough associations, etc. Fourth, theories' *functions* include references to users, purposes, and values. Finally, theorizing does not stand in a dichotomy with *practice* but is rather continuous with it.

Winther (2020a) argues that the standard views clash with these five aspects and that the pragmatic view owns them all. With regard to theorizing as a complex activity with multifarious parts and styles, Winther is very persuasive that these five features are often present. Yet one might wonder whether the semantic and the syntactic views were ever meant to describe or capture the activity of theorizing. Most authors in that tradition seem to address themselves more narrowly, to the products of theorizing only.[38] From this perspective, a suitably modified version of the semantic view can do all the work required by Winther, without falling prey to any inappropriate conflation. True, when conjoined with deflationism about representation, the RSV fuses Winther's semantic and pragmatic views on the nature of scientific theory. It no longer is the case that there is a contrast class since the representational version of the semantic view is pragmatic too! However, I urge that this conflation is coherent, fitting in well with Winther's desiderata, and providing a hospitable environment for a deflationary account of representation.

I also agree wholeheartedly with Winther that the different accounts of theory all have some merit. Each applies to some instance of theoretical work, and each is illuminating in its proper context. Winther (2020a) applies all three conceptions to a case study in population genetics, concluding: "The Syntactic, Semantic, and Pragmatic views are often taken to be mutually exclusive and, thus, to be in competition with one another. They indeed make distinct claims about the anatomy of scientific theories. But one can also imagine them to be complementary, focusing on different aspects and questions of the structure of scientific theories and the process of scientific theorizing. For instance, in exploring nonformal and implicit components of theory, the Pragmatic View accepts that scientific theories often include mathematical parts but tends to be less interested in these components. Moreover, there is overlap in questions—e.g., Syntactic and Semantic Views share an interest in formalizing theory; the Semantic and Pragmatic Views both exhibit concern for scientific practice." This meta-pluralism provides grist for my mill since the inferential conception of representation is naturally compatible with all views of scientific theory, including the semantic view, when properly understood. And ultimately this serves my purposes best since my aim is not to defend any view on the nature of scientific theory. It is rather to defend a pragmatist and deflationary approach to representation, namely, the inferential conception. And no account of scientific theory can get in the way of that.

Chapter Summary

This chapter begins the defense of a pragmatist and deflationary account of scientific representation as the more viable approach to the practical inquiry into the nature of scientific modeling. An important part of this defense involves showing that the inferential conception is compatible with any extant account of scientific theory. This has not always been made clear in the literature, where a thought persists that the semantic conception of representation is essentially linked to a substantive structuralist account of representation. The first part of the chapter is devoted to showing that this is not the case since there is a representational version of the semantic conception of scientific theories that is perfectly compatible with deflationary accounts of representation, including the inferential conception. The second part of the chapter turns to deflationism per se. It presents in some detail the first candidate for such a theory, R. I. G. Hughes's DDI account, and it argues that it is deficient in a few respects. It then extends this account by developing suitably deflationary versions of denotation and interpretation, respectively. The resulting account is deflationary and compatible with any extant account of scientific theory, even though narrowly applicable only to those representational sources capable of denoting.

7

Representation as Inference

In this chapter, I finally develop the inferential conception as a refinement of the DFDIF account developed in the previous chapter. More precisely, the three-part DFDIF account is reduced to just two essential elements (representational force and inferential capacity), which, it is argued, encapsulate summarily and more effectively all its deflationary virtues. The division into just two parts is further justified historically by grounding it on insights learned from both Hertz and Boltzmann in part 1 of the book. Specifically, I show that the inferential conception can be understood as a natural consequence of Hertz's emphasis on a minimal criterion of conformity, together with Boltzmann's defense of the informative nature of models. The second section of the chapter then provides the bulk of the defense for the inferential conception by showing how it effectively responds to those objections that, as we saw in chapter 5, make its competitors untenable. In the third section, I apply the inferential conception to the case studies and models discussed in chapter 3. In all cases, it is argued, the inferential conception respects established distinctions and illuminates the workings of models.

The Inferential Conception of Representation

The last chapter advanced arguments for an RSV of theories (according to which theories are representations) together with a deflationary account of representation (according to which representations are features of modeling practice). It is now time to put things together and simplify by distilling the essential elements of these views into a streamlined account of representation and theorizing called the *inferential conception* (Suárez 2004, 2010b, 2016). Deflationism recommends that we focus on the elements of practice and patterns of use that give rise to scientific representations, while the semantic view invites us to understand model building as the central pillar of any theoretical science. The main practical role of theories in science is to generate informative inferences in the way of either explanations or predictions of relevant phenomena. And, while not all models are or become part of theory, those that do make up theory are primary vehicles for inference. This suggests that the elementary or basic features of scientific representation must be conducive to inference in practice. The deflationary instinct then recommends raising such elementary or basic features in practice to the category of essential or necessary features. This allows a further simplification of the DFDIF account of representation.

SURFACE FEATURES IN MODELING

The striking result derived in the previous chapter is that both the demonstration and the interpretation steps in Hughes's original DDI account are features of practice that require us to perform a set of tasks on the source system and the target system, whether real or imagined. The interpretation step is better understood as the application of models comprising properly two sets of distinct activities: partitioning and mapping. Although the partitioning step is conceptually prior to either of the other two steps, in practice all three activities (demonstrating, partitioning, and mapping) intermingle in complex ways. Thus, in the Galileo example discussed by Hughes, once the convention "area ~ time traversed" is in place, any demonstration carried out on the geometric source immediately carries over to conclusions about the kinematic target system. However, the partitioning conventions are often changed in the course of modeling activity and often very much in response to the consequences of the demonstration itself. Thus, if area had not corresponded accurately with observed time traversed, Galileo would presumably have implemented a different convention and so on. So it is best to think of these three activities in terms of their effective upshot, namely, the licensing of some surrogative inferences from features of the source, as partitioned in its prescribed description, to features of the target, again under some prescribed description.

The inferential conception takes this licensing of surrogative inferences to be at the heart of representation and hence also at the heart of scientific modeling. And, while the capacity to implement surrogative inferences is not a sufficient condition for any source A to represent any target B, it is arguably necessary. The carrying out of surrogative inferences does not require that the inferences be to felicitous conclusions, never mind true conclusions.

What is more minimally required is the enforcement of rules of inference that distinguish the properly licensed conclusions from those that are not so licensed. This is indeed very minimal and perfectly compatible with the conclusions and the premises in such reasonings being false, improbable, highly implausible, or even absurd. What is at stake in licensing is the distinction between valid and invalid inferences from representational sources to their targets—not in any way whether such inferences are sound or cogent. It is thus important to distinguish different normative layers in the modeling practice. What I am referring to as *licensing* applies only to the most basic standards of validity.[1] There will be additional norms at any point in the modeling process to distinguish, among the licensed inferences, those that are also sound and cogent.

It is hard to see how the mere licensing of some surrogative inferences can be lacking in any instance of representation. Where such licensing is lacking entirely and there are no inferences whatever one can reliably draw from source to target, the strong intuition is that there is no representation at work at all. Thus, consider, in the most minimal model available, the map of a terrain with only two spots of different size on it, denoting merely two distinct villages with their relative populations. A map must provide a basic set of valid inferences that the user is invited to draw, such as, minimally, that there are indeed two villages, standing at a certain distance apart, one situated to the east, west, south, or north of the other. This requires only some very basic conventions and partitions to be in place: the identification of the dots, the relative sizes indicating populations, the usual conventions regarding depicting direction. It is a matter of course that all these inferences are defeasible: it may well be that the spots are interchanged (east is west or north is south and so on) or that they are misrepresentations of the sizes of the villages or their distances. The map can be a very erroneous and misleading one. However, we could hardly even call it a *map* if it did not invite the mistaken inferences at all. There is in fact no way to assess the propriety or accuracy of a map in the absence of such inferences. A piece of paper with just two arbitrary dots on it is not a representation of anything (it is not a map) unless it invites such inferences with respect to some domain that it purports to represent. Our first slogan can be this: *Nothing represents unless it has the capacity to license some surrogative inferences*, however minimal, inappropriate, misleading, or incorrect these inferences might be.

Several observations and caveats apply to the slogan, and there are significant informational constraints on the notion of inferential capacity. The first thing to note is that, so construed, inferential capacity is a property of a representational source in its relation to the target of a representation. Thus, we can say that Galileo's geometric model source has inferential capacities vis-à-vis the kinematics of rectilinear motion. We can also say of the Forth Rail Bridge diagrams, maps, and charts that they possess inferential capacities toward the bridge itself. Billiard balls, suitably construed within the kinetic theory, have inferential capacities toward the dynamics of molecules in a gas. And the highly idealized stellar structure models have powerful inferential capacities toward the observable quantities in stellar astrophysics. In all these cases, we are ostensibly ascribing capacities to source systems within a system of practice that uses them as representations of their corresponding targets. These model sources have inferential capacities toward their intended targets; while they may lack any such capacities toward other targets for which they are not intended in practice. In line with many other relational properties, while the inferential capacity of a model is inherently relational, it is nonetheless properly understood as a property of the model source system.[2]

Second, inferential capacities are contextual in ways that go beyond the mere statement that they are relational properties. Unlike other relational properties, they also require competent and knowledgeable agents to use them in the way they are intended in practice. What is for a well-trained and competent scientist evidently a representational source for a particular target need not appear to be so to the uninitiated. And, since, on the inferential conception, representation is a matter of use, the inability in principle of any agents to make proper representational use of a source also deprives it of representational status. A source can hardly be said to have any inferential capacities if there are no agents that can possibly master them and employ them for their own purposes in inquiry. If no agent can possibly draw the required licensed inferences, then the source system is arguably not in practice a genuine representational source. There can be, on this conception, no scientific model to speak of if whatever is presented cannot in any way be employed by some agent as the source of surrogative inferences toward the target.

However, we should be wary of the modality here. For a representation there must be a possible agent that is sufficiently competent and knowledgeable to draw the licensed inferences. But that possible agent need not be any actual agent. So a model can lie dormant and unused for any specified period. It matters not so much that it is not actively used as such as that it *could* be used as such by a possible agent. While this statement leaves open the definition of *possible* agency, nothing much will hinge on how this modality is filled in. All the scientific models and case studies that we have so far reviewed make it clear that the inferential capacities of the models are available to anyone sufficiently well versed in the technicalities involved. It is also evident that, in all those cases, the pedagogical aspects—the instruction in the proper use of the model—are an important and relevant part of the scientific context in which the models get used.

Finally, and most importantly, the sort of inferences that can be derived from a model source with respect to its target cannot trivially boil down to or follow strictly from the mere statement that the source represents the target. Such trivial inferences can always be drawn about any source object, namely, that is a representation of its target. Similarly, we can always safely assert about a name that it denotes whatever it names. This is an inference that is always trivially both valid and sound. If *Galileo* is employed as a name for Galileo (the seventeenth-century Italian scientist), it cannot fail to be the case that the statement "the name *Galileo* denotes Galileo" is true, and this is indeed a true statement about the name *Galileo* in relation to its object. But to rely on any such trivial inference to establish that there is indeed representation would be blatantly circular and grossly uninformative. It says nothing about its object other than it is so named, and it follows strictly from the very representational relation that obtains and nothing else. So it cannot be what explicates the relation that obtains. We must therefore insist that the inferences that are licensed about the target from a representational source be minimally informative: they must make substantial statements about the target that go beyond the mere fact that the target is represented by the source. I call these inferences *specific* (Suárez 2004), for lack of a better name. They explicitly rule out the establishment of representation by mere stipulation. A case of baptism or pure conventional stipulation does not by itself amount to any cognitive representational success since it cannot provide meaningful information about the specific target chosen other than the act of baptism itself. On the contrary, the stipulation would hold regardless of which object or system is chosen as the target.[3]

The inferential capacity of a representational source toward a target is not the only necessary condition on representation. A source can have extraordinary inferential capacity toward a certain target, even including to conclusions that are correct, true, appropriate, and salient, but it can fail to be a representation because it is inert. It lies unnoticed, entirely outside any representational practice. The fusing of the demonstration, ascription, and mapping activities involved in the last two steps in the DFDIF account, in terms of an overarching kind of inferential capacity, does not preclude the additional requirement that the source be endowed with some generalized denotative function, or, as I shall call it, *representational force*. Thus, the piece of paper with the two spots on it might very accurately represent the relative positions and populations of two villages in the vicinity if it is indeed used as a representation. However, if no one has ever paid any attention to the piece of paper, or if it has not been made part of any representational practice, then arguably it is not a representation at all. It is at best a *potential* or *dormant* representation. Our second slogan can thus be: *Nothing represents unless it is used as such by someone in some context*, however convoluted, obscure, or esoteric that use may be. An inert object that stands alone is not a representation until or unless it is introduced into a normative representational practice (see Suárez 2004, 768).[4]

The notion of representational force, or *force* for short, was introduced in past publications of mine (Suárez 2004, 2010b, 2016) and is meant as a generalization of denotative function to all representational devices, as follows. *Force* includes denotative function for those symbols that may acquire or develop such a function; but it also covers cases of representational sources that do not deploy denotational function yet point toward some target in an appropriate representational manner. Thus, physical objects and names typically have denotative function, but it is not so clear that conceptual systems or mathematical equations have such functions. Still, as we have seen, there is no question that both conceptual and mathematical models can and often do represent. So, arguably, we need a more general notion capturing all cases beyond denotative function. The *force* of a representational source therefore refers to its capacity to lead a competent and informed user to its representational target, regardless of what kind of device it is and how it instructs us toward its targets. Like denotational function (of which it is a generalization), force is a "relational and contextual property of the source, fixed and maintained in part by the intended representational uses of the source on the part of agents" (Suárez 2004, 768). This means that there is no need to refer to any self-standing antecedent relation between source and target in order to determine the force of the source toward its target. The force of a representational source is not per se a relation but rather—just like denotative function—a feature of the modeling practice, which explains how and why sources are used in practice in the way they are.[5]

These two features (representational force—or, more narrowly, denotative function—and inferential capacity) are the two most general necessary conditions one can provide for cognitive representation.[6] They nicely generalize and simplify the fully deflationary version of Hughes's DDI account, the DFDIF account developed in the previous section. Together, they provide the set of necessary conditions for any object, system, artifact, conceptual schema, mathematical entity, or item of any sort to represent any other such item. In some cases, particularly those that employ explicit symbols, representational force can crystallize as denotative function, but this need not always be so. In all cases, some specific surrogative inference toward the target should be sanctioned or licensed by the modeling practice, however minimal or uninformative such inference might be. The drawing of inferences requires some ascription of a system description to both source and target and a correlative partition of both into their parts and properties; a piece of demonstration carried out on the source, typically when considered dynamically, in its time evolution; and, finally, a mapping of claims about the source to claims about the target, where the validity of such mapping is akin to inferential validity, normatively leading from premises regarding the source to conclusions about the target. So, to sum up, on the inferential conception of representation [inf], we can say of any putative source object A that it represents some putative target B only if (i) A's representational force is B and (ii) A has specific inferential capacities toward B. These conditions still call for some unpacking, and they deserve a full exploration in terms of their import and implications, both for scientific modeling and for other areas in aesthetics and epistemology in which a consideration of scientific representation turns out to be relevant. However, a conclusion that we can draw at this stage is that no more general necessary conditions are forthcoming, meaning that—in the spirit of deflationism—there is no possible completion of this characterization in terms of any set of necessary and sufficient conditions.

This strikes me as the reasonable conclusion to draw from the debate over the last two decades. Thus, in some of my early writing (e.g., Suárez 2003), I posed several challenges to the analysis of representation in terms of substantive necessary and sufficient conditions. French (2003) attempted to close the definition of *isomorphism* logically, Bartels (2006) attempted something similar for *homomorphism*, and other proposals have been discussed to the same effect for a variety of measures of similarity (Giere 2004; Poznic 2016; Weisberg 2012). None of these attempts to provide a substantive analysis of representation strike me as successful. All of them suffer from counterexamples, and the arguments themselves often presuppose what they set out to prove (Suárez 2015, 2016). The only closing down of the definition that seems available is the equivalent of Wright's abstract minimalism, discussed in chapter 4. But, as we saw then, it does not allow for representation to play any substantive explanatory role.[7]

VERTICAL AND HORIZONTAL LICENSING: CONFORMITY AND INFORMATION

It is now time to come full circle to our historical account of the modeling attitude. For the inferential conception recapitulates, in an economic and elegant summary, some main features of the modeling attitude as we saw them emerge in their historical context. Chapter 2 reviewed Hertz's five basic conditions on representation: conformity, permissibility, correctness, distinctness, and appropriateness. Only conformity is necessary for a scientific representation. And conformity, recall, is the claim that "the necessary consequents of the images in thought are always the images of the necessary consequents in nature of the things pictured" (Hertz 1894/1956, 3). But, as we already noted, the necessity referred to at the level of our images cannot generally be the same necessity that operates at the level of the objects pictured. The latter is typically physical necessity and ruled by the laws of nature, while the former is the kind of logical necessity that follows from the rules of inference that operate on the *Bilder*, or, in our terminology, the rules of inference in the handling of our representational sources. We are now able to make it clearer what Hertz's conformity amounts to. Consider the diagram in figure 6.1 above with the provisos that—as has now been established—denotative function is to replace denotation and interpretation is best thought of as application. The intermediate demonstration step consists in the application of the rules that operate within the representation itself. These are rules of inference licensed by the modeling practice for the conceptual handling of the elements and features of the source, as partitioned in accordance with our description. In Galileo's example, this involves, for instance, the counting of the units of distance in the geometric diagram, and any comparisons between areas of enclosed surface must be carried out entirely at the level of this geometric representation.

Following the diagram in figure 6.1, I shall refer to these rules of inference—which operate at the level of the representation in and by itself and allow us to carry out demonstrations on the source object or system—as the *vertical* rules of inference. (The terminology nicely overlaps with Mary Hesse's [1963/1966] and is in line with contemporary distinctions in Bartha [2010, 2019].) The sort of licensing that the practice imparts to distinguish and apply such rules I shall correspondingly call *vertical* licensing. The overarching requirement of conformity then merely entails that the state descriptions in the source must always be applicable back to the system—that is, that they can always be rendered as descriptions of the state of the system described in the target. This is best envisaged dynamically. The state of the system modeled evolves according to its own dynamic natural laws, while the representational source's state description evolves in accordance with its own dynamic rules. It is then possible to infer a later state description ($S_{t'}$) from an earlier one (S_t), where $t < t'$. The requirement of conformity states that $S_{t'}$ can apply back to the system, thus rendering a description of the later state of the system in nature.

Note that, as was revealed in chapter 2, Hertz emphasizes that conformity does not require any of the other conditions. *Permissibility* is a requirement of consistency: the vertical rules of inference licensed must work coherently with other principles and laws of thought, particularly all those principles in classical propositional logic and the predicate calculus. A conformed representation need not be permissible, but it remains a representation (a *Bild* in Hertz's terminology) and a candidate for a scientific model if it conforms. In other words, a representation can include contradictions and impossibilities (as fictional representations commonly do) without precluding it from acting as a representation, but only on the condition that it conforms. It is undoubtedly desirable that a model be permissible, but this cannot be taken for granted. Instead, as was noted by Vaihinger (1924), many fictional models contain inconsistencies, but this does not generate a failure of validity of inference because all horizontal inferences are materially constrained by the targets to which they get applied (see Suárez 2009a). The rules of inference from a model source to its target go way beyond mere logical consistency and also answer to the material constraints at the point of application (Tan 2021). Thus, for example, a point particle is an inconsistent idealization that can be informatively applied to the physics of motion and collision in classical mechanics.

The remaining conditions apply to the horizontal relations that obtain at the point of application. *Correctness* is a requirement of accuracy. A representation is correct when its state descriptions effectively match those that occur in the target system. Again, this is not a necessary requirement for a representation. Thus, a conformed and legitimate representation can err in its ascription of states to the system: either the initial state S_t or the final state S'_t can be misdescriptions. What matters to conformity is the possibility of ascribing such states to the system, not that they be ascribed correctly. This is reflected in the emphasis in the inferential conception of representation on the validity of the inferences—both vertical and horizontal—as opposed to their soundness or cogency. In other words, a correct representation invites inferences that are sound in addition to being valid. Not only are those inferences licensed by the modeling practice, but they are also fortunate in being sound as well as licensed. And, while this is a desirable feature of a model, it is evidently not to be taken for granted.

Distinctness is a kind of correctness about the central aspects of the target system. Hertz writes that the most distinct representation "is the one which pictures more of the essential relations of the object" (Hertz 1894/1956, 2). Thus, distinctness is a feature of our horizontal inferences back onto the target system. It critically relies on the partition that has been applied to the target system, which may well have been chosen to privilege certain relations or properties of the target as essential. Finally, *appropriateness* is a form of simplicity. It is a requirement on the source itself, as partitioned, in relation to the target. The simplest representation is the one that makes fewer superfluous claims because it contains fewer features about which we can make any claims that have no possible mapping onto claims made about the target. Once again, simplicity is desirable yet hard to achieve in practice, but, at any rate, it is certainly not necessary for representation. Scientific models are not always streamlined and elegant in their simplicity. Sometimes they are convoluted, complex, and very difficult to handle. The inferential conception makes clear why these additional conditions are indeed not necessary for representation. Recall that, according to the inferential conception, only two minimal conditions are necessary: representational force and inferential capacity. They can jointly help us prove only conformity since they entail a minimal capacity of the source A to license inferences appropriately about the target. Any additional feature of the representation, however desirable, is a bonus but in no way necessary for a source to work as a representation of its target.

Nevertheless, correctness, distinctness, and appropriateness all point to one valuable feature of representations, namely, their informativeness. In chapter 2, we found that Boltzmann added an information-gain requirement to Hertz's conformity. The requirement goes further than simply prescribing, as was done in the previous section, that the inferences be specific, that is, minimally informative beyond the very representation at stake. But it does not go much further. The sort of information that needs to be conveyed by a representational source about its target can be very minimal, and it need not be veridical. In terms of the contemporary discussions, this entails a suspension of the so-called veridicality thesis for information, namely, the view that all information needs to be factual.[8] It is best to think of it as Shannon information, the sort that can be quantified in the terms of communication theory (Shannon 1948/1963). Here, an elementary analogy with a communication channel can be employed according to which the representational target (system or phenomenon studied) is akin to an information source, with the user of the model as the receiver of the information and the model a complex channel of communication that requires coding at the source (partitioning of the target) and decoding at the receiver (demonstration and further partitioning at the representational source). While this is schematic and merely suggestive, it agrees with the overall picture of a communication channel rather well (fig. 7.1).

Hertz's three additional conditions can then all be seen as adding something to the minimal amount of information conveyed by a merely conform-

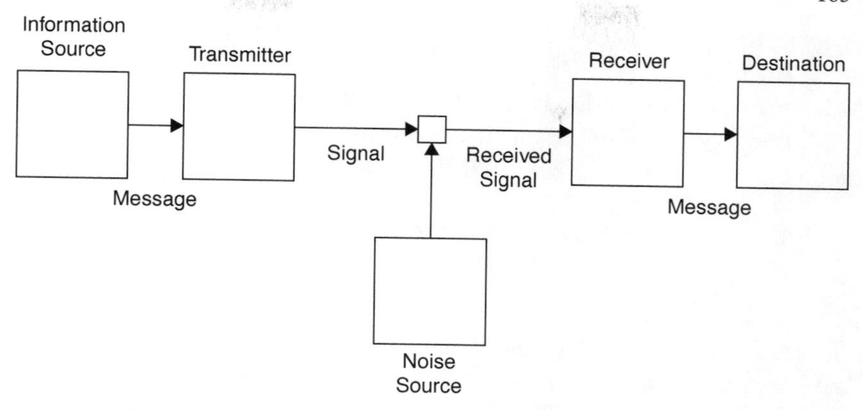

FIGURE 7.1. Shannon's mathematical theory of communication (see Shannon 1948/1963).

ing representation. *Distinctness* can be said to reduce the analogue of noise in the channel. A maximally distinct model is one that conveys all the essential information generated at the information source (i.e., the representational target), thus suffering from no noise. By contrast, *simplicity* can be said to reduce the analogue of equivocation in the channel: the simplest model is the one that contains no superfluous information, thus conveying no misleading information about the representational target. These conditions dispense with the veridicality thesis since they increase the communication channel information transmission, but not in any way that requires such information to be true. *Correctness* is the one condition that adds to the efficient communication channel a fair amount of veridical information about the target. On the present analogy, the veridicality thesis about information applies to representational sources only to the extent that they are the correct *Bilder* for their designated targets.[9] This all seems of a piece with Boltzmann's own intuitions about information, which he too thought quantifiable as a measure of what the model reveals about the source, without requiring the model to be entirely faithful to the target system.

The thought is that models are "information tools," that is, devices constructed with the aim of providing us with information (in the generic sense of communication theory) regarding their target systems. Note that this view chimes well with the current trend to view models as epistemic artifacts, except for the emphasis on the epistemology (Knuuttila 2011, 2021).[10] Models are indeed cognitive artifacts for *information gain* (in the not necessarily veridical or epistemic sense of the term), and this explains much of the methodology of model building that goes on in science.

The Inferential Conception Overcomes the Objections

The inferential conception is meant to provide the most minimal account of scientific representation. In line with previous chapters, we can refer to it as follows:

> *The inferential conception of representation* [inf]: A represents B only if (i) A's representational force is B and (ii) A has specific inferential capacities toward B.

That such representations are capable of transmitting (generic, not necessarily veridical) information is already built into the prescription that the inferential capacities are specific as defined. Note also that this statement stops well short of a full analysis of the constituent of representation. It establishes only necessary conditions. This is a strength when it comes to confronting the objections to the more substantive accounts that we studied in chapter 5. For most of the objections traded on the stronger conditions imposed by the similarity or the isomorphism accounts. By contrast, the inferential conception is quite content to accept that its account of the *constituent* is compatible with many different *means* of representation. For recall that, in Maxwell's colorful terminology in chapter 2, the means of a representation are merely the conductors by means of which we can study targets by reasoning about sources. So the means already provide all the required inferential capacities of sources relative to their targets. And we are thus now able to run through the objections, one at a time, analyzing how the inferential conception provides ways to confront each of them, either directly or by avoiding commitment to their presuppositions.

(a) *The argument from variety*: [sim], [iso] do not apply to all representational devices.

In agreement with the sorts of deflationism that we have canvassed in previous chapters, [inf] leaves open the possibility that different varieties of similarity and isomorphism appear as the means of different representations, each in its own context. This provides ammunition to overcome the objection from variety to the substantive accounts. For, since it is so minimal, [inf] can now apply to all representational devices while leaving open the possibility that different cases of modeling can have different means of the representation in each context. In running through the objection, we analyzed four different cases. In scale models, such as toy bridges (the San Francisco Bay/Delta models or the Forth Rail Bridge), similarity relations seem to be acting as the means of the representation while not constituting it. This is perfectly compatible with [inf] since it adds only the proviso that the inferential capacities of those model sources vis-à-vis their targets are exercised via similarity relations.

In some cases, as was noted, denotation in fact obtains over and above denotative function, and the bay model does denote the bay, the delta model the delta, and the Forth Rail Bridge diagrams the bridge once it is erected. And the mapping between claims about the source and claims about the target that we explored in the last chapter does coincide in fact with a genuine relation between the objects so denoted. There is no a priori reason why denotative function and mapping must be made as concrete as means, but it is certainly consistent with [inf] that they be as concrete. Similarly, as we saw, analogue models are often supported by constrained similarity relations. This is certainly the case for the Phillips-Newlyn machine model of the economy. It is perhaps not so evident for the billiard ball model, in which the analogy is essentially conceptual. But, as we saw, there are ways to lay down similarity relations between the dynamic properties of the systems involved. These again are means for the representation, and they are perfectly compatible with the minimal constituents as defined by [inf].

The ether models, such as Maxwell's (1861–62/1890), fail to denote, but they do possess denotative function, as we saw, and representational force. Nor can the inferential capacities of the models be questioned since they provide such an extremely illuminating and enriching set of conclusions, including the discovery of the displacement current and the postulate of Maxwell's equations of the electromagnetic field. On the contrary, the inferential conception has been expressly built to accommodate fictional representation (Suárez 2004). Whether similarity can be extended to cover cases of fictional representation is unclear, and this may be a case where the means of representation are the very rules of inference enacted for this model. At any rate, whatever means there are to distinguish in practice those inferences that are licensed from those that are unlicensed will be sufficient for the unproblematic application of the inferential conception to these cases.

We saw that, while mathematical models—such as the Lotka-Volterra model or statistical models of data—are prima facie candidates for isomorphic relations, they turn out in more complex ways to involve nested surrogative reasonings between their different layers. These models are again no objection to [inf] since any means to promote surrogative reasoning from source to target is compatible with the source's inferential capacities. No matter how the role is fulfilled, the only minimal condition imposed by [inf] is that it indeed be fulfilled. This is also compatible with those theoretical models, such as isomorphism, that appear to employ structural relations as the means of their representation. And, while the [iso] conception cannot provide for the constituent of representation, isomorphisms between the source's structure and the target's structure are certainly the means of many a theoretical representation. This again is fully in agreement with [inf], which leaves open what the means are that carry out or exercise the source's inferential capacities with respect to the target. The flexibility allows [inf] to capture any means of representation via models since the means are the *conductors*, in Maxwell's felicitous phrase, of the reasoning that take modelers from considerations about the source to considerations about the target, and vice versa.

Next:

(b) *The logical argument*: [sim], [iso] do not possess the logical properties of representation.

The second argument against substantive theories, recall, proceeded from the premise that scientific representation must share the logical properties of representation in general. It must be nonreflexive, nonsymmetrical, and nontransitive. This is hard for any matching account such as [iso] or [sim] to comply with. But it is easy for the inferential conception since both representational force and inferential capacities are stipulated as directional notions. They are meant to be conditions on the source relative to the target. So, according to [inf], a representational source will typically not be a representation of itself because representational force—or its instantiation in denotative function for those symbols that carry it—is rarely, if ever, reflexive. While the source A may have inferential capacities on itself, it does not typically have representational force on itself. And, indeed, with the very rare exceptions of self-referential models, we never use a model to represent itself. Representational force is nonreflexive. Also, by parallel reasoning, according to [inf], the fact that A represents B in no way requires B to represent A. That A has representational force toward B or that A has denotative function with respect to B in no way requires that B also has representational force toward A or denotative function with respect to A. Unicorn pictures have denotative function toward unicorns, but it does not follow that unicorns have denotative function back toward unicorn pictures. The Lotka-Volterra model represents the populations of two species in the Adriatic, but those populations certainly do not represent it. The essential directionality of representational force—and denotative function—prevents easy symmetries. And that is just what is needed for cognitive representation in general.

The argument goes through for transitivity mutatis mutandis. Representation is not generally transitive, and to this end I employed an example from art, the series of canvases that Francis Bacon produced around Velázquez's famous portrait of Pope Innocent X. They are properly representations of the Velázquez portrait and do not thereby also become representations of the pope. Representation in general is very rarely transitive. [inf] accommodates this feature easily since representational force—and its special case in denotative function—is directional. If A has representational force toward B and B has force toward C, it does not follow that A has force toward C. We can see this easily for the special case of denotative function. A unicorn picture has denotative function toward unicorns, and a description of a unicorn picture (including its frame, texture, color palette, etc.) has denotative function toward the unicorn picture. It does not follow that the description has denotative function toward unicorns. Indeed, it patently does not. When it comes to transitivity, even the second necessary condition in the inferential conception, namely, inferential capacity, can fail. For A can have the appropriate inferential capacities toward B in the right context, and B can have inferential capacities of its own toward C in its own proper context. However, it does not follow that there is any context in which A has inferential capacities toward C. This would depend on, among other things, the nesting of the relevant properties in A, B, and C, and there is no guarantee that they would be appropriately ordered for transitivity to obtain. We can conclude that [inf] does not fall prey to the logical argument in any of its varieties.

Now:

(c) *The argument from misrepresentation*: [sim], [iso] do not make room for the ubiquitous phenomena of mistargeting and/or inaccuracy.

Recall that mistargeting is the mistaken ascription of a target, often based on misleading similarity or isomorphism, to a source that does not actually possess representational force toward it (because it represents something else). The friend disguised as Pope Innocent X can be taken to be the representational target of Velázquez's portrait, but he really is not. This is an objective matter of fact about the painting in the right context of interpretation. If anything, we could perhaps, within the context, suppose that the friend represents the portrait. That is, it is possible in the context that the person denotes the portrait (say, if someone turns up at a party disguised as "Velázquez's famous portrait") or even the pope (if that someone turns up disguised as "Pope Innocent X as represented by Velázquez"). Notwithstanding, there is certainly no denotative function that the Velázquez portrait has toward the person so disguised; it would always be a mistake to ascribe that target to the portrait. True, this is clearly an easy mistake to make in the presence of the friend and on first inspection of the portrait if one is in complete and blissful ignorance of its provenance. But that such a mistake is likely does not make it less of a mistake.

Mutatis mutandis for cases of scientific representation. The quantum state diffusion equation can be taken by someone uninformed about its role in quantum mechanics to represent instead a classical diffusion process, such as Brownian motion. The essential directionality is patent in the quantum states involved, but they too can be mistaken for representations of classical states. The mistaken ascriptions would perhaps be understandable and might even be useful in the context of some inquiry. But they are mistakes nonetheless, which also gives the lie to the notion that representation is mere similarity or isomorphism.

The inferential conception gets around the objection in the simplest way possible by simply stressing its first, necessary part. The portrait has representational force (denotative function) toward the pope and nothing else, and the equation has representational force toward the quantum diffusion process and nothing else. Whether similarities obtain with other objects, persons, or processes is immaterial unless such similarity gets mobilized within some articulate practice to determine some new representational force. Recall the fundamental point stressed throughout the book: the setting of the representation highlights the appropriate and salient relations, not the other way around. Regardless of their inferential capacities toward other systems, objects, or phenomena, it remains the case that the portrait is directed toward the pope and not the friend and that the equation is directed toward the quantum diffusion process and not the classical particle in Brownian motion. This already dispenses with the argument from misrepresentation based on mistargeting.

Yet we still need to confront the part of the argument based on inaccuracy. Recall that models are often inaccurate (sometimes they are even grossly so) and that it is sometimes useful to know and calculate how far from the mean or the norm those divergences are. These are issues that [iso] and [sim] are unable to address or provide any guidance whatever on. Can [inf] do better? I argue that it can. First, inaccuracy as a phenomenon is very easy to fit into the inferential conception. For recall that the inferential capacities of sources toward targets built into the account are normative only with respect to validity. There is no requirement that the source be capable of yielding sound inferences to true conclusions about the target. (This of course might be so and an important virtue of any model, one that we would overall prefer, but it is not a condition for it to be a representation.) Hence, on the inferential conception, a model might be extremely inaccurate—even off the mark—but this does not preclude its being a model and properly a representation. Of course, being off the mark and being inaccurate are good reasons to dismiss the model as an unhelpful one in practice. Yet there is nothing in the model qua representation, beyond its inaccuracy, that renders it objectionable.

How about quantifying the degree of divergence? This of course will be a matter to be judged for each model in its context of application, for which a lot more concrete detail regarding the case would be required for us to be able to say anything informative. But the application of [inf] can certainly be part of such an informative answer. For the inferential capacities of the model include, among other things, a measure of what Hertz called *distinctness*. True, some of the inferences the model licenses toward the target will be false, but some quantitative inferences may well be true, and a weighted average of such inferences could in principle be constructed to offer some guidance on the question of accuracy. A crude measure, for instance, would simply compute the ratio of sound inferences within the larger set of all valid inferences. A not so crude measure would appropriately weight each inference in accordance with its importance within the specific inquiry. For instance, just to go back to the Galileo example, sound inferences regarding the times taken to transverse would be critical to the model; while inferences regarding the physical workings of the kinematic system are irrelevant, and there would presumably be others lying at other points on the spectrum. A not so crude measure would then apportion weights $\{w_1, w_2, \ldots, w_n\}$ to the divergence of each quantitative feature $\{f_1, f_2, \ldots, f_n\}$ and calculate the weighted ratio of those features accurately predicted over all features predicted $\{f_1, \ldots, f_n, \ldots, f_m\}$ thus: $(\Sigma_{i=1}^{i=n} w_i f_i)/(\Sigma_{i=1}^{i=m} w_i f_i)$. While this is still crude, it provides the beginning of a procedure to quantify inaccuracy. This shows that [inf] has a potential to handle the objection from inaccuracy, something the other proposals cannot do from the start.

Next:

(d) *The nonnecessity argument*: [sim], [iso] are not necessary for representation. Representation can obtain even if [sim], [iso] fail.

First, note that, unlike trivial similarity (the sharing of any properties, however artificially concocted), the conditions on representation according to [inf] are in no way universally satisfiable. It is not the case that any source has any given inferential capacity toward its target. It is also certainly not the case that any source has representational force toward any target. On the contrary, while both are set within a practice, the practice certainly does not condone any arbitrary association. There are objective matters of fact about both inferential capacities and representational force. It just so happens that these are objective matters of fact regarding the features of a practice rather than any self-standing relation between objects. Within the modeling practice of classical mechanics, for instance, friction is modeled as a linear term of velocity; any other proposal is not acceptable modeling practice. In quantum mechanics, probabilities are computed as the square modulus of the wave-function amplitude, and there is no scope for other choices. In the Lotka-Volterra model, prey and predator populations are modeled by a nonlinear pair of intermingled equations, and there is no scope to resolve equations further into a set of linear equations. And so on. In sum, the charge of triviality patently does not apply to [inf], which does appropriately distinguish representational sources and targets from inane objects or systems that play no representational role.

The nonnecessity argument has no bite against [inf] since the inferential conception has been constructed precisely to capture the most minimal conditions that cognitive representations in general meet. I already argued that the conditions are so minimal as to be met by any instance of a "re-presentation," however ductile, mistaken, or grossly inaccurate. Note that they are conditions relevant to a practice only, thus imposing no requirements whatever on the nature of the objects involved as source or target or their relation. The inferential conception thus overcomes the nonnecessity objection by construction.

Finally:

(e) *The nonsufficiency argument*: [sim], [iso] are not sufficient for representation. Representation can fail to obtain even if [sim], [iso] hold.

The easiest argument for [inf] to meet is the nonsufficiency argument since the inferential conception simply does not advance a set of jointly sufficient conditions for representation. Recall the distinctions presented in chapters 4 and 5. The constituent of representation is as fully described as it can be by means of these two necessary conditions, but they will typically not suffice in any instance of scientific representation. Other conditions will apply in addition since any application of a representation requires some means to be in place. These can include varieties of isomorphism and similarity relations or other relations between source and target[11] articulating mappings between claims about the source and claims about the target. There is a lot of leeway here to establish additional conditions if they serve to advance the fundamental aim of surrogative licensed inference. It is even conceivable that cases of representation arise where no additional condition or relation between representational sources and targets exists. While in these cases the means of the representation may be surrogative inference drawing itself (as suggested above for Maxwell's model), the cases remain pathological or exceptional. It is rarely the case that representational force and inferential capacities are sufficient in general for representation. Some effective means must obtain in addition.[12]

Reviewing the Case Studies: Inference and Information Transmission

The inferential conception [inf] overcomes the objections that make its competitors untenable. But is it able to shed new light on our modeling case studies? I argue that, for an array of cases of different types, there are aspects of the practice of model building that make a difference to an inferential approach. In running through these case studies, I focus particularly on four issues of relevance to the defense and development of the inferential conception, as follows:

- The ostensible aim, sometimes even explicitly so, of these models is surrogative inference drawing. The assumption that this aim drives modeling activity explains the design and development of the models.
- Much of the modeling activity, particularly in the publication and dissemination stages, concerns what Boesch (2017, 2019a) refers to as *licensing*: the establishing and conveying of the horizontal and vertical rules of inference that are effectively operative in the handling of the model.
- Many of the considerations that go into the development and refinement of the model are directed at achieving a virtuous dynamic equilibrium between the two parts in the inferential conception's "fork": representative force and inferential capacities. This involves sometimes modifying a model's inferential capacities (the licensed set of rules) to account for changes in the representational force or, conversely, its intended force to accommodate new inferential capacities.
- An upshot of this activity—not merely a by-product since it is often intended—is a refinement in the model understood as an efficient communication channel that can convey (not necessarily veridical) information about the target.

The three main case studies intended as benchmarks will be commented on in turn, with comments interspersed about some of the other models reviewed at different points in previous chapters of the book.

THE FORTH RAIL BRIDGE

Consider two of the models discussed in chapter 3: Benjamin Baker's anthropomorphic human model of the cantilever (fig. 3.3 above) and his diagrams and plans for the Forth Rail Bridge (fig. 3.4 above) as detailed in Westhofen (1890). The point of the first model is to exemplify the cantilever principle that the bridge is built on. We are invited to imagine the setting depicted in the photograph, where two men hold the lower bars in compression with their arms in tension, thus supporting the weight of the suspended man in the middle. But there is a lot of detail that is truly irrelevant to the purpose of the model, including the actual people involved, their physical appearance, and their clothes; not to mention their names, relative social positions, and personal ethical and political values. None of these aspects of the model matter, even though one could draw myriad inferences concerning them. But these are not inferences licensed in the context of use of the model. The inferences that we are invited to draw concern exclusively the balance of forces and strengths—specifically, the combination of compression in the lower bars and tension in the men's arms. These are licensed in context because they are the ones that truly serve to exemplify the principle of any cantilever bridge. There are then several comparative inferences from the model as enacted to the diagram that sits above the figures in the photograph, which in turn represents the bridge.[13]

The relevant licensing applies to both vertical and horizontal rules of inference. On the one hand, some vertical inferences are licensed within the model—mainly concerning the balance of forces and the stability of the man in the middle position as well as the weight at both ends, which must be proportionate to the tension forces. Otherwise, the arms would go up, and the man in the middle would fall. On the other hand, there are horizontal inferences licensed since the cantilever principle that applies in the human anthropomorphic model applies to the bridge as well. One is thus invited to infer that the upper arm of the bridge is in tension while the lower arm is in compression as well as that the weight at both ends of the bridge must be proportionate to the tension experienced since otherwise the arms would blow up and the middle span of the bridge would collapse under its own weight. These are legitimate inferences to make from a model that is intended to exemplify the cantilever principle that is embodied in the bridge. While the model is a curiosity of sorts, it is remarkably illustrative, and it serves wonderfully as an illustration of the appropriate licensing of both vertical and horizontal inferences. It is also notably successful in simultaneously conveying in a very succinct and intuitive manner general information regarding the principle of the cantilever and specific information about the nature of the concrete forces operating in the actual Forth Rail Bridge.

Baker's diagrams and plans for the bridge depicted in figure 3.4 depict the physical and geometric features of the bridge, including its array of different parts, lengths and widths of the different tubes and tubular structures and so on. As in the anthropomorphic model in figure 3.3, there are several licensed inferences among all those inferences one could arbitrarily draw from the graphs toward the bridge. For instance, the graphs are informative regarding the relative widths of the upper and lower parts of the bridge's arms, the relative positions of its different parts, and their comparative shapes. This includes the L-shaped girders with maximal tension strength for the upper spans and the compression-resistant tubular-shaped lower spans. These are the inferences that are licensed. Those regarding other accidental features (such as the depth of the Forth's basin, the shape of the boats underneath the bridge, the color of the bridge's parts, etc.) are not so licensed in the context.

The whole series of bridge plans also serve to exemplify the dynamic aspects of model building. As reasoning regarding the different features of the bridge advances and is refined, the diagrams are increasingly tailored to the task in hand. What begins as a rough sketch for the general principle of the cantilever is progressively refined into increasingly complex structures of arms, piers, and girders, each responding to the overall need for a robust construction able to overcome extreme side-wind strengths. The representational force in figure 3.3 is toward the general principle, but, by the time the implementation has been achieved in practice, the plans in figure 3.4 point toward not the general principle but distinct parts of an actual bridge. There is productive tension between the representational force of a model and its inferential capacities, a tension that lies at the heart of its ongoing development as a model.

THE BILLIARD BALL MODEL IN THE KINETIC THEORY OF GASES

The four key aspects of representation in the inferential conception are, if anything, exemplified even more clearly in the billiard ball model for a gas. Chapter 3 emphasized that the billiard ball model is in fact Boltzmann's perfected version of an earlier ideal spheres model in James Clerk Maxwell. The ideal spheres model is highly idealized. Gas molecules are represented as perfectly elastic, perfectly spherical small particles with no internal structure. They move randomly, collide freely, and dissipate no energy, so their interactions are entirely conservative. As a matter of fact, gas molecules are more like billiard balls in that the latter are subject to friction, possess some elasticity, and dissipate energy in their interactions with each other (and with the billiard table). Yet there are evidently important differences (Suárez and Pero 2019, 361ff.). First, billiard balls are shiny, opaque, hard, etc. None of these properties credibly apply to gas molecules. At best the model pretends that these properties obtain in the target system of gas molecules, but it cannot seriously postulate them. In other words, no inferences are licensed to these properties in the gas molecules, as a matter of course, in the practice of building and applying the model.

Second, there are properties of the gas molecules that are ignored or denied in the billiard ball model, such as viscosity, thermal conductivity, and free expansion. The model simply abstracts such properties away. It is hard to see how it would not since some of these properties are collective properties of the gas as a whole and there are no analogous properties of the system of billiard balls, collectively considered. Thus, viscosity, for example, is a physical consequence of the density and temperature of the whole gas. But a system of billiard balls lacks any such connection between the balls' density and temperature and their "viscosity." Even if one could define the temperature and density of the set of billiard balls collectively considered (which one cannot, except as averages), there would be no corresponding property of viscosity. And beware the nature of this denial. It is not that we do not know or cannot determine the viscosity of billiard balls but rather that billiard balls have no viscosity at all. It is impossible to use the model to learn about the viscosity of the gas molecules because there is simply no analogue property obtaining in the model. And this is something we know fully and with complete certainty. Such inferences are not possible, and it would be erroneous to pretend to infer to any conclusion regarding the viscosity of a gas on the basis of the billiard ball model. There is just no feature in the practice of using the model that can possibly license such an inference.

The inferences that are licensed are those relative to the dynamic motion of the molecules, and they are licensed within limits. For instance, billiard balls experience gravitational forces and friction in their movement that gas molecules do not. Yet, in terms of their random motions and near-elastic collisions, gas molecules are somewhat like billiard balls. Those are the only inferences that are licensed by the model. Of course, once the full kinetic theory is set up properly, there is no need to investigate the analogies further. However, as we saw in part 1 of the book, the model provides some useful heuristics for the development of the theory and a good illustration of those dynamic properties of gas molecules.

The inferential conception cuts to the heart of this distinction between licensed and unlicensed inferences, without further detour. The licensing of the vertical inferences—or rules of demonstration—within the system of billiard balls is provided by the classical mechanics of collisions. The licensing of the horizontal inferences—the interpretive or mapping aspects of the model reviewed above—is provided in the building of the model and its application in the community of physicists in late nineteenth-century Europe. Those inferences are never licensed simply as a matter of mere stipulation. Rather, they grow organically as the model develops and is moreover applied in the further development of the kinetic theory (Brush 2003; Darrigol 2018). This development occurs in space and time. It is conditioned partly by the natural world that gas molecules and billiard balls inhabit and partly by the social world where scientists operate. The licensing of the relevant inferences is not therefore a mere act of volition or intention on the part of any given agent or scientist. It requires setting up intended uses of the model and making sure that all its future applications obey the norms of application in that use. In other words, the legitimate uses of a model do not answer to arbitrary intentions or designs. They are, rather, sanctioned within the relevant practices of the community of model users.

This takes care of inference drawing and licensing. It also exemplifies the ways in which a billiard ball model can be said to be informative regarding the properties of gas molecules. We do learn something—however limited and conditional—about the molecules that make up a gas when we compare them with billiard balls. Namely, we learn that they are dynamically like billiard balls; they exhibit nearly completely elastic collision forces and displacements. This enlightens us regarding the nature of the motions of molecules in the gas, and we can build on these assumptions to create a more comprehensive theory of the behavior of the gas. But what about the dynamic equilibrium mentioned earlier in this section between representational force and inferential capacities? Where is this to be found as a productive force in developing the model?

A more detailed review of the historical development of the kinetic theory of gases provides some enlightenment about this specific point.[14] The theoretical background is provided by the assumption (originating in Bernoulli) that gases are composed of minute molecules traveling at great speed in all directions. The greater the speed, the greater the pressure the gas exerts on the walls of its container, and the greater its temperature. By 1859, when James Clerk Maxwell turned his attention to gas dynamics, Rudolf Clausius had already established the mean free path for gas molecules, and he was working on a program to develop a theory to explain the properties of gas compounds by appealing to the dynamic properties of the compounding molecules. However, he was not focused on figuring out the distribution of molecules' velocities within the gas, and certainly he was not assuming that the distribution function can express a probability. Maxwell's contributions to the kinetic theory of gases were key advancements in both respects, and they gave rise to what is now known as the Maxwell-Boltzmann distribution.

The historiography of the last twenty or so years has, in addition, revealed that Maxwell's introduction of the distribution function for the velocities of gas molecules took direct inspiration from the then nascent science of statistics. Already at Edinburgh, around 1850, Maxwell was exposed to the dispute between his mentor Forbes and John Herschel regarding the use of probability arguments to determine the distribution of stars (Harman 1991, 125). Herschel published an extended review of Quetelet's works in that year's *Edinburgh Review* (see Herschel 1850/2014), which Maxwell may have read then. He certainly studied it closely later, in the winter of 1857–58, when the essay appeared reprinted as part of Herschel's essays (see Harman 1991, 124–34; and Porter 1981).[15]

The key lesson that Maxwell derived from Herschel was the ubiquity of the error distribution function, which both authors associated with underlying uncertainty. As has been accurately pointed out, Maxwell's derivation of the velocity distribution function for gas molecules strikingly parallels Herschel's (1850/2014) derivation of the measurement error distribution function (Gyenis 2017; Harman 1991, 124–34; Porter 1981, 1986). The critical assumptions required to derive the Gaussian form of the molecular velocity distribution along three different axes *x*, *y*, and *z* are essentially transpositions from the equivalent assumptions in the theory of errors. Thus, it seems plausible that Maxwell was from the start inclined to regard the properties of gas molecules as subject to the uncertainty that is at the root of measurement error. Following Gyenis (2017, 54), the first assumption (A1) is that there is indeed a unique and stationary distribution function for the velocities of all molecules in a gas. This is not a trivial assumption. Molecules' velocities could be unevenly distributed in different media, and they could be unstable, ever evolving and changing in regular patterns depending on regularities within their environment or in irregular patterns out of irreducible stochasticity or noise. After all, Clausius did not even think of ascribing a distribution function to the velocities of the molecules in a gas, never mind a stationary one for every gas.

The second (A2) and third (A3) assumptions that Maxwell took from Herschel are well-known and widely recognized (Harman 1991, 126–28). A2 stipulates that the joint distribution function over velocities factorizes into separate distribution functions for each of the velocity components along *x*, *y*, and *z* directions. In other words, the velocity components are independent of each other. A3 states that the magnitude of the distribution depends on the particle's velocity magnitude only, that is, on its distance from the origin of the velocity vector and not at all on its direction or the orientation of the velocity vector. In other words, the distribution is isotropic. Herschel (1850/2014, 394ff.) found a justification for these assumptions in the theory of errors. Given our lack of knowledge of the actual causes of measurement error, error is as likely in any given direction in space, regardless of the axes of rectangular coordinates with respect to which it is measured. He illustrated this with an example in two-dimensional space with *x*, *y* coordinates (Herschel 1850/2014, 394): "Suppose a ball dropped from a given height, with the intention that it shall fall on a given mark. Fall as it may, its deviation from the mark is *error*, and the probability of that error is the unknown function of its square, i.e., of the sum of the squares of its deviations in any two rectangular directions." This postulate is, according to Herschel, the consequence of "*complete* ignorance regarding the sources of error, and their mode of action," or, in contemporary terminology of its proximate and ultimate causes and causal paths. Given such ignorance, the distribution of errors is just as likely in any given direction, depending only on magnitude. The larger the distance from the mark, the less probable the error becomes. This is best modeled as "an *even* function, or a function of the square of the error, so as to be alike for positive and negative values" (Herschel 1850/2014, 397).

While demonstrating proposition 4 in his paper, Maxwell applied a parallel reasoning in his derivation of the distribution function for the velocities of the molecules in a gas. He first assumed that "the existence of the velocity [component] x does not in any way affect that of the velocities y or z, since these are all at right angles to each other and [hence] independent" (Maxwell 1860–90, 380). And then he immediately supposed in addition that "the directions of the coordinates are perfectly arbitrary, and therefore [the number of particles within the element of volume (*dx*, *dy*, *dz*) after unit of time] must depend on the distance from the origin alone." Together, these two assumptions yield—as in the theory of errors—that the distribution function must satisfy the spherically symmetrical independence assumption equation: $f(x) f(y) f(z) = \phi(x^2 + y^2 + z^2)$. The only function that satisfies this equation is the Gaussian:

$$N(v, v+dv) = N\frac{4}{a^3\sqrt{\pi}}v^2 \cdot e^{-(v^2/a^2)} \cdot dv,$$

where N is the total number of particles and $N(v, v + dv)$ is the number of particles within each unit velocity interval $[v, v + dv]$. This is the famous proposition 4, and, on its basis, Maxwell goes on to derive a number of striking results regarding the phenomenological properties of gases that can, he claims, all be experimentally tested.

In terms of the inferential conception of representation, these results constitute the inferential capacities of Maxwell's (1860/1890) model, the first and founding model in the kinetic theory of gases. Thus, proposition 6 establishes an earlier version of Boltzmann's notorious H-theorem (see Gyenis 2017), namely, that, when two systems of particles are brought together (when two gases mix in the same vessel), the mean vis viva of the particles will become the same in the two systems. Then, in proposition 12, Maxwell proves a version of Avogadro's hypothesis, namely, that the number of particles per unit volume *N* is the same for all gases at the same pressure and temperature. The key assumptions (A1) regarding the existence of a stationary distribution and (A2) its spherical symmetry follow from the model assumptions employed by Maxwell, namely, the decision to model gas molecules as perfectly elastic spheres with minute but finite radius *r*. This was not, however, the only possible choice. Maxwell himself discusses two alternatives, and he goes on to employ one of them in part 2 of his paper. The alternatives considered by Maxwell license different and novel inferences. They thus build up alternative inferential capacities.

First, there is the option to model the particles as "centers of force" around a one-dimensional point, with the forces being negligible everywhere except at a certain small distance from the central point, identical to the radius *r* of the perfect sphere in the original model. Within that radius, the force "suddenly appears as repulsive force of very great intensity" (Maxwell 1860/1890, 378). Maxwell (1860/1890) does not employ this model assumption at any point, however, since he claims it to be "evident" that either model will lead to the same results. Yet, as we shall see, for good reasons he turns to it in Maxwell (1866–67/1890). The second alternative is, of course, to model the gas molecules as not perfectly elastic or not perfectly spherical, and, in Maxwell (1860/1890, pt. 3), he turns to just this assumption and abandons spherical symmetry. In the terminology of the inferential conception of representation, the description of the modeling target as perfectly elastic, minute, but perfect spheres colliding elastically is abandoned in favor of a less idealized conception. Why? The short answer is: Because it increases the inferential capacities of the model source toward its target, allowing Maxwell to infer certain features of gases that he would be unable to predict on the old model. Not only that. He feels dissatisfied with the original model of perfect spheres because it leads him to counterintuitive results regarding the viscosity, or internal friction, of gases. Thus, he calculates the ordinary coefficient of internal friction as $\mu = \frac{1}{3}\rho l v = 1/(3\sqrt{2})(Mv/\pi s^2)$, where ρ is the density, l is the mean length of path, v is the mean velocity, s is the distance between the centers at collision, and M is the mass of the gas. In other words, Maxwell assumes that the effect of density on viscosity is canceled out. By contrast, Stokes had found out by conducting experiments on air that $\sqrt{(\mu/\rho)} = 0.116$, that is, that the viscosity or internal friction of a gas is inversely proportional to its density (Maxwell 1860/1890, 391).

In part 3 of the 1860 paper, Maxwell consequently extends his account to the case of collisions between perfectly elastic bodies of *any* form, including nonspherical forms. The critical difference is that collisions between nonspherical bodies do impress rotational forces on them since the force of impact will not generally be along the line joining their centers of gravity. Instead: "The force of impact will depend both on the motion of the centers and the motions of rotation before impact, and it will affect both these motions after impact" (Maxwell 1860/1890, 405). The inferential capacities of the model shift accordingly, and Maxwell is now able to calculate ratios of translation and rotational velocities, which allows him to derive a host of new results, including the equipartition theorem (see Maxwell 1860/1890, 408–9 [proposition 23]). Further inconsistencies with known experimental results lead him to develop yet another model in which he adopts the first alternative and moves on to conceive gas molecules as centers of force (see Maxwell 1866–67/1890). There is a heuristic in the progressive development of these models according to which the different inferential capacities of the model are tested and lead to new proposals regarding the appropriate sources for reconceptualized targets of gas molecules. In each instance, the description of the target changes in response to results derived in view of the inferential capacities of the sources. The overall aim can be described as one of maximizing the amount of positive (not necessarily veridical) information that the source yields about the target. The ever-developing models progressively adjust their representational force in view of their own inferential capacities, which are in turn revealed in the process, and vice versa.

STELLAR STRUCTURE MODELS IN ASTROPHYSICS

The four stellar structure equations reviewed in chapter 3 (the hydrostatic equilibrium, continuity, radiative transfer, and thermal equilibrium equations) jointly display formidable inferential capacities. Together, they enable astrophysicists to infer the values of observational, or empirical, quantities such as luminosity (L) or effective surface temperature (T_{eff}). The models input values for the unobservable or theoretical quantities such as initial chemical composition, pressure, and temperature at the core, heat flow, and rate of nuclear energy release per unit mass in addition to the initial size of the cloud of gas. They then go on to output temperature and luminosity at every layer, including the outer photosphere. The main purpose of such models is to facilitate inference to those observable quantities that are then plotted on an HR diagram. This then allows astrophysicists to plot the star on the diagram, classify it accordingly (whether as a main sequence star, a white dwarf, or a red giant), and consequently rather accurately predict its life cycle. The inferential capacities of the model are transparently apparent in its use, and the diverse assumptions built into the model are meant for their advancement.

In their standard interpretation, the equations that make up the model (really a family of models) license just the appropriate inferences to the relevant observable properties. That is, an appropriate use of the model sanctions inferences to the effective temperature and luminosity of the star. While such inferences need not be to correct conclusions regarding any given star (whether they are depends sensitively on the inputs), they are licensed by the model under the standard understanding of its terms. The vertical rules of inference applicable are those pertaining to demonstrations within the models that anyone with a mastery of the mathematical equations can deploy; they rely essentially on a proper understanding of the symbols that appear in the equations, including all the mathematical symbols, such as differential variables and operators. Computer simulations are often run for this purpose since they mainly involve calculations starting at the core of the star and rising through the layers. So these rules of inference are explicit to the point that they can be automatized. The licensing of these vertical rules of inference is therefore embodied in the mathematics and in the computing techniques that can be brought to bear on resolving the equations for different sets of initial and boundary conditions. These techniques evidently constitute a large part of the training that astrophysicists undertake, often in their earliest stages of instruction.

How about the horizontal rules of inference that allow for surrogative reasoning? These rules involve an understanding of representational force, both in its denotative function role and in its mapping role in application, not just coarsely but in the finely grained details too. The proper use of the horizontal rules of inference requires first some understanding of what the model stands for. Besides an acquaintance with the conventional denotation of the physical symbols (for pressure, temperature, heat flow, etc.), one must know precisely what the conjunction of the equations stands for, namely, the interrelated dynamic evolution of the critical quantities that propel a star's life cycle. The equations make sense conjointly only as the equations for stellar interiors, and they work together in extracting conclusions about the dynamic evolution of stars given that they apply to those properties of stars' interiors. Claims about the model properties thus translate into claims about a star's life cycle. But, as usual, one must beware the translation keys, which map claims about certain properties over some but not all such claims. In the case of stellar structure models, claims about the internal properties of the inner, unobservable layers are not mapped over for correctness, while claims about the surface properties of the outer layers are. The licensed inferences work in all cases, but they do not always require correctness for the acceptance of the models.

The key distinction is between those claims in the models that purport to carry over to actual stars and those that act instrumentally in generating expedient inferences to the former descriptive claims. It underwrites both the sense in which stellar structure models can be said to be informative and the sorts of heuristic dynamics that inform their development by modelers. This involves critically an understanding of the role that model assumptions play in these inferences. Thus, as we saw in chapter 3, the central assumptions in stellar structure models—assumptions that ground and justify the application of the four equation template models to stars—are not strictly true. Some are openly untrue. For example, recall that the EA holds at best locally at the radiative core of the star. In no other star layer are convective forces entirely negligible, and in the outer layers they can even become dominant. But the assumption is conducive to extracting estimates for the surface observable surface properties, particularly surface temperature (T_{eff}). That is where the justification for the model assumptions lies, and it is incumbent on the model's users to know this, on pain of drawing the wrong kinds of inferences. In chapter 3, it was noted that at least two other assumptions required to justify the mathematical equations of stellar structure have the same character, namely, the IA and the SSA. Neither of these assumptions can be said to be approximately true or true at the edges, but, together with the UCA, they jointly conspire to make sure the evolution of the stars depends only on initial mass. This allows for the construction of a model with just the four main stellar structure equations and only one free parameter, adjusting for the remaining initial and boundary conditions.

The models thus channel information regarding the values of the different quantities and their interrelations, hence enabling astrophysicists to plot these quantities onto an HR diagram. This is turn allows them to predict the past and the future of the star's life cycle. The information thus channeled is both veridical (the observable quantities) and nonveridical (the values of properties at the inner core and the various assumptions that are meant to hold there). The combination is felicitous in as much as the veridical information and the nonveridical information jointly generate successful and compact predictions for a large set of stellar objects. Finally, different corrections introduced into the models allow for an application of the same representational sources to different target systems (e.g., anomalous stars lying outside the main sequence, white dwarfs, red giants, etc.). The corrections are introduced into the model equations via refinements and adjustments carried out on the assumptions, such as modulating the UCA to the different chemical compositions registered or experimentally observed. In other words, the inferential capacities of the models lead to a more precise stipulation of their targets, often involving extension to previously excluded objects or systems, and vice versa. The adjustment in the representational force that comes with the extension of the domain of application of the model to previously unsuspected systems also brings out new inferential capacities of the model source (the mathematical equations in this case). These often in turn come hand in hand with refinements or adjustments to the central assumptions.

The Modeling Attitude Revisited

In Maxwell's terminology, as we saw in chapter 2, models are conductors that lead modelers from considerations regarding representational sources to corresponding considerations regarding their targets. Boltzmann adds the idea that such considerations are informative about aspects of the target subject to inquiry. Astrophysicists are thus led from a study of highly idealized mathematical objects and structures, in the four key modeling equations in stellar structure models, to the genuine properties of stars and their relations. Ecologists are led from the study of the abstract structures instantiated in the Lotka-Volterra model to an appreciation of genuine properties of prey and predator populations in complex ecological systems. Architects and engineers use maps and graphs for buildings and bridges as blueprints that allow them to anticipate, predict, and control for actual features of newly erected buildings and bridges. And Maxwell and Boltzmann themselves employed this informative inferential approach in developing the kinetic theory of gases. All these models, reviewed through this book, are representations; and the qualities that Maxwell and Boltzmann ascribe to them are the (only) essential qualities of representation itself. Instead of looking for an abstract relation of representation, one that metaphysically constitutes it, the modeling attitude invites us rather to focus on modeling practice and its thickly normative layers of licensed rules of application.[16]

The relevant licensed rules of inference divide into two kinds, which I have called *vertical rules* and *horizontal rules*. Vertical rules apply to the demonstrations carried out within the models, a part of the modeling activity in which commonly the models gain a life of their own.[17] They apply to the internal workings of the sources considered as self-standing objects, and licensing can there proceed in a relatively unconstrained manner, attending to the features of the models only. But a model qua representation is not so separate from the object of its study. It is rather essentially linked to its purposes in surrogative reasoning. Therefore, if we wish to understand the representational nature of modeling fully, the horizontal rules of inference must be considered too. These are the rules of inference that allow modelers to walk the route from the target under study to the model source and then back, informatively, from the source to the target at the point of application. Licensing there takes a different form since what is at stake is the use that the source might have as a representation of the target. This step requires a considerable amount of judgment as to what the appropriate moves are, a judgment typically instituted by instruction in the use of the model. The key distinction here is between those inferences that are meant to be informative in the veridical sense (they provide us with information regarding the actual properties of observed or controlled for aspects of the phenomena elicited by the target) and those that are informative only in a nonveridical sense. These latter inferences inform us rather about a regime of possibilities underlying the model, including plausible causal mechanisms that may underpin the phenomena, but also—and here critically judgments must be made—about implausible, wild, and off the mark simplifications and distortions introduced for instrumental purposes, mainly for the sake of greater inferential expediency and economy.

A suitably deflationary account of representation would want to build nothing else into the notion of representation other than these two surface features in the practice of modeling, features that I have here referred to as *representational force* and *inferential capacities*. Yet such a thin account has much to recommend it: first, in terms of doing justice to the history of the modeling attitude that suggests it and, second, in terms of serving modeling practice right, allowing us thus to understand the processes that lead to the construction of models better and enlightening us about their use in practice thereafter. I have in fact suggested that the productive tension generated by the two-forked inferential conception—between the establishment of representational force and the study of inferential capacities—is often at the heart of what drives model construction and development in practice. I have also illustrated this in a variety of examples extracted from both the contemporary practice of model building and the history of modeling that accompanies the birth of the modeling attitude. The contours of a thinly defined yet correspondingly very widely applicable conception of representation should have emerged. The overriding aim of modeling practice in this conception is surrogative inference drawing. This attitude, which finds sources in the relativity of knowledge thesis so prevalent in Scottish commonsense philosophy but develops into a much larger and multifarious practice over the course of the twentieth century, requires no further support or grounding in any metaphysics. Rather, the scientific modeling practice itself grounds and articulates any further metaphysical or epistemological distinctions that might be made.

The inferential conception is thus metaphysically as thin as it can be. Yet it illuminates many aspects of the practice of model building. It has additional applications beyond that practice, however, and in part 3 of this book I would like to suggest some ways in which the inferential conception can illuminate recent debates in aesthetics and epistemology as well. This should not be surprising since representation is as ubiquitous in the arts and humanities as it is in scientific practice. The inferential conception is meant to apply to the larger class of cognitive representations, of which—as Maxwell himself was able to comprehend—scientific representation is but a kind.

Chapter Summary

This chapter outlines and defends the core theoretical commitment in the book, the so-called inferential conception of representation, or [inf]. This deflationary and pragmatic conception of representation arises as a refinement of the extended account developed in the previous chapter. It postulates two merely necessary conditions on representation, namely, representational force and inferential capacity. Both are strictly relational and contextual properties of representational sources that can be understood only as part of a thickly normative set of modeling practices. They are essentially elements in a practice and do not hark back to some undefined ontology of models or their target systems. The relation between sources and targets must thus be understood in a suitably deflationary spirit as a functional feature of our use of sources, not as a heavily constrained metaphysical relation that would require the existence of its relata. It is then shown that [inf] overcomes the objections that make its substantive competitors untenable. It is also argued that [inf] has roots in the historical emergence of the modeling attitude, with its minimal conditions aligning well with Hertz's conformity criterion and Boltzmann's information insight. Finally, it is shown that [inf] provides an insightful account of the dynamics of model building in the three benchmark case studies developed in chapter 3.

PART III

Implications

8

Lessons from the Philosophy of Art

The outlines of a new deflationary and pragmatist conception of scientific representation are now within our purview. The view incorporates insights from the history of what I call the *modeling attitude*, and it can thus be seen as the result of a long historical process that crystalizes around a minimal set of two requirements for any given thing (object, system, description, equation, process, or complex entity—the *representational source*) to be taken as a representation of anything else (the *representational target*). While these requirements are not generally sufficient for representation and must be supplemented with additional conditions in different domains, they are as general as they can be for representation across domains to obtain. To the basic insight that a representational source must possess some primitive and unanalyzable *representational force*, we need to add the requirement proper to cognitive representations that they are capable of transmitting information. In other words, the source must also be capable of delivering inferences toward its target.

In being so minimal—in laying down such spare requirements for what counts as representation—the inferential conception is also able to furnish an appropriate account of the representational qualities of a large variety of scientific models. Yet the two conditions are illuminating, not just in terms of what must obtain in any given case of a scientific representation, but also in terms of the practical need to maintain each other in check, as it were, through what I have referred in the last chapter as the dynamic equilibrium between *force* and *inference*. It is the productive tension between both elements in the inferential conception's twofold fork that propels model building forward and explains the different strategies and methodologies adopted by scientists in their development. Part 2 of the book presented arguments in favor of the inferential conception as well as those that militate against alternative, more substantive conceptions. In part 1, what I call the *modeling attitude* was presented in detail; and a number of templates in contemporary scientific modeling were provided that underpin the extent and reach of such an attitude in present-day scientific modeling practice. Yet, as I have characterized it, the inferential conception presents itself as an account of *cognitive representation* more generally, not circumscribed to the sciences only. On the contrary, the idea is that the account can generalize to all those areas where cognitive gain—the acquisition of knowledge, information (veritable or otherwise), or merely a preliminary or enhanced description—is of practical interest. I have been at pains to emphasize how artistic representation often provides a benchmark against which to judge the different accounts of representation in science. All representations must, after all, have something in common by virtue of which they are indeed representations, and there is no better or more paradigmatic benchmark than art.

It is now time to confront the issue head on. Are scientific representations just another kind of cognitive representation, and, if so, how? What marks them out specifically, and what do they have in common with other types? How in particular are they related to our most paradigmatic cases of artistic representations? And what, specifically, can philosophers of science learn from philosophers of art about representation as a cognitive achievement, and vice versa? My answers to these questions in this chapter will by and large be very positive. I shall argue that indeed representations in art and science are just varieties of a more general nexus of cognitive representation. The differences between them turn out to be conceptually minimal and to affect mainly ways of assessing their appropriateness rather than any constitutive or defining principle in their makeup. Thus, empirical assessments based on evidence are critical in the acceptance of scientific representations or models but do not normally play any significant role in the suitability of artistic representation. And, while the aim of accuracy is one of the most cherished in the application of many models, it plays virtually no role in the assessment of artistic representation. By contrast, artistic representations are subject to aesthetic criteria that are not operative in the sciences.

Nevertheless, I have argued that these are all issues that regard the assessment and evaluation of the success of our representations, for whatever purposes we set them out, whether epistemic or aesthetic, but do not affect their being representations per se. In other words, they regard what Wendy Parker (2010, 2020) and Anna Alexandrova (2010) refer to as the *adequacy conditions* that allow us to evaluate or assess our models. These conditions include but are not exhausted by issues of representational accuracy and empirical adequacy. And the prior question that is at stake here is precisely what makes a representation a representation, however inaccurate or inadequate. The distinction suitably mirrors that introduced in part 2 of this book between the means and the constituent of a representation. Issues of model adequacy concern the functions and roles of representational means and how effective those means are toward the adequacy/accuracy of models. Questions about models qua representations instead concern issues regarding the constituent of representation, namely, whatever conditions obtain for cognitive representation in general. About such matters of constitution, I argue in this chapter that there are no significant differences between artistic representation and scientific representation, that the inferential conception applies in both cases.

There is another reason to broach the discussion of artistic representation in this book, which is that there are important lessons for both sides in the debate. Most importantly for my purposes, it turns out that many of the arguments that philosophers of science have been debating over the last decade are in fact versions of arguments that philosophers of art have handled and known for a much longer time. While philosophers of art have sometimes taken inspiration from scientific modeling, particularly in linguistics or cognitive science, the inspiration has rarely gone the other way. Yet I take in this chapter precisely the rather unusual view that the philosophy of art can appropriately inform and improve our understanding of scientific representation, and vice versa. There are some subtle turns and nuances in the inferential conception of representation (at which I can only gesture here) that can be put to good use in discussions of aesthetics and the philosophy of art.

Thus, one of the points that I have been emphasizing as essential, at key stages throughout the book, is the fact that the recognition of a representation is conceptually prior to and often in practice temporally synchronous with the recognition of any salient relation between source and target that might be employed for surrogative inference. In other words, it is not the case that the recognition of some salient existing relation between source and target gives rise to or enables representation. It is rather the other way round. The normative activity of representing is what enables agents to select the salient aspects or properties of source and target as well as the relations between those properties that might serve the purpose of surrogative reasoning. This insight I have ascribed to first Putnam (1981) and then van Fraassen (1994, 2008), although I doubt that they would put it in these terms. The most congenial treatment I have found is rather in the philosophy of art. When suitably generalized, Richard Wollheim's (1968/1980, 1987) notion of seeing-in comes closest to the right description of the phenomenon and its implications for the notion of cognitive representation in general.

Thus, I intend in this chapter to take the implications of some insights in the philosophy of art yet further and suggest that they generalize beyond the plastic arts. I do so partly to shed light on the practical workings and functional characterization of that otherwise inevitably elusive notion (inevitably since it is meant as an unanalyzed and necessary primitive in the inferential conception) introduced in previous chapters, namely, force. But I do so also to show that the same considerations that we saw militating against the similarity account of representation in science also militate against resemblance theories of pictorial representation. Furthermore, these arguments point toward similar solutions in terms of a twofold account of cognitive representation more generally. It hence makes didactic and logical sense to start with a consideration of resemblance theories of pictorial representation and their shortcomings.

From Figurative to Abstract Art and Beyond

In the assessment of paintings, there is a critical difference between issues of figurative degree, which can be judged on a spectrum, and bivalent issues of representation, which must receive a yes/no answer. As we shall see, the distinction fits well with some ideas that we have already used in previous chapters with respect to scientific models. I shall focus on famous paintings through history to illustrate the first issues of degree. The chronology goes along roughly from the most evidently figurative paintings in the sixteenth and seventeenth centuries all the way to the development of abstract art in the twentieth century. And, indeed, the distinctions that I intend to discuss gradually decrease along that chronological line, a decrease eventuating in a sharp or abrupt change in the assessment of the works discussed with the sudden emergence of nonrepresentational art in the twentieth century. This is the second distinction, which is rather abrupt and not at all a matter of degree but a contrast between those paintings that represent and those that do not. Hence, the spectrum would seem to go from replica to figurative to impressionist, cubist, and the like to the sort of very distinct abstract art that ends in what Mondrian famously called *neoplasticism* or *pure plastic art* (Mondrian 1920/2017, 1926/2017, 1937/1971). And this can be seen to correspond roughly with a spectrum of degrees of the figurative: from full or complete, to partial or symbolic, to indirect or evocative, to openly nonfigurative but still representational, and then to ultimately purely plastic or nonrepresentational art.

Hence, we have a nice set of two distinct questions. What makes one painting more clearly figurative than another? What is at the heart of the distinction between representational and nonrepresentational art? The former is a question of degree, while the latter is really a binary issue, even though it appears as a limiting case at the sharp end of the spectrum of figurative representation. And, while the first one concerns the vehicle of representation and its qualities (as a means for representation), the latter concerns the definition (or the *constituent*, in the terminology of past chapters) of representation *tout court*. The difference between the two issues can best be illustrated by means of four outstanding pieces of art from the history of painting.[1] In exhibiting them, I aim to consolidate some intuitions that I think we all prima facie share in response to both questions regarding both the order that seems intuitive in the rank of representational paintings and as the nature of the visual experience that seems to characterize nonrepresentational painting. In due course, these intuitions shall be put to work in a larger theoretical schema that encompasses scientific representations as well, thus showing them to be operative in the larger nexus of cognitive representations.

DIEGO VELÁZQUEZ'S *POPE INNOCENT X*

Let us first turn to an example already considered in chapter 5, namely, Diego Velázquez's 1650 portrait of Pope Innocent X (of which fig. 5.1 is a reproduction), which hangs in the Doria Pamphilj Gallery in Rome. This "masterpiece amongst all portraits"[2] is a figurative piece of representational art if there is any. We know whom it represents, in what role of command, and the precise dates and circumstances in which it was painted.[3] More generally, the category of portraits (both paintings and photographs) stands at the extreme end of unambiguously representational work. By its nature, a portrait is a representational source that stands for its target, often to the point that it will take its place, symbolically and sometimes even legally (as, e.g., the photographs or portraits of the head of state in official pronouncements or ceremonies in many countries). When it comes to Velázquez's portrait, there is no scope for reasonable doubt regarding its intended target—namely, Innocent X—in its acting role and function as head of the Catholic Church at the time.

It is also a remarkably *accurate* portrait (in addition to being for many reasons a very remarkable painting by an extraordinary painter). It is reported that, on first seeing it, the pope himself pronounced the words *troppo vero* (too true) on account of the remarkable veracity of the portrait—its faithful depiction of not just his physical appearance but, we can suppose, his personality as well. Justi describes the main traits in the pope's personality as follows: "Although of saturnine temperament, and often prey to moody thoughts, Innocent freely unbent and indulged in playful or caustic badinage with those who enjoyed his full confidence." And the portrait captures these traits astonishingly well and alluringly so too: "There are few portraits, few paintings of any kind, that have at all times so instantaneously taken possession of all classes of observers. Someone remarked that if he gazed any longer at the head of Velázquez's portrait, the man would haunt him in his dreams. . . . The glance, drawn from the deepest recesses of a character at once suspicious and reserved, concentrates in itself the whole being of the aged statesman" (Justi 1889, 356–57). Together with other works that Velázquez produced while in Rome, the portrait cemented his international reputation, word passed on quickly among the main patrons of art in Europe, and ever since the "painter amongst painters" (Édouard Manet's dictum quoted in Carr 2006, 79) has enjoyed a prominent place of honor in conventional histories of art.[4]

HANS HOLBEIN'S *THE AMBASSADORS*

The second example I should like to present has also been debated at length in the literature on both artistic and scientific representation. Hans Holbein's *Ambassadors* (fig. 8.1) is a major work of sixteenth-century art.[5] Although on the face of it looks as representational as the Velázquez portrait (in fact, it seems to portray simply two standing characters in front of several scientific instruments), there are other aspects of the painting that turn out to be key. These go well beyond the merely figurative and stretch into the metaphoric and symbolic. Holbein depicts an anamorphic skull in the front part of the stage, an object that was typically taken in art of that time to represent mortality. The custom at the time also involved cleverly disguising this object, often by representing it in the back of canvases or in all kinds of elliptical ways. The contrast of mortality with wealth was a constant at the time. Death was the great equalizer that brought all differences in fortune, property, and distinction to one and the same level in the afterlife. Holbein characteristically captures the contrast by means of the depiction of a crucifix in the upper-left-hand side of the portrait, barely unveiled behind the sumptuous green tapestry that fills the background.[6]

Thus, the main characters' ostentatious wealth contrasts with the oblique and elusive—yet overwhelmingly present—skull representing mortality, and the promise of an afterlife is symbolized in the crucifix. We know all this because these were established representational conventions at the time, used often before by Holbein himself (Foister, Roy, and Wyld 1977, 44–58). And these are not the only symbolic references. For instance, many of the scientific instruments and tools on display in the back of the room are charged with symbolism. The inevitable sense of the passage of time is apparent in the

FIGURE 8.1. Hans Holbein's *The Ambassadors* (1533). Image courtesy of The National Gallery, United Kingdom.

many instruments for the measurement of time on the first shelf. The fact that they are poorly arranged and not set to the location of the characters in London at the time is often taken to symbolize the discord and disorder of the epoch, marked by religious division. Georges de Selve was intensely involved in Counter-Reformation attempts at pacifying the Protestant uprising, and even Jean de Dinteville's own melancholic state during his stay in London has been suggested to be represented in some of the chosen objects (Foister, Roy, and Wyld 1977, 30–43). All in all, Holbein's portrait is, no doubt, a figurative and representational painting, but it contains a large range of objects that stand in symbolic and conventional relations to certain aspects of the intended target. We can thus see it as one step up in the move toward abstraction.

This move toward abstraction becomes undeniable in much twentieth-century art, such as late impressionism, symbolism, expressivism, abstractionism, cubism, and so on. In this sort of art, the symbolic elements become dominant and are not obviously conventional. They are typically much looser and subjective. The interpretation of twentieth-century art correspondingly requires greater use of both hermeneutics and careful historical scholarship to unearth a multiplicity of meanings and intentions—not just those of the artists but also those of the people who commission or help build up the paintings, curate exhibitions, and arrange for exhibits as well as those of the critics, commentators, and public who in their own ways add to the complex and multifarious meaning of a work of art in the twentieth century. We shall come back to this, but it should be obvious that the representational content[7] of a painting does not necessarily look much like its target, at least in the way most figurative art does.

PABLO PICASSO'S *GUERNICA*

An instructive example, particularly since it is so well-known and widely discussed, is Picasso's *Guernica* (fig. 8.2).[8] The famous painting was produced in 1937 for the Paris World exhibition, and its commission, production, and subsequent history are extensively chronicled. We have thus excellent insight into all the various elements required for a correct appreciation of the painting and its various representational targets. The complex social and historical background can be briefly summarized as follows.[9] The painting was originally commissioned by the Spanish Republican government during the Spanish Civil War. Picasso had been living in Paris since the early years of the twentieth century, and in September 1936, just a few months after the war started, the Republican government made Picasso the director of the Prado Museum in absentia. He took up the post with relish, employing it to order the eviction of many of the Prado's treasured contents to Geneva in order to preserve them. Then, in January 1937, the government commissioned him to produce a major work for the Spanish pavilion at the Paris World Exhibition in 1937. Picasso accepted but hesitated over the subject, for a while simply playing with the motive of an artist in his studio, and procrastinating. Then, on April 26, 1937, an event took place that shocked him out of his creative hiatus, the bombing of the Basque city of Guernica. The bombing—perpetrated by Hitler's Condor Legion and Mussolini's Aviazione legionaria, with Franco's consent (Preston 1993, chap. 9)—was powerfully chronicled by George Steer in the *New York Times* and, across the Atlantic, in *The Times* a few days later (see Steer 1937a, 1937b). Picasso entered a state of creative frenzy over the course of May 1937 in response to these events, completed several preparatory sketches, started working on the actual canvas on May 11, and finished the huge painting (3.5 × 7.5 meters in size) on June 4. It was thereafter unveiled

FIGURE 8.2. Pablo Picasso's *Guernica* (1937).

in July as expected, but it was not immediately a significant success, criticized by both sides of the political spectrum as being unduly pessimistic. The small Spanish pavilion was dwarfed in comparison with the monumental and uplifting yet confrontational German and Soviet pavilions, which attracted most of the attention.

The subsequent history of the painting confirms its provenance, the manner of its production, and the charged political events that lie at its origin. *Guernica* was first sent on a tour of European—mainly Scandinavian—nations, where it began to gain critical attention. It was on display at London's Whitechapel Art Gallery in September 1938, at the time of the signing of the Munich appeasement agreement between Chamberlain and Hitler. It picked up tremendous support mainly from the British Labour Party: Clement Attlee attended the unveiling at the Whitechapel, at which he gave a rousing speech, and the painting toured Leeds, Manchester, and Liverpool, the main centers of recruitment for the international brigades fighting in Spain. It subsequently toured the United States, initially helping raise public support for the cause of US intervention in World War II, and thereafter, through 1951, being shown in most of the major American galleries. After further shows in South America and Europe, its state deteriorated, and it was finally deposited at the Museum of Modern Art in New York in accordance with Picasso's express wishes. It remained there until it was finally transferred back to Spain in 1981, also in compliance with the spirit of Picasso's stipulations.

It is undeniable that the painting's target is the bombing of the town of Guernica. But to say only this is to do very scant justice to the history, import, meaning, and, ultimately, the representational force of the painting.

Admittedly, its representational force has changed somewhat over the years, starting off as merely a description of the unbearable cruelty of what may be the first carpet-bombing of a civilian population, and thereafter progressively picking up political significance. Thus, *Guernica* was initially read as a warning of the rising threat of fascism and a call to arms against it and sometimes used as a piece of propaganda. But thereafter it came to be seen as a more abstract yet brutal indictment of war and a cry for world peace. And, even if we focus only on the narrowly construed target, there are all kinds of questions that can be raised. Does *Guernica* represent the town of Guernica? Or does it represent the people of Guernica dying at a particular time on a particular day? Or does it rather more abstractly represent the event of the bombing of Guernica itself, presumably not an instant in time but the three and a quarter hours during which the town was reportedly repeatedly bombed? But, in this context, what can the ungulates possibly represent? The crazed horse has in the past been identified with the Spanish progressive republican government and the bull with fascism. While both interpretations were rejected by Picasso, it would be naive to take them to represent merely a horse and a bull present at the scene. Other objects are also open to meaningful interpretation. Thus, the broken sword has been taken to represent the broken spirit of humanity, disharmony, and defeat. The crying mother holding the baby is often taken to represent the victims and their pain (most of the men in Guernica were out of town that day, as reported, so this is representative in not merely metaphoric ways). The broken glass and buildings symbolize generic ongoing destruction, and the overall frame can be easily taken to stand for the claustrophobic space the bombing took place in. There was literally no escaping devastation and death.

Guernica is most definitely a piece of representational art, but its targets are multiple, changing, and evolving and permanently open to interpretation as they evolve. The painting as we have come to know it truly represents all those things mentioned above at once, each of them properly in its right context. The targets are not conventionally stipulated by any passerby or individual observer. They are instead the outcomes of complex, eminently social processes of refinement, selection, and judicious interpretation. They are determined by judgments that surround the diverse appreciations and uses of the painting. They start perhaps with the intentions of the artist himself, and they are often highly considerate of his stated aims. But they always inevitably go way beyond his original intentions, and they tend to coalesce around stabilized community judgments. While the judgments are relative to communities, they are bound by the properties of the canvas itself, its history, evolving uses, and the artist's stated intentions. And, while these boundary conditions are essential if we are to fix the representational target of a piece of art, they do not exhaust it or fix it independently. There is still need for normative conventions to be stated, appreciated, and respected in the context of some communal use. True, Picasso's *Guernica* achieved nearly universal appreciation and a remarkable consensus throughout at least the Western world in terms of what its targets signify. This is extremely rare for a painting, but it does not refute the insight that the determination of the target of a painting requires stable communal judgments to be in place. It just shows that those communal judgments (and their underlying values) can be extrapolated and imported in different contexts and cultures, to the point of reaching an extraordinary near-universal consensus. The community of interpreters of *Guernica* is by now arguably the entire world.[10]

This example more generally shows that the targets of paintings are typically abstract, dynamically evolving in time, often subject to the requirements of context, and highly sensitive to critical reception, which can vary across cultural contexts. Even those targets that can be said to be most definitely concrete are not to be regarded as instantaneously described objects or their properties. They are better thought of as exemplars, types, or tropes evolving dynamically and thus capturing the essence of processes, events enduring in time, or generally a course of such processes or events of a given sort over a period. The same is in truth applicable to portraits as well, as becomes clear when one reflects carefully on the historical origin, uses, and reception of both the Velázquez and the Holbein portraits through the ages. Even photographs can be uninteresting when simply depicting things as they happen to present themselves to the camera. Great photographers are typically those who can transcend the instant and thus display some enduring characteristic, scene, or property that is significant in terms of its reception or the overwhelming response to it.

While scientific models will rarely exhibit such extremes of plasticity and malleability in their use and are rarely so context or response dependent, they are nonetheless not entirely dissimilar. A thorough understanding of a model requires us to probe sufficiently into its history, uses, and reception within a community of inquirers. Without any appreciation of these external social factors, it becomes impossible in turn to understand and appreciate the representational features of a model (its representational force and inferential capacities). Let it be recalled, however, that the factors are external only in the sense that they are not physically built into the object that plays the role of representational source, just as the uses and history of a painting are not built into the canvas itself. But on the inferential conception of representation—or any other credible account for that matter—a representation in art or science alike does not just comprise the source object (in fact, the source object on its own is, on the account defended in this book, no representation of anything). Rather, it critically incorporates the established norms of use within a particular community of inquirers, that is, those norms that determine both force and inferential capacities. These norms are often the outcome of complex social processes that originate or at any rate involve the act of creation of the model and therefore the creator's intentions and goals. But they necessarily go way beyond these initial intentions, particularly for long-lasting models that achieve significant scientific success. In all these respects, scientific modeling is rather akin to artistic representation, which is another reason why the comparison in this chapter is apposite.

PIET MONDRIAN'S *LOZENGE COMPOSITION WITH YELLOW, BLACK, BLUE, RED, AND GRAY*

There is yet another dimension of helpful comparison, one that concerns the distinction between representational and nonrepresentational art and is best illustrated by considering one of Piet Mondrian's celebrated colorful compositions (such as the one depicted in fig. 8.3). Mondrian famously defended *nonfigurative* art,[11] by which he meant roughly what we now mean by *nonrepresentational* art. And, indeed, it does make sense to draw the distinction between representational and nonrepresentational art just where he does since ultimately at the confines of figurative art lies an attempt to deal with the pure plastic form of the painting itself, which involves in no way going beyond the canvas or hankering toward any other possible object (abstract or concrete, real or imaginary). To attempt to read Mondrian's diagonal and straight-line compositions in any other way—as a representation of anything standing beyond the canvas—just seems a mistake.[12] All the painter's efforts are directed against such a reading; and the properly licensed and sanctioned use of the paintings as such does seem to preclude such understandings. (One can evidently use the painting in other ways—by, e.g., purposely matching it to a grid of roads or rail lines—but in doing so one would not be following the licensed use of the painting as such.)

In terms of the inferential conception defended in this book, Mondrian's paintings lack representational force altogether, and whatever inferential capacities they may have toward any putative target are of no representational use—when they are regarded as paintings. To turn a Mondrian painting into a representation of something requires a change in the normative use that is licensed in the artistic context. This involves taking it to be something other than it is—a map or a description, in other words, a model of something.

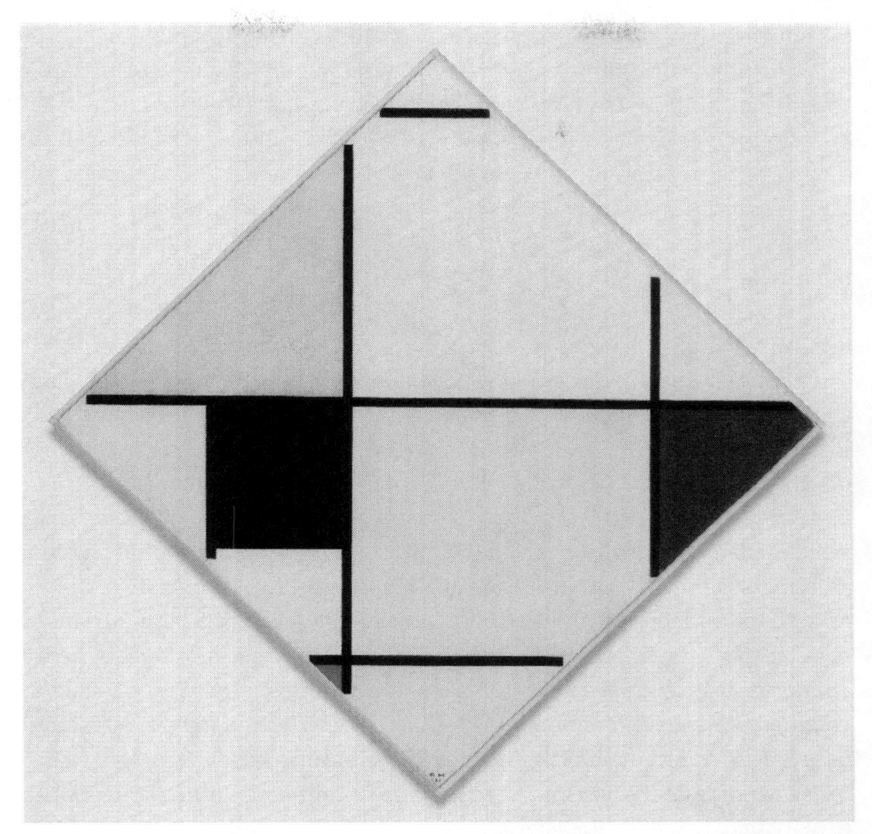

FIGURE 8.3. Piet Mondrian, *Lozenge Composition with Yellow, Black, Blue, Red, and Gray* (1921). The Art Institute of Chicago.

Yet we are not prevented from playing around with the plastic qualities of the painting, and indeed Mondrian advises one to do so for the sake of objectivity.[13]

Taking this over to the debate over scientific models has some interesting consequences for the artifactual nature of models, always built with very particular aims and goals in mind; as well as for the practice of abstract modeling in the limit where a scientist disengages with the target of the model altogether and instead chooses to focus entirely and exclusively on manipulating features of the model itself. In both cases, it will be pertinent to analyze the distinct conditions that must obtain if representation is to be truly at work—as opposed to constituting, as in the case of Mondrian's paintings, a mere expression, investigation, or more complex game with the pure plastic properties of the model source. But, first, we need to answer some questions within the philosophy of art itself regarding the nature of representation. What distinguishes representational from nonrepresentational art? And what can be the grounds of our discriminating judgments regarding degrees of abstraction/figurativeness? It turns out—or so I shall argue—that the relevant views and positions in the philosophy of art correlate neatly with some of the views we have already canvassed and discussed in earlier chapters regarding scientific representation.

Resemblance and Its Discontents

We can now consider the two questions that I raised at the start of this chapter with the four cases of art that I introduced in the previous section in sight. Recall that the two questions are as follows: what explains the decreasing degree of figurativeness (which appears to be inversely correlated with increasing degrees of abstractness), and what explains the abrupt change in representational quality that appears in the final stage, when a nonrepresentational painting such as Mondrian's *Diagonal Composition* is considered? A theory or account of depiction ought to provide enlightenment on both questions. And, in line with the discussions in previous sections, a substantive theory or account will attempt to provide the answers on the basis of some relation obtaining between representational sources (in this case, the canvases) and targets ranging from people in their settings (Pope Innocent X, Jean de Dinteville, Georges de Selve), to conventional symbols and allusions (the pope's psychological profile, other symbols of mortality and resurrection in *The Ambassadors*), to the much more multifarious, abstract, and diverse targets of a piece of abstract art such as *Guernica*. What could this relation be?

The obvious candidate is the visual similarity in physical appearance between the canvas—as it presents itself to us—and the diverse targets, in other words, the *resemblance* that paintings hold to their targets. The suggestion that paintings depict via resemblance relations and that this is the feature that explains differences in figurativeness or abstraction is of course as old as philosophical reflection on art itself, originating in Plato, and has received a great deal of critical attention in the philosophy of art through the years. There are undeniable intuitions that back it up, particularly as an account of the degrees of accuracy, faithfulness, or figurativeness of a painting. Thus, the Velázquez portrait is, we are told, as accurate as any painting can be, while Holbein's *Ambassadors* is more complex, incorporating a degree of symbolism and purported inaccuracy, and *Guernica* is yet more ambiguous and multifarious. Still, they are all representational paintings, and the various degrees of resemblance to their targets indicate only a difference in their abstractness, a difference that can pertain to the mode of representation as much as to the nature of the targets. These are clear intuitions that favor the resemblance theory. It seems undeniable that paintings—portraits in particular—tend to resemble their targets and that they are often judged in accordance with how well they capture some of the essential characteristics of the physical appearance of their targets.

Thus, resemblance may come to the rescue in relation to our first question of degree regarding accuracy, figurativeness, and/or abstractness. However, it cannot possibly do as an answer to the second question, the binary question as to what makes the first three paintings representational (in their varying degrees of accuracy and abstractness). It is obvious that there are no more salient resemblances between *Guernica* and its targets than there are between *Lozenge Composition* and a closely matching array of roads or rail lines. And, even among those paintings that undeniably play representational roles, the different forms (the means) of their representation are not always easily subsumed under any kind of resemblance relations. The psychological characteristics of Pope Innocent X find no resembling elements in *The Ambassadors*, other than vaguely symbolically, perhaps, in the prominence of the color red. The strong symbolism implicit cannot be explicated by means of any meaningful resemblance. If one did not know the cultural connotations of the different objects, one would be lost when it came to its actual targets. Quite clearly a lot of what makes a painting a valuable representation of its target cannot in any straightforward manner be subsumed under any resemblance relations. The last forty or so years, in particular, have seen severe criticism raised against this account of pictorial depiction as well as a number of attempts at resurrection in the form of warmed-up and modified resemblance approaches.

The theory of depiction as resemblance came under scrutiny in the 1970s when the main defenders of both alternative conventional and phenomenological approaches to depiction coalesced around their rejection of the resemblance theory. In his landmark *Languages of Art* (1968/1976), Nelson Goodman advanced powerful arguments against resemblance as part of a defense of his linguistic theory of art. Roughly at the same time, in his seminal *Art and Its Objects* (1968/1980), Richard Wollheim advanced objections to resemblance theories, in the context of his larger criticisms of the illusion theory championed earlier by Ernst Gombrich (1959/2002). Goodman's main arguments were logical in character, and we reviewed them in chapter 5. Their application to art is straightforward, and we already considered them in relation to Velázquez's *Pope Innocent X* (fig. 5.1) and Bacon's *Study after Velázquez* (fig. 5.2). By contrast, Wollheim concentrated on the implausible phenomenology for the experience of art involved in resemblance theories, and he advanced a different proposal that grounds depiction on the essential hallmarks of the phenomenology of perception, most prominently what has come to be known as *twofoldness*. Together, these arguments make a formidable case against any view that takes resemblance as the *constituent* of depiction or artistic representation. As we shall see, however, neither set of arguments necessarily compromises the role of resemblance as representational *means*, that is, the kind of relation between a painting and its target that (once the representational role of the former toward the latter has been established) allows any informed and competent agent to draw conclusions regarding the target on account of properties of the source.[14]

In the philosophy of art, two different theories have been traditionally discussed: those that appeal to objective similarity or resemblance and those that appeal to subjective or experienced similarity or resemblance. The arguments against these similarity theories in the philosophy of art parallel those against different kinds of similarity theories of scientific representation presented in chapter 5. However, just as then, not all arguments work equally well against both types of theories (subjective theories are more recent and enjoy some measure of success), so it pays to devote a bit of time to their differences. The traditional view takes it that "there are objective similarities between pictures' design properties and properties of their subjects" and that "it is in recognizing these similarities that their subjects are identified" (Lopes 1996, 18). This is equivalent to the substantive similarity conception of representation [sim] considered in chapter 5 and fails owing to the same logical objection (namely, the essential irreflexive, asymmetrical, and intransitive character of representation). The objective approach can be finessed, however, in ways that again parallel some of the moves that we studied in chapter 5. One could, for example, add a proviso that the resemblance in question be salient or intended in some generally identifiable way. This would circumvent the problems by adding a property of intention or salience that is not reflexive, symmetrical, or transitive, just as required. However, as already noted in chapter 5, there is no general and independent notion of intended salience available. The only condition general enough is precisely the one that picks out whatever properties are in fact salient to the representational work done, which can hardly help any theory of the *constituent* of representation.[15]

On a subjective resemblance theory, by contrast, depiction is not constituted by anything in those things depicted and depicting or in their relation. In the terms of the inferential conception, it is not constituted by any correspondence of properties between the source object (the canvas) and the target system (the scene or character depicted). It is rather a feature or quality of our experience of those things.[16] Yet, for the notion not to be entirely an arbitrary or subjective feeling or judgment or the result of a private language, it must be the case that there are objective features in our conventional visual experiences that can be put into some sort of resembling relation. Peacocke (1987) suggests that the relevant features pertain to what he calls our *visual fields*, aspects of the sensational contents of our experiences in front of both objects depicting and objects depicted that can then lay down the appropriate field relations.[17] Thus, consider Constable's *Salisbury Cathedral from the Meadows* (1831). It represents Salisbury's cathedral as seen from a certain angle, across the meadows, on a gray, overcast day. The relevant visual field is the two-dimensional intercalated plane that would project the scene, on a particular overcast day, as perceived by an observer placed at the right position across the meadows (and that we can only conjecture Constable to have occupied while producing the painting). This application of projective geometry to paintings known as Alberti's rule (Alberti 1450/1966, 51) is not unusual since the Quattrocento.[18] It directs artists "to treat every picture as a transparent surface on which the outlines of objects seen through it may be inscribed" (Lopes 1996, 24). Much figurative painting follows this rule, but not all representational art does, as is made clear by a quick comparison with the four cases of representational art so far discussed.

The one case that most clearly adopts something akin to Alberti's rule (or, at any rate, that can be interpreted or understood in this way) is, of course, Velázquez's *Pope Innocent X*. There is a possible point of view regarding what we might take to have been the scene at the time of the painting's making such that projecting the scene onto an interposed plane would generate a visual field that corresponds approximately to the visual field that would be experienced by a perpendicularly positioned observer of the painting as it hangs in the museum today. The representation of the objects depicted in Holbein's *Ambassadors* (i.e., discounting for the strong conventional symbolism) can also be recast in accordance with Alberti's rule. The anamorphic skull would correspond to the visual field experienced from a position at a highly oblique angle. This is contrary to the position the viewer would need to adopt to re-create the visual field that (we can suppose) the rest of the figures depicted would exhibit in fact.[19] While this requires contradictory visual fields for any actual observer, it can still be supposed that a collage need not adopt Alberti's rule univocally and with respect to only one point of view throughout the whole painting. Instead, it might have distinct points of view corresponding to each of its parts or batches.[20]

It would thus seem that an approach of this kind, in terms of objectively experienced resemblance, could apply to all figurative painting. Nevertheless, there are some probing questions that shed doubts on the objectivity of the visual fields involved and the putative fields relation. These questions particularly concern the reconstruction of the experienced visual field at the time the painting was made—or, at any rate, the field that could conjecturally be reconstructed to have been the visual field. In other words, these questions concern the putative target of the representational relation.

The objections do not trade on the fact that the scene depicted no longer exists or is no longer available, which is clearly the case in both *Pope Innocent X* and *The Ambassadors*. They rather turn on the fact that the scene as depicted was never actually in existence and could not have been. The scenario is counterfactual and perhaps even impossible. We noted above, in connection with *The Ambassadors*, that paintings are often actually produced not in one setting but in several. This is clear from the large number of elements taken from different and independent scenarios that are then put together in ways often anachronistic or simply impossible at the time. The robes and other clothes in *The Ambassadors* are pastiches or inventions, as are, recall, the instruments depicted by Holbein in unnatural and distorted positions. There is therefore no real visual field that one could experience naturally as the target. There is not even a field that the sensational visual field of the canvas could possibly resemble. Here again, as we saw in chapter 5, the problem is that, if resemblance is a relation (and what else can it be?), then it will fail to obtain in cases of artistic and scientific representation of fictional targets. Consequently, even merely experienced resemblance does not *constitute* representation.[21]

Yet, as was pointed out earlier, there is no denying that pictures often do resemble their subjects. But, while this is widely acknowledged to be true, it is also widely thought that the resemblances that do obtain between a painting and its subject cannot explain why the painting is a representation in the first place. In the terms that we are employing here, resemblance can settle only issues of accuracy or abstractness. It cannot help to settle the prior question regarding representation. It is thus not the case that paintings are representations *because* they resemble their subjects. Rather, it is the other way round. Our recognition that a painting (such as *Pope Innocent X*, *The Ambassadors*, or *Guernica*) resembles its subjects is a consequence of our recognition that it represents them.[22] This is perhaps most conspicuous in the most abstract cases of art, like *Guernica*, where it is patently clear that we could not possibly fathom how the painting resembles its diverse targets before we acquired some understanding of what it represents. But it really applies across the board, including to most figurative paintings, as soon as we notice the complexities involved in the judicious selection and construction of the targets themselves. It is down to Holbein's great talent to have understood and put into existence the likenesses—physical and symbolic—that *The Ambassadors* holds to its target subjects. And it is Velázquez's great genius to emphasize aspects of the pope's personality, thus bringing into being likenesses that are not, as it were, lying out there for us to pick out but are rather created ab initio. In other words, once what I call *representational forces* are determined, distinct likenesses and resemblances are brought into existence that can then be employed as means of the representation—as the conveyers of information about the subject targets so resembled. The logical order is here also the order of practice and the order given in experience, and the question remains what, if anything, constitutes representation in the arts.

Conventionalist and Phenomenological Approaches to Depiction

In this book, I focus only on two answers to the constitutional question regarding artistic representation. This is partly because they are the central answers discussed in the philosophy of art literature but also because they happen to be most plausibly generalized to representation at large. They are sometimes referred to as *linguistic* and *perceptual*, but, for my purposes, it is best to think of them as the *conventionalist* and *phenomenological* approaches, respectively. They were principally championed by Nelson Goodman and Richard Wollheim, respectively, and, since they are both well-known, they are described only briefly in this section. In the following section, I generalize with due caveats to a balanced combination of the phenomenological and the conventional aspects that integrate into a (thin and suitably deflationary) notion of representational force.

CONVENTIONALIST APPROACHES

The rejection of resemblance views can lead naturally to the supposition that depiction is a form of conventional representation by signs, in the way of semiotics. It led Goodman (1968/1976) to suppose that the model of linguistic representation—according to which different names and symbols conventionally stand for different objects and their properties—is also applicable to artistic representation, including depiction. Goodman's thought is that of course pictures resemble their targets but that this is because resemblance is trivial since anything resembles anything else in myriad ways. It cannot therefore suffice for representation or depiction that the source resembles the target since there are all kinds of resembling objects that do not depict or represent each other. This is a version of an argument against similarity theories for representation more generally, one that we rehearsed earlier (in chap. 5). The upshot in this context is that there must be a stronger notion that hooks the depicting object to its target independently of any resemblances. Goodman's suggestion is the following: "A picture that represents—like a passage that describes—an object refers to and, more particularly, denotes it. Denotation is the core of representation and is independent of resemblance" (Goodman 1968/1976, 5).

Denotation is a variety of reference, the sort of relation that a name holds to its bearer, a predicate holds to the set of members in its extension, and, according to Goodman, a portrait holds to its subject. Goodman is a nominalist about all of them, in the sense that he claims that what the name, predicate, or depiction holders have in common is nothing other than the very referential relation they hold to their names, predicates, or depictions. A predicate is defined via its extension, and it does not apply to the set of members in this extension by virtue of any property that those members share or hold in common. On the contrary, the only property they hold in common is precisely that they fall under the predicate.[23] Goodman's theory thus assimilates pictorial representation (and by extension all artistic representation) to the sort of conventional symbol systems that characterize linguistic representation. Most philosophers of art are skeptical that this assimilation can be carried out, and they claim that there are at least some essential experiences in painting that transcend the merely conventional (see Abell and Bantinaki 2010; Hopkins 1998; Hyman 2006; Hyman and Bantinaki 2017; Lopes 1996, 2005; and Newall 2011).[24] I agree. The tenor of this book has been to emphasize the different ways in which models in science—and representations more generally—emancipate themselves from language. In this chapter and elsewhere, I am seeking parallels with artistic representation precisely because art is paradigmatic of object-to-object representation. It would be self-defeating for the overall project and the sense of the analogy that I am pursuing if artistic representation turned out to be just a case of linguistic or word-to-object representation after all.

Nevertheless, even Goodman accepted that depictions differ from linguistic descriptions in definite ways, so, even if both could be understood as denotative symbolic systems, there are distinct features of depiction that distinguish it from other symbolic systems. Goodman (1968/1976, chap. 4) discussed two types of requirements on a symbolic system: syntactic and semantic.[25] Broadly speaking, syntactic requirements apply to the formal features of the symbolic system itself, while semantic requirements apply to its capacities as a conveyor of representations, thus roughly the features in the targets that they depict. A pictorial symbolic system differs from other symbolic systems in being *analogue* and *relatively replete*. It is analogue because it is both syntactically and semantically *dense*. It is *syntactically dense* because "it provides for infinitely many characters so ordered that between each two there is a third" (Goodman 1968/1976, 136). And it is *semantically dense* because it is not possible to classify every object in the target system as complying with one character or set of characters in the symbolic representation but not another (Goodman 1968/1976, 152)—or, in plainer English, because "it provides for a dense set of classes of denotata" (Hyman and Bantinaki 2017, 14). Moreover, a pictorial system is relatively replete because it requires "delicacy of discrimination" (Lopes 2005, 148). Its interpretation is sensitive to many different properties of the representational source: "Any thickening of the line, its color, its contrasts with the background, its size, even the qualities of the paper, none can be ignored" (Goodman 1968/1976, 229).

Perhaps then the right analogy is between scientific models and specifically depictive representational systems, those that are analogue and relatively replete. There is one particularly attractive feature of the analogy that deserves to be considered. Many diagrams (including scientific diagrams) would seem to fit in with Goodman's prescriptions for depiction: they are syntactically and semantically dense as well as relatively replete. If we consider the classification of models offered in chapter 3, most scale and analogue models have these features. A good range of mathematical models do too, insofar as there are real-number valued parameters that need to be fixed (syntactic density) and that often exhibit "delicacy of discrimination" (semantic density and relative repleteness). Consider in this respect the Lotka-Volterra model, in which the parameters can be fine-tuned in ways that can sensitively affect the ratio of prey and predator populations. It is unclear that theoretical but nonmathematical models are symbolic systems of this sort, and certainly most qualitative models are not. But they can be regarded as either exceptional or falling under other systems of symbols. In other words, Goodman's approach has the significant virtue to turn depiction into just one kind of representation, one in which the larger notion of denotation within a symbol system—of any sort—unifies the diverse types of representation.[26] Undoubtedly, this is one reason why some philosophers of science have felt inclined to employ denotation as the answer to our second binary question regarding the confines of representation (Elgin 1983, 1997, 2009, 2017; Hughes 1997, 2010; Frigg and Nguyen 2020).

Yet there are other problems with Goodman's approach that make even this restricted analogy unsuitable (see, e.g., Hyman and Bantinaki 2017; and Lopes 1996, chap. 3). First, a set is dense only relative to an ordering, and Goodman (1968/1976, 136) cashes the ordering out in terms of discriminating properties of the denoting symbols: "The ordering in question is understood to be such that any element lying between two others is less discriminable." It is doubtful that pictures, diagrams, or scientific models can be so ordered. But, moreover, "discriminable" seems to make sense only as a notion of perceived resemblance (between the symbols themselves, for syntactic density, and between elements of the target, for semantic density). It is then hard to make this compatible with Goodman's nominalism. There is in addition tension with some essential phenomenological features of our experience of pictures and with what is known as their *natural generativity* (Hyman and Bantinaki 2017, 16; Lopes 1996, 71). This is the capacity of pictures to instruct us regarding the world, not merely symbolize it. As Wollheim puts it (1987, 77): "If I can recognize a picture of a cat, and I know what a dog looks like, then I can be expected to recognize a picture of a dog. But on the semiotic view this ought to be baffling. It should be as baffling as if, knowing that the French word 'chat' means a cat, and knowing what dogs look like I should, on hearing it, be able to understand what the word 'chien' means." Natural generativity (or *transfer*, as Wollheim puts it) has an equivalent for scientific models too, so this is a particularly troubling feature of Goodman's account.[27]

Finally, there are the perennial objections regarding the critical central role that denotation plays in Goodman's account of representation. For, as we have seen, in Goodman's account, denotation is the core of representation, including depiction, and it is just as arbitrary or conventional as description: "Descriptions are distinguished from depictions not through being more arbitrary but through belonging to articulate rather than to dense schemes; and words are more conventional than pictures only if conventionality is construed in terms of differentiation rather than of artificiality" (Goodman 1968/1976, 230–31). Yet, as we saw in chapters 5 and 6, *denotation* is a success term and requires existence. A denoting symbol denotes an object only if the object exists, and, if that object is identical to another, then the symbol necessarily denotes the latter too (Lopes 1996, 59). This is problematic for our analogy since, as we saw in previous chapters, the targets of many scientific models or their parts are nonexistent, fictitious, or unreal.[28]

PHENOMENOLOGICAL APPROACHES

It is not a new thought that any understanding of depiction must fundamentally say something about whatever human experience is involved in the perception of depiction.[29] For our purposes, the idea becomes central to Anglophone discussions of the nature of depiction in the 1960s, and those discussions have never really waned, the central insight still enjoying distinguished prominence.[30]

As we shall see, the central insight in several ways aligns well with an inferential conception of representation in general, but the comparison must be handled with care since the phenomenological experience involved in depiction turns out to be a rather peculiar one from a more general standpoint that considers all cognitive representations.

The most influential among contemporary phenomenological approaches is undoubtedly Richard Wollheim's (1968/1980, 1987), which developed in response to Gombrich's (1959/2002) illusion theory. It pays to summarize the key differences between these two phenomenological approaches briefly since they have corollaries for representation more generally, including scientific modeling.[31] Gombrich's illusion theory was based on the three conjoint ideas that observers of pictures are often aware of the surfaces of the pictures (and can pay attention to brushstrokes, marks, and techniques on display on the surfaces), are often aware of the scenes represented in those pictures (or, more generally, of the picture's representational contents, i.e., the images of their targets that they convey), but are never aware of both surface and content simultaneously. Instead, Gombrich defended the notion that the perceptual experience of paintings has the flipping characteristic of the duck-rabbit figure so often explored in gestalt psychology. One might initially perceive the figure as a duck, but, once one manages also to perceive it as rabbit, it becomes possible to switch back and forth between perceptions. Yet it is impossible to perceive both simultaneously. Thus, regarding a painting of a battle horse, according to Gombrich (1959/2002, 236): "To understand the battle horse is for a moment to disregard the plane surface. We cannot have it both ways. . . . [T]he better the illusion the more we see a picture as if it were a mirror." On the illusion theory, depiction is a form of representation in which the surface of the painting elicits the perceptual response characteristic of seeing the scene depicted. The painting does re-create in the mind the very experience one would have in the presence of the target *in the flesh*, as it were. One can of course get close to the painting and observe its brushstrokes and technique instead.[32] But, according to Gombrich, one can never experience both surface and content simultaneously.

Wollheim criticized Gombrich's illusion theory and instead defended the position that the characteristic mode of perception that distinguishes depiction is twofold. It allows the experience of the surface and the experience of the content of the painting simultaneously. He took inspiration from Leonardo da Vinci, who trained his students to see objects and images in all kinds of surfaces, such that, "if you look at any walls soiled with a variety of stains, or stones with variegated patterns . . . , you will therein be able to see a resemblance to landscapes graced with mountains, rivers, rocks, trees, plains, great valleys, and hills in many combinations . . . and an endless number of things which you can distil into finely-rendered forms" (da Vinci 1989, 222). The general phenomenon, which we are all seemingly able to experience from a very early age and appears to be cross-cultural and widely spread through the epochs (seemingly beginning with Neolithic painting), was famously branded *seeing-in* by Wollheim (1968/1980, 209).[33]

There is here a helpful contrast to be drawn with the phenomenon of seeing-as, which Wollheim initially thought central but is perfectly amenable to a conventionalist treatment. Thus, in Goodman's (1968/1976, 27–31) system, seeing-as is simply representing-as. We represent *x* as *y* if we represent *x* by means of a *y* representation (a caricature that, say, portrays Winston Churchill as an infant, is an infant representation of Churchill). By contrast, seeing-in is fundamentally a kind of unanalyzable experience, one that cannot be conventionally defined as any set of signs within a symbol system. It can be studied only through the consequences it has in our cognitive and aesthetic preferences, habits, uses, or practices. According to Wollheim (1968/1980, 210ff.), three considerations favor seeing-in as the appropriate phenomenological experience involved in depiction.

First, one can see events and states of affairs *into* a painting; but one cannot see a painting *as* a state of affairs or *as* an event. I can see the characters in Holbein's *Ambassadors* as wealthy mortal beings, Velázquez's Pope Innocent X as a commanding man, or the bombing of Guernica as carnage. But I can only *see in* the Holbein canvas death itself being the great equalizer or the contrast of mortality and death. Mutatis mutandis for the saturnine temperament *in* the Velázquez portrait and the rising threat of fascism *in Guernica*. I cannot see any of the targets portrayed *as* any of these things because seeing-as necessarily requires reference to objects and their properties (or their parts). Yet paintings evidently represent affairs, events, and processes, not just objects. As noted earlier, this is true even of most portraits and artistic photography.

Second, there is no localization requirement on seeing-in, which is rather holistic (Wollheim 1968/1980, 211): "If I see x as y, then there is always some part (up to the whole) of x that I see as y. . . . Seeing-as has to meet the requirement of localization, whereas no such requirement is placed on seeing-in. I may see x in y without there being any answer to the question whereabouts in x I can see y." Velázquez's portrait represents the commanding power of the pope as head of the Catholic Church, but there is no specific part of the painting that does so. Holbein's *Ambassadors* represents the contrast of mortality and wealth and the disharmony of conflicting times, but—other than the symbolic skull, and this only precisely because it is a conventional symbol—there is no part of the painting that does so. Picasso's *Guernica* represents the rising threat of fascism, the pain and destruction brought by war, and the suffering of the people of the city of Guernica. But, on pain of tergiversation, there is no part of the painting that looks *as* just any of that.

Third, and finally, "seeing-in permits unlimited simultaneous attention to what is seen and to the features of the medium" (Wollheim 1968/1980, 212). In other words, seeing-in accounts for the *twofold* nature of pictorial experience. One can see the commanding power of the church when and if one looks at the Velázquez portrait, or see mortality in Holbein's *Ambassadors* when and while one looks into it, or see violence in *Guernica* while contemplating it. By contrast, seeing-as retains the gestalt feature. One cannot simultaneously see some x as a y and be visually aware of the features of x that sustain this perception. I can gain a sense of how, say, Churchill is depicted as an infant only by fixing my attention on the part of the painting that so depicts him. There is debate in the literature regarding the nature of twofoldness, but the phenomenon itself is well established. It is not seeing-as, and it does not sit easily within conventionalist approaches to representation.

Generalized Seeing-In: Inflection, Normativity, Artificiality

I claim that these three advantageous features in Wollheim's account transcend depiction. They can be seen to be at work in the generation of representational force in cognitive representation more generally, including scientific modeling. Undoubtedly, the nature of the experiences differs greatly since handling models rarely boils down to merely observing a picture. But the overall architecture of the typical experience in the use of models shares all the abstract features that distinguish seeing-in. There is an unavoidably speculative dimension to the comparison (and hence this final section of the chapter is more adventurous and less scholarly in style). But it can be established phenomenologically through a detailed study of the features of the relevant cases in their phenomenological use.

Thus, for instance, take the first requirement imposed by seeing-as, the objectual requirement, from which seeing-in frees us. Only objects can be *seen as* something else in a painting. Events and states of affairs can only be *seen into* a painting. This evidently lines up well with any account of scientific representation since, as we have seen, scientific models rarely just represent static properties of objects at an instant but instead normally represent events, states of affairs, effects, and processes whether they are understood as successions of events or as self-standing dynamic entities. This is undoubtedly true for both the billiard ball and the stellar structure models that we studied in chapter 3 since they represent dynamic processes. But, even in the case of the Forth Rail Bridge, it can be argued that the engineer's diagrams are meant to represent not the bridge as it stands at any given instant but rather the relations between different parts of the bridge expressed abstractly as quantities of torsion, tension, and so on. In this sense, they too can be taken to represent abstract states of affairs.

Similarly for the localization requirement. The phenomenological experience of seeing-in is not localized in the sense that it cannot be ascribed to just one or another part of the work of art. When we consider most scientific models, particularly analogue, mathematical, or theoretical ones, we get the same sense that no aspect of them specifically carries the character of their representation of the target. Rather, the full sense of weights and counterweights in the Forth Rail Bridge model is carried by the whole graph, including the subsidiary appendixes and depictions of each of its parts. The billiard ball model, as a whole, represents the thermodynamic aspects of a gas arising out of the kinematics of its molecules. No individual billiard ball or set thereof can be taken to represent thermodynamic phenomena or the underlying kinematic properties. The structure of a stellar interior is not a property of any one of the four equations or variables that describe the dynamic behavior of every one of its layers, parts, and quantities. And the prey-predator dynamics in the Lotka-Volterra model is not localized in any given part of the model. It is rather a comprehensive dynamic feature of the *working* of the whole model, one that is exhibited only when the model is run as a simulation over a given parameterization of the data. Obviously, the experiences that are called forth in all these cases involve not *perception* but only *use*, and their normative import is given not by any sense of what we ought to perceive visually but by a more comprehensive account of our experience in using them. Nonetheless, the phenomenology of their use is not very different from the point of view of this localization requirement. Scientific representation is thus best rendered as a generalized form of seeing-in.

Finally, we come to the central feature of Wollheim's account, twofoldness. Again, there is no denying that this is meant to be, if anything, a universal feature of the characteristic perceptual experience of specifically pictorial art. Thus, it would apply in full only to diagrams or graphs that represent in ways akin to those in which pictures do. There are of course such diagrams in science, including many of the scale models that we studied in chapter 3, and understanding what they stand for as models requires a twofold perceptual experience like that characteristic of pictorial experience. We need to be simultaneously aware of the features of the San Francisco Bay model and those of the San Francisco Bay that it represents to operate on it *as a model*. We can of course instrumentally operate on the material aspects of the model without any knowledge of or regard for the Bay Area or indeed without any knowledge that it is in fact meant to represent the Bay Area, but it is questionable that in that case we are *operating on it as a model.*

A more interesting question still is whether some version of twofoldness applies to other cognitive representations. For instance, is there a sense in which I can use the Lotka-Volterra model as a model only if I am somehow implicitly applying it to a population of prey and predators, however abstractly considered? Obviously, I can just ignore the target of the model or be completely unaware of it and nonetheless operate on it formally as a set of mathematical equations whose symbols I know how to operate on syntactically. But, again, even Goodman would agree that it is questionable that this use of the model is a representational one. It is questionable that I am using the equation *as a model of anything*. So there is a sense in which, in order to operate on a model *qua model*, I need implicitly to conceive of its target system behaving as the model says it does. The twofoldness thesis seems to apply too at this level in the practice of the use of the model. It is again not to be considered a feature of our *perceptual* abilities. Rather, it applies to our implicit conceptions, those that are presupposed in our use of the model. This is a conceptual version of twofold experience requiring us to be aware of the formal features of the model while simultaneously holding onto our conception of the target modeled.

This generalization of *seeing-in* to models that are not merely depictive has interesting consequences, for many equivalents of the relevant features that have been ascribed to it in the pictorial setting obtain too in the larger setting of all our cognitive representations. They thus turn out to be illuminating of the features of representational force in the practice of modeling in general. In other words, the analogy with the philosophy of art at this point becomes genuinely enriching for our conception of representational force more generally and for what is consequently involved in the practice of scientific modeling. I shall distinguish three aspects that deserve discussion: the nature of what is known as *inflection* in *seeing-in*, as applied to depiction; the relevant normative dimension in the appropriate perception of paintings; and the implications for nonrepresentational paintings such as Mondrian's *Lozenge Composition*.

INFLECTION AND THE DYNAMICS OF MODEL BUILDING

An objection to Wollheim's (1968/1980) earliest account of *seeing-in* is that it is surreptitiously calling for two different experiences: the experience of the surface of the painting and the separate experience of what the painting depicts, that is, its target or content. In his later writings, however, Wollheim makes clear that there is just one experience with two different aspects within a single "fold" that can thus interact and relate in interesting ways (Wollheim 1987, 1998; Wollheim and Hopkins 2003).[34] As Lopes (2005, 123–24) puts it: "Features of the design may inflect illustrative content, so that the scene is experienced as having properties it could only be seen to have in pictures." One has an inflected pictorial experience when one ascribes to the content of the painting some features or properties that are induced by features or properties of the surface, for example, when one sees a rugged landscape in a painting only by virtue of the rugged and rough brushstrokes in the canvas. The converse can occur too. The features of the scene depicted, as conceived in thought, can affect how different features of the surface are perceived or experienced. One can come to perceive soft and gentle colors and strokes in a painting that represents a soft and gentle scene. This is in part because of what Wollheim (1998, 224) calls the *conceptual permeability* of seeing-in, the fact that it can be affected by thought. How we think about each aspect of our perceptual pictorial experience can affect the nature of the corresponding aspect of that integral experience. It is hard to explain or expect this if the experiences are double or divided, but it is only natural to think that different aspects of one integral experience would so depend on and affect each other.

While the rich and involved debate over the nature of inflection continues and there are different accounts of the unitary nature of seeing-in on Wollheim's account (e.g., Hopkins 2010, 2012; Lopes 2005; Nanay 2010; Newall 2015; and Wollheim and Hopkins 2003), what matters to us here is only the fact that inflection very obviously carries over to representational force in cognitive representation more generally. A model source can affect how we see and conceive its target phenomenon, system, or object. Think of Baker's anthropomorphic model of the Forth Rail Bridge described in chapter 3 (fig. 3.3 above). Operating this concrete model can give a sense of the very precise balance of weights that must obtain in the bridge. The billiard ball model evidently affects how we see gas molecules, particularly if we think of both concretely as statistical systems of independent entities. But, conversely too, as we explore (whether experimentally or merely in our thoughts) the features of the system modeled, we often come to see certain aspects of the model source in a different light. When we realize that the prey-predator population in the Adriatic does not always behave as predicted, we come to see the limitations of the Lotka-Volterra model. We now perceive the model to be not exact but only an approximate idealization. While, in most scientific models (or, more generally in most cognitive representations), this sort of feedback back and forth is conceptual rather than perceptive, it certainly seems to adopt the generalized form of inflection.

In fact, inflection is one way of understanding the dynamics of modeling practice that I discussed at the end of part 2. For it was there suggested that scientific models are constructed in response to certain dynamics whereby force and inferential capacities constantly modify each other. But if force is—as I am indeed suggesting—an essentially twofold experience, one that takes in both source system and target system simultaneously, it is bound to be the case that our apprehending of the inferential capacities of the model can be affected by what we understand to be the main features of the target phenomenon; and vice versa. Something very akin to conceptual inflection is at the heart of our ability to modify models' inferential capacities and forces in response to our understanding of each and their relation.

THE NORMATIVITY OF REPRESENTATION

At the end of chapter 7, I emphasized the normative aspects of the inferential conception. Both representational force and inferential capacities are normative requirements. They are subject to norms, including the norms of valid logical reasoning, but also more specifically tailored norms that apply to the selection of targets (representational force) and the inferences that are licensed (inferential capacities) within the context. The latter are divided into vertical and horizontal rules of inference, which apply to the demonstrations within models and their application to their targets, respectively. These rules are evidently tailored to the specific representational sources and targets involved, they emerge in the context of a richly layered practice, and they are ultimately socially sanctioned and maintained by communities of practitioners. Their subsequent application is therefore grounded on expert judgment, which in turn reinforces the rules. They are then passed on in diverse pedagogical and teaching contexts by knowledgeable instructors, often—in the model of Warwick (2003)—by means of what Thomas Kuhn (1962/1996) referred to as *exemplars*. The slogan, recall, is that representation is located not "in the objects," or "in the mind," but in the social world, as a part of richly textured model-building practice.

It is instructive at this point to see that, perhaps despite first appearances, phenomenological accounts of pictorial representation agree with this view of the sources of "representational normativity." I say *despite first appearances* because there is an obvious difference in how the different set of rules for licensing inferences as well as the norms that select and maintain force are established in the first place and how loose they can be. It is evident that there is more leeway in the setting of these norms and the interpretation of a piece of art than there is in a scientific model. Some works of art are in fact exceedingly open to interpretation, while others are less so, and often this goes along with the degree of figurativeness discussed earlier in this chapter. Thus, *Guernica* is notoriously open to interpretation in a way that Velázquez's *Pope Innocent X* and Holbein's *Ambassadors* are not. A creative artist also has more freedom in setting up new conventions and rules of inference than a scientist ever does. Scientists are not entirely corseted or constrained, but, on the contrary, they can and often do introduce new conventions and rules to go along with their models.[35] In the long run, the representational uses typical of artworks and scientific models tend to coalesce around the same set of underlying normative social commitments and structures and in a very similar fashion too. In both cases, it takes a sufficiently knowledgeable and competent agent to apply those norms correctly, and this kind of context-dependent expertise cannot be substituted by any algorithm or abstract set of universal rules.

Wollheim's most famous essay, "Seeing-as, Seeing-in, and Pictorial Representation," precisely begins with a discussion of the standards of correctness in the interpretation of painting: "What is unique to the seeing appropriate to [pictorial] representations is this: that a standard of correctness applies to it and this standard derives from the intention of the maker of the representation." This initially appears to Wollheim to be unique, separating pictorial representation "from other species of the same representational genus, i.e., representational seeing," since the intention of the maker does not necessarily fix the correct interpretation in other nonpictorial representations, such as test diagrams or photographs. But one might wonder what the difference here can be when the standards of interpretation are after all fixed in advance for the artist to comply with: "What the standard does is to select the correct perception of a representation out of possible perceptions of it, where possible perceptions are those open to spectators in possession of all the relevant skills and beliefs. If, through incompetence, ignorance, or bad luck of the artist, the possible perceptions of a given representation do not include one that matches the artist's intention, . . . nothing or no-one is represented" (Wollheim 1968/1980, 205–6). I invite the reader to substitute in this paragraph *scientist* for *artist* and *interpretations* for *perceptions*. The resulting paragraph is one that, I believe, we would all be inclined to agree with. There is a normative background against which both artists and scientists work when they produce a new work of art or introduce a new scientific model. Naturally, that background is richer and much more stringent for a scientist (particularly as regards the vertical inferential capacities of the model). But, essentially, unless the artist/scientist complies with the normative background that is shared by the audience, nothing would result that can be correctly interpreted as a *representation*. Given the looser set of norms and the greater leeway, this is always bound to happen more often with artworks than with scientific models. Nevertheless, entirely incomprehensible models that are unable to capture any phenomena at all have also been introduced often in the history of science. They just tend not to appear in the textbooks.

This all brings us to the social or communal nature of the normative background that is at work in both artistic representation and scientific representation. It is through the enforcement of (often tacit and implicit) shared values, interpretations, and standards (and particularly the licensed norms for the application of force and vertical and horizontal rules of inference) that artworks and models alike gain their representational purchase. This is how both their *representational force* and their *inferential capacities* are socially sanctioned. It seems to me that most philosophers of art now appreciate this (see, e.g., the essays in Kemp and Mras 2016), and Wollheim himself came to a clear understanding of it in time, for instance, when he brilliantly wrote (Wollheim 1987, 48): "In a community where seeing-in is firmly established, some member of the community . . . sets about marking a surface with the intention of getting others around him to see some definite thing in it: say, a bison. If the artist's intention is successful to the extent that a bison can be seen in the surface as he has marked it, then the community closes ranks in that someone who does indeed see a bison in it is now held to see the surface correctly, and anyone is held to see it incorrectly if he sees, as he might, something else in it, or nothing at all. Now the marked surface represents a bison." This strikes me as a rather apt description of the similar processes employed by scientific communities to fix the representational force of their models.

MODELS AS ARTIFACTS WITH LIVES OF THEIR OWN

So far, I have been emphasizing the different ways in which the philosophy of art can enlighten philosophers of science about the nature of representation. There is, however, a final set of issues at the crossroads of artistic and scientific representation where the enlightenment can go both ways. It concerns the answer to the first binary question raised at the start of this chapter: Where is the frontier between representational and nonrepresentational art? How do we distinguish the Mondrian composition from the other three paintings?

The analogy with seeing-in allows us a better grasp of the claim that, when a model ceases to be a model, it becomes mere formal theory. What this means is that its use ceases to have any twofold nature or structure. A nonrepresentational use of a model is an experience that entirely lacks the dual aspect of the perception of the representational target, just like a nonrepresentational work cannot be perceived as a unitary experience with two aspects—the surface and its content. It provides merely the experience of its surface. So, just as in the limit the Mondrian painting has no purchase beyond itself and invites no perception of any scene thereby depicted, a nonrepresentational model is not really a model since it invites the consideration of no target object or system beyond itself. The model source gains a life of its own when the model loses its representational force, thereby ceasing to invite any licensed informative inferences regarding any ulterior object, system, or phenomenon. It is of course not the case that the source loses its internal structure; rather, there is nothing in the practice to warrant any licensed inferences from the source toward any target whatever. All horizontal rules of inference are suspended and put on hold, and the only rules of inference remaining are vertical. These rules allow us to manipulate the model source itself but lack any consequence for a target since there is none.[36]

There is a further thought that deserves exploring but, in a more speculative spirit, can be sketched here only briefly. One growing view in the philosophy of science emphasizes the artifactual nature of models—the different ways in which models are constructed to work as tools toward a range of purposes. Tarja Knuuttila (2011, 2017) has forcefully argued for a hybrid representational view, one in which the artifactual nature of models is best revealed in the mediation that models carry out between theory and data.[37] But the view has interesting links with the methodology for the evaluation for models' adequacy for purpose (Alexandrova 2010; Parker 2010, 2020) that I touched on lightly at the start of this chapter. For, on the adequacy-for-purpose view, models are essentially always *means* toward achieving some desired value, virtue, or outcome, whether it be, say, explanatory power, empirical adequacy, unified phenomena, covering laws, experimental tests, interventions, or manipulation of causal variables. So the natural view of models is accordingly instrumental. Models are constructed as means for something else, and, in that sense, they are artifacts built for a purpose.

This instrumentalist thought lends itself to a more radical form of artifactualism, one in which models are studiously constructed in response *only* to the needs of our practice, without regard for their fit with their putative targets. I suggest that one way of thinking about models in this instrumentalist guise is to conceive of them as initially merely flat surfaces—that is, not as genuine models or representations—that can nonetheless become representations (or not, depending on needs) at a later stage of inquiry. This radical artifactual view (Sanches de Oliveira 2022) thus dispenses with the twofold experience altogether. In dealing with models in this guise, scientists merely experience them as surfaces and have no simultaneous experience whatever of their contents. This is a form of extreme nonrepresentationalism that takes models as always being born in the state analogous to Mondrian's *Lozenge Composition*, bereft of any representational force. I do not believe that it applies to many models (since on my view most models fully merit their name and are born as twofold representations from the word go); nor do I think it is sustainable for very long in the history of any model (on pain of the term *model* losing its handle altogether). But it is a view that makes sense of some concrete objects that are specifically built with a purpose and can thereafter become models in the full sense of the term I use here, such as model organisms in synthetic biology (Knuuttila and Loettgers 2013; see also Ankeny and Leonelli 2021). On the analogy of a Mondrian painting, the inferential conception affords an intriguing understanding of such objects as lacking representational force and therefore having no licensed horizontal rules of inference. Yet these objects can be fully normatively constrained in the vertical rules of inference they deploy and are therefore always susceptible of becoming fully models as and when they are employed to explore other systems of interest.

The Phenomenology of Force

The previous section argued for an extension of Wollheim's notion of seeing-in to cases of representation outside the perceptual arts. It is necessary to depart from the phenomenology of perception that characterizes painting since scientific models are not objects of perception, nor are they often even visual entities. We must instead move toward a more general phenomenology of the experienced use of representations as well as the features of the practice of building such representations. Nevertheless, some of the hallmarks of seeing-in remain in this new guise. There is therefore a sense in which a generalization of seeing-in outside the visual arts has some purchase over the general conditions of cognitive representation more generally. The phenomenological approach is, I believe, at the heart of any deflationary account of scientific representation that would focus on the practice and the experience of using representations, as opposed to any more essentialist leanings or attempts to probe into their natures.[38]

Obviously, there is no denying the evident dissimilarities between both phenomenologies. The phenomenology typical of the experience of the visual arts is—at best—a case or instance of the more general phenomenology involved in setting up representational forces for any cognitive gain. Yet this already opens new avenues of research into the nature of pictorial experience since seeing-in is now understood to be a concrete application of representational force to the special case of depiction. And, while the establishment of force in pictorial experience has been studied by philosophers of art, the other fork in the inferential conception, which ought to apply to all cognitive representation and hence also to most cases of depiction, has been hardly even touched on. Yet one can ask many questions about the inferences that are normatively licensed, within the appropriate contexts, by a painting such as Velázquez's *Pope Innocent X* or Holbein's *Ambassadors*, not to mention *Guernica*. Can one infer to the dominant power of the church, and, if so, how exactly? What more can be said about disharmony of the times from a consideration of the instruments depicted? How do the different licensed inferences from *Guernica* change in accordance with cultural and political context, and how exactly are these normatively implemented? The application of the inferential conception to artistic representation opens a Pandora's box of new questions that philosophers of art have not considered in detail.[39]

There is also no denying that conventional symbolic elements play a more prominent role in scientific models than they play in art. This is clear from a mere inspection of the typical media involved, which in scientific models—as we saw in previous chapters—can take many forms beyond merely visual depiction, including highly abstract formal symbolic systems in the language of mathematics. Nonetheless, the issue here strikes me as essentially one of degree and not in any way requiring a fundamental characterization of different types of representation involved. Even Wollheim of course agreed there are conventional and symbolic elements involved in the visual arts. The symbolic and conventional meanings imported into Holbein's *Ambassadors* are undeniable, and there is much in modern abstract art that blurs the lines between conventional symbolism and visual perception (an egregious example is Magritte's introduction of the sentence "Ceci N'est Pas une Pipe" in his celebrated 1929 painting *La trahison des images*, but there are many others). But the issue at stake here is whether the conventional elements define the nature (the constituent) of representation and whether they are thus sufficient for an understanding of what is involved in the practice that sets up representational forces. Wollheim's answer is that, as concerns depiction, they do not suffice, partly because nothing can define representation, which can be approached only in a deflationary spirit through its phenomenology. I thoroughly agree with the spirit of this answer, and I believe that it generalizes fully to cognitive representation in general. There is undoubtedly much philosophical work to be done in this area since the conceptual generalization of twofold experience that I am proposing is new and original, as far as I know, and requires further unpacking. At least I hope to have provided the outlines of how such a research program could proceed and why the analogy between scientific representation and artistic representation continues to hold considerable promise in the elucidation of both.

Chapter Summary

This chapter embraces the analogy between artistic and scientific representation and defends the view that there is much they share. It first delves into four cases in the representational arts that mobilize some of our intuitions regarding degrees of figurativeness and the sharper distinction between representational and nonrepresentational art. It then proceeds to discuss conventionalist and phenomenological theories of depiction in order to illuminate the nature of representational force. The exposition then focuses on Wollheim's (1968/1980, 1987) phenomenological account. This is not only because Wollheim's views are very influential in the philosophy of art but also because his characterization of seeing-in exhibits a notable fit with aspects of the notion of representational force that I introduced in previous chapters. Both force and seeing-in are deflationary in spirit and aim at surface characterizations of phenomena that admit no deeper theoretical illumination. Yet the surface features can be described, and there is much in common. I do not go as far as claiming that representational force and seeing-in are identical, even for depiction. But the twofold nature of pictorial experience can be thought of as the concrete application of representational force to the specific case of depiction since their consequences are so similar.

9

Scientific Epistemology Transformed

Scientific representation has been claimed to have relevant overlap with—and important implications for—metaphysics, the philosophies of mind and language, and epistemology as well as the philosophy of art. In this book, I focus only on the implications for the philosophy of art and scientific epistemology, which I believe to be genuine and of considerable interest for philosophers of science, epistemologists, and philosophers of art alike. The last chapter acknowledged the relevance of philosophical aesthetics, and it discussed in depth the connections between the inferential conception of representation defended in this book and the phenomenology of artistic perception. In this chapter, I shall embark on a similarly detailed study of the relevance of recent discussions of scientific representation for the epistemology of science. These are the two main areas of philosophical research—outside the philosophy of science and the study of modeling—where I claim there to be significant implications for or from the inferential conception of scientific representation developed in earlier parts of the book. They come to show that the inferential conception is not just an account of scientific models. Rather, it serves as an account of cognitive representation more generally.

The attentive reader already knows that I regard the implications for metaphysics to be somewhat overblown, and one of the aims in the second part of the book was to minimize or deflate the lofty ambition to read off any metaphysics of relations from scientific models. Instead, I have focused on the practice of model building, which must play a central role in any philosophical study of scientific representation. This has consequences for what are cognate areas of shared interest and how those shared interests can best be cashed out. For instance, artistic representation—painting in particular—is also a deeply representational practice, so the links are evident and must be acknowledged. I shall argue in this chapter that an understanding of representation as inferential practice also has consequences for any practice-based epistemology of scientific knowledge. In some sense, the links here are deeper since there are not merely nominal connections between the varieties of cognitive representation (as is the case with art). Rather, scientific representations themselves are the typical subject of such evaluative practices. Here, as elsewhere, I argue that there are significant payoffs in going deflationary about some of the central notions in epistemology while remaining pluralistic about science's main epistemic goals.

As for the overlap with the philosophy of language, I do accept that there is a genuine connection with semiotics and the practice of sign making that did turn up already in chapters 4 and 5. That connection plausibly stands to be further probed and studied (to the point, perhaps, of deserving a separate book). However, as repeatedly noted, the notion of reference is not here placed at the heart of scientific representation at all. So referential semantics and much of traditional philosophy of language finds no place of honor in this account. I doubt that I stand alone now in this regard. The semantic turn in the philosophy of science discussed in chapter 6 already pointed a route to the philosophical analysis of science away from any purely linguistic relations. Even closer to home, there is now an extensive movement in the philosophy of science away from truth or reference as the critical relations for evaluative purposes as well as a very large contemporary literature on the role of scientific fictions within scientific models.[1] Assumptions within models can be characterized as fictive or fictional, depending on failure of reference. In the latter case, there is outright failure of reference, as exemplified in some of the models that we analyzed in the first part of the book. The thought that such fictions play a critical central role in scientific representation thus already implicitly jettisons any constitutive connection between representation and the linguistic notion of reference (see, e.g., Elgin 2017; Suárez 2009a; and Toon 2012a).

Finally, as regards the philosophy of mind, I have rejected the view that scientific representations are merely mental states, and I continue to emphasize intended uses rather than intentions at the heart of any articulate account of scientific representation. It is a matter of course that a lot of the practice of cognitive representation generally relies on thinking activity, in particular, inferential activity. In that sense, representation requires the development of certain mental activities and skills and, consequently, can be said to involve mental states (at least on some central accounts of what thinking or mental activity generally involves). This can entail some limits on what cognitive science can say informatively about scientific representation more generally. Yet this is not to say that the implications of a deflationary or an inferential account of scientific representation for cognitive science will be minor or insignificant. But that would really be the topic for another book.[2]

Hence, all that remains is to broach the significant implications for scientific epistemology, and a good place to start is Ian Hacking's *Representing and Intervening* (1983). The claim at the root of Hacking's book is that representation is linked to scientific theorizing and presupposes notions of fit, truth, or reference between theoretical terms and the objects purportedly represented. Representation is thus a relation that stands between bits of our science and the world; and as such, according to Hacking, it turns out to be both unassailable and impossible to evaluate from any representation-independent standpoint. This gives rise to the notorious issue of the incommensurability of paradigms or theoretical choices, and it demonstrates the futility of representation as a hallmark of either progress or realism in science. No criterion of realism for either the entities putatively referred to by our best science or their properties can be achieved through representation, according to Hacking. There is an inevitable sense of self-reference in any representational relation, in the sense that representations ultimately refer only to further representations or that they are mediated by them. And this feature turns out to be self-defeating in any attempt to provide a representational relation with the appropriate epistemic credentials to sustain a scientific realism worth its name.[3]

By contrast, in a marked and notorious dichotomy, Hacking defended the view that experimental intervention can provide apposite criteria for both scientific progress and the reality of the entities hypothesized by science and/or their properties. Experimental intervention in the world is a type of action that creates its own material conditions. Since this type of action is already in the world, the global questions as to its relation to this world, its putative reference to bits of it, or the truth conditions in this world for the corresponding statements do not arise (or at any rate have no purchase on the type of activities involved and their significance). Underlying Hacking's contrast, there is the thought that representation can provide only the sort of knowledge derided in John Dewey's critique of the spectator theory of knowledge (Dewey 1943).[4] By contrast, experimental intervention has the potential to provide genuine knowledge of reality.

In other words, experimental realism is born out of a contrast between a rather passive and substantive concept of representation, on the one hand, and an interventionist account of the activity that characterizes experimental life, on the other. Yet, on the view defended in this book, as we have seen representation is also an activity, one moreover with significant interventionist practices at its heart. We have seen that models are built in stages and in constant response to their perceived matches with our ever-evolving understanding and conceptualizations of phenomena. These interventionist practices are thoroughly social, in that they centrally involve some important normative constraints that are social or communal in character. And they can gain an autonomous life of their own too.[5]

It does not then seem possible neatly to distinguish representation from experimentation on such grounds. Both are similarly autonomous, embodying their own community constraints within their own often-intermingling practices. And both require us to think carefully about the role of human interventions in carrying them out. Representation thus construed seems to be involved in both types of theoretical and experimental work, in similar ways. I shall go on to argue below that a deflationary conception of representation belies the thought that there is any significant cognitive difference between theoretical work and experimental work in the sciences. Both involve the same sorts of representational practices. This is in keeping with the defense of an extended version of the semantic conceptions of theories that places deflationary representation right at the heart of theorizing as well as any experimental work. Whatever differences have been ascribed to these two sorts of activity (theoretical vs. experimental) perhaps have sources in the social classes that traditionally carried them out or in the sorts of practical endeavors and goals that they typically aim for. I submit that they are not to be found in any cognitive ability peculiar to their modes of representation. There are, however, key epistemological differences in the assessment of the accuracy or adequacy for purpose of those representations that are characteristic of theoretical and experimental work, and these can be argued to ground some experimental realist claims still appropriately.[6]

Thus, in the first section of this final chapter, I consider Ian Hacking's and Nancy Cartwright's arguments for experimental realism, and I argue that no attempt to make the position cogent can escape the ubiquitous use of a deflationary inferential conception of representation. The key distinctions are indeed epistemological, and they are best understood against the unified background of a deflationary inferential conception. The hope here is to give experimental realism a second airing, thus allowing us to confront some of the old objections raised against it.

The remaining sections of this chapter consider alternatives to this form of experimental realism within the work of some of the main authors that are regarded as constituting the contemporary canon in scientific epistemology.

Bas van Fraassen's constructive empiricism is discussed first, and it is argued that its early forms fall prey to objections deriving from its use of a substantive conception of representation. This gives constructive empiricism the form that has been the object of so much criticism in the literature and that many of us now regard as untenable. Constructive empiricism, I urge, fares significantly better when framed within a deflationary inferential conception of representation.[7] At any rate, van Fraassen's favored semantic conception of theories can be made compatible with a deflationary account, as already pointed out in chapters 5 and 6. While not savaging the whole of constructive empiricism, this does show that a certain form of theoretical instrumentalism, one inviting a voluntarist epistemology regarding hypothetical posits, is a tenable position.

The next section focuses on some very influential contemporary pluralist epistemologies that emphasize the variety of science's goals. These normative epistemologies put social, moral, or political values at the core of science. Arthur Fine (1987/1996) was perhaps the first to espouse and defend an NOA that directs philosophers to study the actual norms of scientific practice. Helen Longino (1990, 2002) and Philip Kitcher (2001b, 2009) are known for their accounts of the intermingling of values and facts in scientific activity. This is in line with much postpositivist critique of the distinction between the contexts of discovery and justification and the fact/value dichotomy (e.g., Fine 1987; and Putnam 2002). While largely endorsing these views, I find in key passages and parts of these authors' work remnants of a substantive conception of representation that weaken their positions. I argue that the adoption of a natural modeling attitude instead—the contemporary descendant of the implicit modeling attitude of the nineteenth-century modelers studied in the first part of the book—together with an explicitly deflationary inferential conception of representation would strengthen their epistemological standpoints. The ultimate epistemological goals of Fine, Kitcher, and Longino are best served by a deflationary account of scientific representation.

The final section, which closes both chapter and book, suggests that the natural modeling attitude offers some enlightenment on the currently much debated notion of scientific understanding. On a deflationary account of scientific representation and its concomitant modeling attitude, a degree of understanding is readily available to modelers of whatever persuasion and practice and whatever tools they employ for the task in hand. This stands in great contrast with the deeper and harder to achieve explanations that theoreticians seek.[8] The circle closes with a renewed defense of an epistemology that respects the critical role played by our experimental and modeling practices in the acquisition of scientific knowledge.

Experimental Realism Revisited

The deeply influential *Representing and Intervening* set a marked contrast between theoretical representation and experimental intervention in the world, and Hacking used the contrast to launch an argument for realism at the level of experimental intervention. The famous slogan "if you can spray them, then they are real" (Hacking 1983, 24), was meant to guarantee the existence of those unobservable entities whose causal properties we know sufficiently well to employ them effectively in our probing into the nature of further objects. But how exactly are we to understand the thesis that underlies the slogan?

It is widely believed that, when first introducing experimental or entity realism, Ian Hacking and Nancy Cartwright mainly had in mind a metaphysical thesis about what is or can be real. Thus, Hacking's manipulability criterion is often thought to express a commitment to the reality of entities endowed with causal properties that can be tinkered with by agents to probe into further systems. As Margaret Morrison (1990, 1) rightly put it: "Hacking contrasts the metaphysical questions concerning scientific realism with those that deal with rationality, the epistemological questions. . . . In arguing for entity realism, Hacking takes himself to be addressing only the metaphysical questions."

Certainly, much of Hacking's and Cartwright's work in the last two and a half decades lends itself to this interpretation. The slogans that popularize the position and that we all use to teach our undergraduate students reinforce it ("if you can spray them, they are real," "whatever can be manipulated is real," etc.) since they explicitly are conditions on reality. Hacking (1983, 146) at least seems to go along happily: "Reality has to do with causation and our notions of reality are formed from our ability to change the world. . . . We shall count as real what we can use to intervene in the world to affect something else, or what the world can use to affect us."

By contrast, Cartwright initially defended a causal inference account of experimental realism that is prima facie at least compatible with an epistemology first position, a position to which I shall return in due course. Yet, for Hacking (1983, 28), the epistemology of causal inference is there only because otherwise the metaphysics would be "idle," and Cartwright followed suit in her later works, many of which are rife with metaphysical claims regarding what nature or the world must be like for our scientific methods to work as they do.[9] The metaphysical hold since then seems strong since, whatever views Cartwright went on to espouse on methodology and epistemology, they are always there to bring out the content of antecedent metaphysical beliefs in a patchy or dappled world: the ontology grounds the methodology of causal inference wherever this is applicable. Her most recent book on these topics contains a chapter devoted to the notion of "causal warrant," signaling a significant change at least in emphasis with respect to previous books. But the change may be rhetorical or cosmetic since the book is essentially a manifesto for the primacy of causal metaphysics over any methods of causal inference (Cartwright 2007).[10]

We shall need greater precision in our definitions, so let us first attempt to summarize this rather common metaphysical form for experimental realism as the *metaphysical experimental realism* (MER) thesis: *Manipulation is a necessary and sufficient condition on reality:* x *is real if and only if* x *can be manipulated.* This definition makes experimental realism prima facie implausible for many areas of science since many of the entities postulated by science are not susceptible to manipulation, at least not if we assume that any object can be manipulated whose properties can in principle be interfered with by a human agent in some carefully contrived and productive manner, typically during an experiment. There are plenty of objects and entities that science takes to be real yet cannot be manipulated in this way. Ian Hacking (1989) himself has made a strong claim along these lines for black holes and gravitational lenses.[11] It seems very contrived that such entities are not real simply because they cannot be manipulated. The very concept of manipulation seems to entail limits on what can be manipulated among all those things that are real. If so, manipulation cannot be a necessary condition on reality since it is not true for every entity *x* that we might suppose is real that *x* can be manipulated.

The defender of MER might weaken the position to a merely sufficient condition, which we can summarize as the *metaphysical condition on experimental realism* (MCER):[12] *Manipulation is a sufficient condition on reality:* x *is real if* x *can be manipulated.* The statement captures appropriately the main metaphysical commitment behind Hacking's and Cartwright's defense of experimental realism. However, it has been the object of rather convincing criticisms, which I claim fall into three different kinds: inadequacy, incoherence, and implausibility.[13] In different ways, they are all attempts to show that manipulation cannot even be a hallmark of reality because it is not possible to classify some entity or type of entity as real independently of the theory that describes its causal powers and our possible manipulations of it. The theoretical representation of the entity plays a critical role in evaluating what in fact is real in all these cases. Yet experimental realism is supposed to provide grounds for existential commitments in the entities postulated by science that are independent of any theoretical representations. Hence, experimental realism fails since it is not able to ground the unobservable entities that science postulates without ipso facto grounding the truth of the theoretical representations that best describe those entities.

The charge of inadequacy has been best voiced by Resnik (1994, 395), who maintains that "the experimental realist can only have knowledge about theoretical entities if she assumes that the theories that describe those entities are at least approximately true." The claim is that in practice theoretical knowledge and experimental knowledge are always intertwined. This is close but nonequivalent to the thesis that all observation is theory laden. So the experimental realist argument for the reality of unobservable entities can be sound only at the cost of major descriptive failure. To put it another way, the experimental realist either fails to provide an accurate description of scientific practice or fails to provide an argument for the reality of unobservable entities that is distinct from the traditional arguments for scientific realism about the entities postulated in our best theoretical representations of the phenomena.[14]

Resnik's argument is widely, although unreflectively, accepted in the literature (e.g., both Massimi [2004] and Morrison [1990] seemingly accept it). Yet it is, in my view, among the weakest arguments against experimental realism. At best, it can directly refute only a *psychological version of experimental realism* (PER) thesis asserting: *Manipulation is a necessary and sufficient condition on the acquisition of existential beliefs: a subject S can acquire the belief that* x *exists if and only if S manipulates* x. Resnik's argument would at least prima facie seem to refute PER since it points out how experimental manipulation and intervention are in fact neither necessary nor sufficient in practice for acquiring existential beliefs. Scientists routinely acquire their beliefs about what sorts of entities are real via theory, testimony from peers, the legacy of some tradition, or a mixture of any of these. This is quite clear, but it is also obvious that no metaphysical or epistemological conclusions are forthcoming from a statement merely about the acquisition of beliefs. The point of experimental realism surely rather involves claims about the content or the warrant of such beliefs, regardless of their provenance. It seems perfectly consistent to accept PER yet deny any of the informative versions of experimental realism that have epistemic bite.

The only way in fact to turn Resnik's argument into an effective weapon against the metaphysical versions of experimental realism so far reviewed is to endow it with some powerful metaphysical underpinnings to any successful theoretical representation. If a theoretical representation is adequate, acceptable, or successful only insofar as it is true in the correspondence sense (so that it corresponds to reality in some substantive way), then the fact that existential beliefs are routinely acquired via some theoretical representation that has these properties entails the existence of those entities since the entities must exist for the representations to be successful or true in this metaphysically substantive way. It follows that, if Resnik thinks the argument holds water against metaphysical variants of experimental realism, he must be assuming a substantive conception of scientific representation. By contrast, on a deflationary account of scientific representation, a theoretical representation is acceptable or adequate to purpose whenever it can be successfully employed in informative inference; there is no requirement that the representational source matches the target in any specified way. Hence, Resnik's argument against experimental realism—even the stronger metaphysical variants—fails on a deflationary account of representation.

A more robust argument against experimental realism charges it with incoherence. Regardless of how scientists acquire their beliefs, it is questionable whether one can coherently separate an unobservable entity from the theory that best describes it. In some cases, particularly as regards unobservable entities, the only concept that we possess of an entity is the one given to us by its theoretical representation. (This is not typically the case for observable entities, for which we have a perceptual as well as a theoretical concept.) Hacking is clear that only a short list of "phenomenological home truths" is required to describe our experimental interaction with an unobservable entity and its properties; we need not believe its full theoretical representation. But the issue is whether we can thus infer the theoretical representation of the entity on the grounds that we can manipulate it. And it appears that we cannot. The phenomenological home truths involve only a subset of the properties of the entity that appear in its full theoretical representation. Hence, what we can infer on the basis of such home truths does not correspond to the full concept of the entity as theoretically represented.

This is at the heart of what I call the *flectron* objection to the metaphysical versions of experimental realism (such as MER or MCER). The entities that we must suppose are real because we manipulate them in the electron microscope are not electrons as we understand them; they are instead particles (call them *flectrons*) that have some but not all the properties of electrons. So MCER does not provide us with any resources to claim that electrons as theoretically represented are real. It forces us to claim instead that flectrons are real. MER goes even further since it positively disallows the inference that there are electrons. On either view, it follows that scientists would be manipulating different entities (flectrons one time, plectrons another, and so on) in each distinct case of a putative manipulation of electrons. The licensed inference would be only to the set of home properties required in the experimental conditions with which we are working, a distinct and different set of all the properties ascribed to it in the full theoretical representation.

But this seems plainly wrong since scientists take themselves to be manipulating the very same entities (electrons) whether they are performing a scattering experiment or operating an electron microscope. (Similarly for historians and philosophers of science who describe the scientists' activities. We all refer to the same entities, namely, electrons.) The only concept we possess is that of the electron, and the only inference that makes sense to ascribe in such cases is to the reality of these particles. Thus, the incoherence charge is a powerful argument against the metaphysical versions of experimental realism—whether in the strong (MER) or in the weak (MCER) senses. On either view, the ontological commitments of scientists ought to dovetail with their experimental practice, always in line with those properties they manipulate, but no further. But this is not what happens in actual practice when the manipulations afford warrant in the full plethora of properties in the theoretical representation. How can this be?

The incoherence charge favors deflated versions of experimental realism that make no strong metaphysical commitments and instead adopt a deflationary account of representation, such as the inferential conception. In other words, the typical objections to experimental realism target only its metaphysical variants (MER, MCER). A purely epistemic reading of experimental realism remains viable. Contrary to Hacking's and Cartwright's early attempts to put ontology and the metaphysics of reality in charge, it benefits experimental realism to put the epistemology and the methodology of model building in the driving seat instead. The claim at the heart of experimental realism ought then to be one concerning warrant and only warrant for our phenomenological home representations, understood in a deflationary spirit.

It is true that experimental realism thus understood aims to align the warrant provided by the methodologies of causal inference in experimental work with the sort of epistemic warrant that operates in ordinary, everyday cognition. It is there not rare to ascribe a stronger form of warrant to antecedent existential commitments in those entities whose properties come to be manipulated in ordinary life activities. Any cursory example can illustrate this. Suppose that someone gives me a model graph of the contents of the kitchen next door. I may come to believe on the basis of this representation that it contains a breakfast table larger than my own, but I have an epistemically strong warrant for this belief only if I am afforded the chance to manipulate it causally, for example, to use it for my breakfast and so on. This is what I referred to in previous work (Suárez 2008) as *causal warrant*, and I employed it in the distinct formulation of the *epistemic experimental realism* (EER) thesis: *Manipulation is a necessary and sufficient condition on causal warrant: our belief in the reality of* x *acquires such warrant if and only if* x *can be manipulated.*

The representational practice that underlies the methodology of inferences in experimental science is indeed the same one that operates in theoretical representational practice. Representing in all these cases boils down to the drawing of licensed inferences that are well grounded and directed with appropriate force. So to go from manipulating an entity under a particular theoretical representation to asserting that the entity is real requires much more than simply assessing that the representation is correct. For the correctness of the representation really says nothing other than that the source has the appropriate inferential capacities, within context and as used, to deliver informative inferences regarding the target, however conceptualized. And that achievement is entirely independent of the reality or otherwise of the target as represented. This inferential practice is primary, and it drives whatever metaphysics there might be in experimental scientific practice. The result is not just a change of emphasis but a fundamental change of philosophical outlook, one that, I argue, allows us to develop a new and more successful defense of experimentalism.

While the representational practices are essentially the same in theoretical work and experimental work—and models are employed in all such areas—the epistemic practices and constraints vary greatly. The version of experimental realism that I defended in the past, following Cartwright, puts causal warrant up front. It assimilates all warrant gained in an experiment to the causal warrant typical of manipulative interventions on causal variables.[15] I now believe that this is too conservative, that something more general is required, a general kind of *experimental warrant* that is accrued through any form of experimental manipulability in a suitably controlled environment. This helps accommodate bona fide simulations and some thought experiments as well as the sort of observational evidence typical of astrophysics and hence allows us to characterize some of the warrant thus acquired as of a stronger type than the one deriving from merely theoretical hypothesizing. While this generalized form of experimental warrant can—and often does—issue in support for causal laws, this need not always be the case. Manipulability need not always be a meaningful path to causal discovery.[16]

There are three different reasons why a specifically causal form of warrant such as Cartwright's provides a poor basis for the epistemic form of experimentalism that I want to defend. First, it unfortunately already presupposes a metaphysical thesis regarding what is real, namely, that only those properties that are causal can be manipulated and only those properties are therefore candidate real properties. But I continue to believe that any metaphysical backup for experimentalism is unnecessary, so this is already a reason to reject Cartwright's causal realism. Second, I have now convinced myself that Cartwright's original argument cannot ultimately do the work demanded to support even an improved version of specifically *causal realism* somewhat weaker than what was originally intended. These two considerations derive to a large degree from Christopher Hitchcock's (1992) critique, and jointly they show that there is no specific form of causal warrant that is not also either theoretical or experimental in the ways described. But there is, finally, a third element. The position that I defended on behalf of Nancy Cartwright builds up a general epistemological *thesis* out of what ought to remain merely a general pragmatic attitude or stance within our epistemic practice. We do not need a general theory of knowledge to appreciate such an experimentalist stance. It really is just the application of the modeling attitude developed by nineteenth-century physicists to experimental contexts.

The form of experimental realism (or perhaps just "experimentalism") that I favor now may be summarized as the following *(epistemic) experimental modeling attitude* (EMA): *Scientific models can possess two types of warrant, experimental and theoretical, and experimental warrant is stronger than theoretical warrant in the sense that no theoretical warrant against the existence of some entity* x *postulated in a model may ceteris paribus defeat experimental warrant for* x.

The notion of experimental warrant is thus linked to what is prima facie manipulated in experimental conditions, regardless of whether it ensues in any form of causal warrant. In an echo of Hacking's claim, it accrues specifically to phenomenological "homely" representations or models of the properties of the entity that are subject to such manipulations. Nevertheless, on a deflationary inferential account of these homely representations, the view stops well short of endorsing Hacking's phenomenological home truths. For, recall, on a deflationary view, nothing follows regarding the truth of the statements from the inferential potency of the representations.

Moreover, there are equally legitimate internalist and externalist versions of this warrant and hence correspondingly internalist and externalist versions of this experimental attitude. An *internal experimental warrant* would require *a belief in manipulation to be a necessary and sufficient condition for experimental warrant.* On this account, someone's belief that *x* exists acquires this special kind of warrant if and only if that person believes that *x* can be manipulated. It is internalist because it stipulates that we acquire experimental warrant for *x* whenever we believe that we manipulate *x*, independently of whether we manipulate it as a matter of fact. Thus, it follows that, like any other form of warrant, experimental warrant is defeasible since our beliefs concerning what we do in fact manipulate often turn out to be mistaken. Thus, the beliefs that acquire this warrant are certainly not incorrigible. The history of science itself shows how such beliefs regarding what is real on account of what is believed to be manipulated have sometimes been defeated and corrected. Hence, our taking ourselves to manipulate *x* cannot be, on this view, a sufficient condition on *x*'s reality. This shows that neither MER nor MCER follow from EMA and therefore that EMA does not have the negative consequences that make those versions of experimental realism untenable.

It might be objected that EMA is too weak, that it makes it too easy to acquire experimental warrant. It is true that EMA would be satisfied by any mistaken belief in the manipulation of a nonexistent entity. There are many historical case studies that make this point. Phlogiston was certainly thought to be manipulated in combustion experiments through the eighteenth century (Chang 2010). And some physicists in the nineteenth century thought that the ether could be manipulated in different ways, at least in principle, if not routinely.[17]

A straightforward way around the objection raised by these case studies is to adopt instead an *externalist version of experimental warrant: Manipulation is a necessary and sufficient condition for experimental warrant: our belief that* x *exists acquires this special kind of warrant if and only if we manipulate* x. This delivers us from the historical counterexamples, but it gives rise to other difficulties regarding how to understand those cases in which the mere assumption that some entity was being manipulated seemed to ensure its acceptance in the community against all kinds of alternative merely theoretical warrant. The ether itself—as was conceptualized in the models that we studied in the first part of the book—provides a cogent example. However, as epistemologists and philosophers of science, we should not, I think, necessarily choose between these two forms of filling in the notion of experimental warrant. Both forms can provide us with insight as to how science proceeds. Both are also ultimately defeasible and therefore show that any purported scientific knowledge of reality is necessarily tentative and fallible.

The take-home message of this discussion is that a deflationary approach to scientific representation enables the defender of experimental realism to avoid the main objections that have been raised against the position. I believe that experimental realism remains a viable epistemology for science, as long as it is cast in a suitably deflationary form, avoiding any metaphysical commitments. That is, it remains the case that the best argument for the reality of electrons is to be found in the routine operations that we are capable of successfully carrying out on electron microscopes. The sort of epistemic warrant that such operations provide for the types of models of the electron typically present in contemporary physics is certainly the best warrant one could have for the existence of any entity lying beyond our perceptual abilities or observational capacities. As we are about to discover, it may well be the best warrant one could have simpliciter for any entity whatever.

Constructive Empiricism Deflated

For an experimental realist, nothing hinges on *observation*—the perceptual capacity of sensorial apprehension in the visual part of the electromagnetic field, which is typically achieved in most mammals and larger animals via the physiology of the optical eye.[18] An experimental modeling attitude does not take perception via the optical eye to be a hallmark of or a criterion for the reality of anything, even though it is typically the case that we can normally see whatever entities we manipulate. But the fact that the terms are almost coextensive should not lead us to think that these properties carry similar epistemic weight. The sort of warrant that accrues to experimental activity in the sciences need not dovetail with our observational experience.

Now, to the extent that the logical positivists' distinction between the observational and the theoretical vocabularies relied on the properties of entities to be susceptible to observation (and this is debatable, as we noted in chap. 6), they were committed to a more passive sort of epistemology. For it is undoubtedly the case that observation is the passive reception of information through a sensorial medium or organ while manipulation—and, more generally, experimentation—requires actions to be carried out by agents intentionally in the material world. If Hacking and the new experimentalists (in the wake of Deweyan pragmatism) are right, then nothing as passive as merely observing a hypothetical entity can possibly provide us with the kind of robust experimental warrant that can grant its reality. There abound mirages and hallucinations and other optical effects, and what is observable seems relative to observational conditions, circumstances, observed capacities, and the like.

In *The Scientific Image* (1980) and later publications, Bas van Fraassen revived the epistemology of logical positivism. He brought back the idea that the domain of the observable precisely delimits the boundaries of what can be known—of all human knowledge. His antirealist philosophy of science, known as *constructive empiricism*, embraces a form of voluntarism as regards the appropriate doxastic attitude toward unobservable entities. Both realist and antirealist attitudes are allowed. While van Fraassen defends the rationality of suspending belief with respect to the unobservable entities of science (the form of agnosticism, e.g., regarding electrons, that constructive empiricism is noted for), he does not really reject realism with respect to those entities either. His liberal account of rationality allows for both. The commitment to the truth of the theory beyond its empirical adequacy is entirely an optional matter that scientific rationality does not rule in or out.[19]

The full development of this variety of antirealism obviously requires great clarity regarding the distinction between the observable and the unobservable. Within the epistemology of constructive empiricism, this dichotomy is a critical tenet since it grounds the distinction between the microscopic entities postulated by science that, according to constructive empiricism, we can suspend judgment about and the objects of our ordinary experience of the macroscopic world in which we are mandated to believe (on either constructive empiricism or scientific realism). Hence, the slogan that popularizes the view is that we are required to believe in tables and chairs regardless of whether we are realists or constructive empiricists. The difference between these two positions kicks in only at the level of the unobservable objects postulated by science. Not surprisingly, the dichotomy—and its fundamental import for epistemology—has been the subject of a lively debate over the years, with most critics voicing skepticism that the domain of the observable can in any way exhaust the limits of human knowledge. The idea that we enjoy privileged epistemic access only to the observable objects postulated by our best science—and that consequently only these objects warrant our belief—is unacceptable to most realists, including, of course, experimental realists (see Bird 2007; Churchland 1985; Gutting 1985; Hacking 1985; Van Dyck 2007; and M. Wilson 1985).

As is well-known, at the core of constructive empiricism there is the difference between the doxastic states of acceptance and belief. For a constructive empiricist, the appropriate attitude toward any scientific theory is acceptance, not belief. And, more specifically, to accept a theory implies only the belief that it is empirically adequate, that is, that what it says about the observable entities in its domain is indeed the case. There is no further requirement to believe in the truth of the theory, that is, no requirement to believe that what it says about the unobservable entities in its domain is similarly the case. Thus, accepting atomic physics carries a requirement to believe in the predicted spectral lines for the emission of different elements but no requirement to believe in the elements themselves, at least not if understood as (kinds of) minute unobservable particles. There are other pragmatic requirements—the sorts of programmatic commitments to a theoretical framework that characterize much scientific work—that typically involve the willingness to see that framework applied to further instances in its domain, stretching the scope and breadth of the theory. But these are, on van Fraassen's account, purely pragmatic and, by and large, devoid of any epistemic import.[20]

We thus come to the crux of constructive empiricism, the definition of a theory's empirical adequacy, and in particular its relation to the theory's truth. The latter must be strictly more stringent than the former since, according to van Fraassen, the truth of a theory requires its empirical adequacy, but not conversely. A theory can be empirically adequate yet false.[21] Van Fraassen worked hard to articulate notions of empirical adequacy and truth satisfying these intuitive conditions within the semantic conception of scientific theories (see van Fraassen 1980, chaps. 1–2; and van Fraassen 1987). In a nutshell, the truth of a theory is understood as an isomorphism to the systems in its domain, while the empirical adequacy of a theory is cashed out in terms of the model-theoretical relation of embedding, as follows: "A theory is empirically adequate if it has some model such that all appearances are isomorphic to empirical substructures of that model" (van Fraassen 1980, 12).

More precisely, an embedding is an isomorphism to a substructure that must be independently specified. A structure, recall, is a domain endowed with a set of n-place relations: $A = \langle D, R_i^j \rangle$. A substructure is a partition of the domain together with all the relations as truncated to that domain: $B = \langle D', R_i'^j \rangle$, where $D' \subseteq D$ and $R_i'^j = R_i^j / D'$ (i.e., those relations in the original domain D that obtain when applied only to those elements in D that are also included in D').[22] For a theory to be true, it is not just the empirical substructures of the theory that stand in an isomorphism to the appearances. Presumably the whole theory must stand in an isomorphism to the structure of the systems in its domain.

The definitions satisfy the intuitive conditions precisely because a full isomorphism is always more stringent than a mere embedding. The former requires a mapping of the whole of a theory's structure onto the structure of whatever system or phenomenon it is intended to represent, while the latter requires only the mapping of a given empirical substructure of the theory or a set of such substructures onto those structures that accurately represent the appearances.[23] Hence, the truth of the theory necessarily requires its empirical adequacy, but not conversely. A constructive empiricist would thus recommend belief only in the empirical adequacy of an acceptable theory. This is the belief that the theory's structure embeds the appearances in its domain, whatever they may happen to be. The belief in the truth of the theory, that is, in the full isomorphism of the theory to whatever systems or objects exist in reality—including those putatively responsible for the appearances—is not required by any canon of rationality that the constructive empiricist might accept. Thus, in van Fraassen's (1980, 73) slogan: "It is not an epistemological principle that one might as well hang for a sheep as for a lamb." The belief in the empirical adequacy of a theory is preferable to the belief in its full truth simpliciter since the latter commits to the entire description, beyond as well as within the observable domain.

This feature of empirical adequacy alone gives constructive empiricism an edge over its scientific realism competitor on at least three additional distinct grounds, all suitably canvassed by Monton and Mohler (2021, 15ff.): the hermeneutical fact that belief in empirical adequacy is the weakest attitude that we can attribute to scientists while making sense of their activities (which almost always involve projection, i.e., future predictions), the capacity to account for apparent violations of this rule of acceptance (i.e., apparent cases of acceptance of theories that do not seem to be empirically adequate) by means of a notion of approximate embedding, and the capacity to make sense of the suggestion that "experimentation is the continuation of theory construction by other means" (van Fraassen 1980, 77). The latter is best exemplified in Millikan's experiment to measure the charge of the electron, which can be nicely cast as a mere attempt to "fill in a value for a quantity, which, in the construction of the theory, was so far left open" (van Fraassen 1980, 77).

Nevertheless, there is also a well-known battery of criticisms raised against constructive empiricism besides those already mentioned regarding the notion of observability and whether it can bear the heavy epistemological weight that van Fraassen puts on it. I would like to focus on the following one here. Van Fraassen's very definition of *empirical adequacy* as embedding of the appearances into substructures is open to scrutiny and can be criticized. For my purposes here, the most pertinent criticisms are those of Lutz (2012, 2014) and Rosen (1994). The former charges van Fraassen's definition with being too limited in comparison to other definitions within the received view. The latter's critique is even more pertinent, and I shall focus on it first. Rosen (1994, 164–69) charges constructive empiricism with a commitment to abstract entities way beyond the nominalism it is intended to embrace, precisely on account of its definition of empirical adequacy.

Certainly, constructive empiricism is an attempt to understand the nature and goals of science abstractly in as sparingly an empiricist mode as possible, and van Fraassen has made much of the commitment to avoid the use of any metaphysically charged modal notions in the account (such as laws, causation, etc.). By contrast, his definition of *empirical adequacy* expresses an implicit commitment to the reality of at least some abstract entities. Thus, as Rosen puts it (1994, 166): "An individual believing a theory to be empirically adequate is thereby committed to at least three sorts of abstract objects: models of the phenomena (data structures), the models that comprise [the theory] T, and functions from the one to the other. To suspend judgement on abstract objects is therefore to suspend judgement on whether any theory is empirically adequate, and this just [is] to give up acceptance altogether." Since abstract objects are unobservable if anything is, the constructive empiricist cannot thereby accept any scientific theories. Rosen considers ways out of this predicament, involving a fictionalist account of the mathematics involved, and finds them wanting, in the sense that they all seem to carry modal commitments that the constructive empiricist would more happily avoid. Nevertheless, it seems that an easier way out for the constructive empiricist is to relinquish the definition of *empirical adequacy* as structural embedding. The definition is independently problematic anyway, as I shall try to show now.

While his early notions were expressed in terms of a substantive theory of scientific representation, van Fraassen clearly became inclined to deflationary accounts later (see esp. van Fraassen 2008). It thus seems appropriate to figure out what the application of a deflationary inferential conception to empirical adequacy would entail in this context. More particularly, I would like to argue that a deflationary rendition of van Fraassen's notions can better respond to some of the criticisms raised against constructive empiricism while respecting and endorsing all its advantages. On an inferential account of representation, the notion of empirical adequacy tracks a representational source's promotion of correct inferences toward the observable aspects of its target. For a theory to be empirically adequate on the RSV, it would need to be the case that it includes models (understood as deflationary representations) that *license inferences to the observable aspects of its (real or fictive) target phenomena or systems.*[24] The definition fits in well with van Fraassen's requirements since only information regarding the observable aspects or parts of the targets is relevant to the assessment of the correctness of the empirical adequacy of the theory on this account—just as desired. In addition, for the theory to be true, it would need to be the case that all inferences licensed by its representations are correct, regardless of whether the target properties are observable. So, again, as required, the truth of a theory entails its empirical adequacy, but not vice versa. And all the epistemic benefits for constructive empiricism follow, including the slogans. It is still disallowed in principle "to hang for a sheep as well as a lamb."

The deflationary definition is also descriptively accurate. It fits in well with the examples of empirically adequate theories discussed in van Fraassen (1980) and a plethora of other cases, such as those false yet empirically adequate theories discussed in the pessimistic meta-induction argument. This can be shown by inspection case by case, although that would require a long and separate treatment. But this treatment is in fact not needed since the structure of the definitions already makes it clear that these false yet empirically adequate theories can hardly fail to fit with the prescriptions. For recall that, on the inferential conception, representation rides on whatever means are available, whether they be isomorphism or similarity. For any of these theories (phlogiston, caloric, ether, etc.), the representations afforded by the theory certainly entail all sorts of conclusions regarding the observable parts of their targets since all these theories were profusely tested in laboratories or in large observational studies. Evidently, testing any empirically accessible predictions of a theory requires one to extract its licensed inferences regarding the observable phenomena, whether in fact already observed or not. Yet the theories are false—or so we now think—in as much as we know that they also license mistaken conclusions regarding the nonobservable aspects—and regarding the putative causes of those observable phenomena (whether they be properties of the nonexistent caloric, phlogiston, or ether).

Nonetheless, all known advantages to constructive empiricism are also forthcoming on the inferential deflationary version of empirical adequacy and truth since it is still true that "the assertion of empirical adequacy is a great deal weaker than the assertion of truth, and the restraint to acceptance delivers us from metaphysics" (van Fraassen 1980, 69). In fact, if anything, the inferential and deflationary account of empirical adequacy goes further in "delivering us from metaphysics." Consider Rosen's argument that van Fraassen is committed to abstract entities corresponding to the theory (including its empirical substructures), the appearances, and the formal functions between the two. None of these in fact obtains on the deflationary version of the truth and empirical adequacy of scientific representations. As we saw in part 2, a scientific theory ought not be reified as a mathematical structure. Nor should the implications toward the observable aspects of its targets be reified. They are merely inferences of the sort referred to in chapter 7 as the *horizontal* kind. Finally, there are no functions that need obtain between theory and appearances or world. The inferential conception certainly requires none, and it sees the relevant connections rather in the practice of use, in what we saw Maxwell recognize as the conductors that logically lead from a system onto its target. An empiricist attitude to scientific knowledge thus instead recommends casting constructive empiricism in the deflationary mold of the inferential conception. And, while this is by no means an endorsement of constructive empiricism, it does recast the whole debate in a different light. It becomes a debate regarding the appropriate licensing of the inferences to the observable part of the targets of the representations within scientific theories.

None of this constitutes a definitive defense of constructive empiricism since all the objections regarding the epistemic import of the observable/unobservable dichotomy remain, including those regarding the modal character of the notion of observability. However, the application of the inferential conception shows that the best defense of constructive empiricism is as a kind of instrumentalism that takes our theories and models to be inference-conducive tools promoting efficient patterns of reasoning from the observable domain to the unobservable, and vice versa. The aim of much scientific activity is perhaps at best a pragmatic type of instrumental reliability (Suárez 1999a). This sometimes requires the empirical adequacy of our theories and models, but more often scientists rest content with the more basic pragmatic achievement of mere reliability: a sustaining confidence that they can continue to apply their theories and models—empirically adequate or not—to new phenomena in their domain successfully.[25]

Reconsidering the Turn to Social Practice

I now turn to three different contemporary attempts to align epistemology with scientific practice in all its normative complexity. These views are therefore close to those defended in this book since the inferential conception also endorses the insight that value judgments play a normative role in representational practices (as part of the rules that serve to set up and maintain both representational force and inferential capacities). While my discussion shall in each case be a bit more cursory, I do hope to provide some pointers as to how the discussion might proceed, particularly how to develop these attempts further to build an epistemology of science that closely tracks the scientific practice. In all cases, my contention is that it is more likely to succeed under the guise of a deflationary and inferential notion of representation.

ARTHUR FINE'S NATURAL ONTOLOGICAL ATTITUDE

The first attempt to bring scientific epistemology normatively close to scientific practice that I want to discuss is also one of the earliest. Arthur Fine launched NOA in a series of papers (e.g., Fine 1984a, 1984b) as well as his celebrated book *The Shaky Game* (Fine 1987/1996).[26] This was an attempt to overcome the realism debate, and it is in this sense closest to the spirit of the inferential conception, which is too an attempt to overcome the debate or at least locate it elsewhere. NOA inveighed against the thought that the epistemic aim of science is easily definable in the abstract. Fine thus criticized realism for attempting to bring the whole of the complex activity that is science under the umbrella of the goal or pursuit of truth. He similarly criticized constructive empiricism for attempting to do the same, this time under the umbrella of the uniform pursuit of empirical adequacy. It is all too facile in this debate to assume that representational success is thus the key to the different positions. The scientific realist would presume that the aim of science—which implicitly defines the activity—is the pursuit of representation. The antirealist, on his part, would presume that something short of full representational success is the aim. The quarrel then concerns how far representation goes.

It ought to be clear that, on a substantive account of scientific representation of the sort that we discussed in chapters 4 and 5, the realist has the upper hand. For, if representation requires significant substantive relations, such as similarity or isomorphism, to obtain successfully between sources and targets, it becomes all by itself a scientific realist goal for science. In other words, the claim that science aims at substantive representational success is in essence a claim on behalf of scientific realism. Van Fraassen's constructive empiricism may seem to come just short of that ideal and hence to define the map of any position contrary to scientific realism. But in fact, as we saw, van Fraassen, at least in his earlier writings, is also committed to a sort of substantive representational success. The requirement that the appearances be shown to be isomorphic to empirical substructures of the theory also trades on a substantive account of representation. The difference between claiming that the aim of science is full substantive representational success and claiming that it is only in part substantive representational success is not in fact that large. Of course, arguably, this is just what van Fraassen needed to mount a formidable defense of his position, particularly in terms of endorsing the appropriate measure of epistemic risk with minimal metaphysical commitments. But it does not seem to be what is required if the aim is just to build a minimal account of representation that can accommodate all parties to this debate.

NOA is indeed clear that scientific realism and constructive empiricism both go beyond what is minimally required to understand science, at least if we are to take it on its own terms. (Thus, the title of both of Fine's papers/chapters: "The Natural Ontological Attitude" and "And Not Anti-Realism Either.") More precisely, NOA wants to toe the safe and homely line that it is possible to accept the truths of science without having to interpret their truth in any way whatever, whether it be as correspondence or coherence or along the lines of any other theory of truth. This is in line with the quietism endorsed in chapter 6. But the affinities go further. NOA, recall, is the "California natural": no additives please. It invites us to take science as it comes, without attempting to impose any grand epistemological narrative on it: "Do not rope [science] in with premade philosophical programs. Trust that science is open and providing all the resources and nourishment that we who study science need" (Fine 1987/1996, 176).

And, indeed, on a deflationary inferential conception of representation, the aims of representation itself involve no commitment to any philosophical program—certainly not to any scientific epistemology. They in fact involve no commitment to any thesis or belief regarding the aim of science beyond a trusting attitude toward the modeling practices of scientists. As we saw earlier, these practices are informed by the normative commitments involved in their communities, which are mainly in place to set up and maintain both representational force and inferential capacities. In other words, just as NOA admonishes, a deflationary approach to representation takes a bottom-up approach to the nature of representation. Rather than imposing any philosophical or metaphysical view from above, it lets the very modeling practice of scientists speak and set the agenda for what counts as successful representation and what does not.

NOA claims that it is possible to accept the core basic truths that scientists accept without endorsing a correspondence theory of truth or any other theory of truth for that matter. We can thus adopt the ontology that is natural to the scientists who accept such pronouncements without adding any metaphysical layer onto these sentences that would explain their reasons for being true. Fine's neutral stance on semantic matters regarding truth has, however, come under some criticism, perhaps most eloquently voiced by Alan Musgrave in his response (Musgrave 1989; see also Psillos 1999, chap. 10). Fine claims that scientists accept at face value certain sentences as true without necessarily endorsing any correspondence relations (and so should we, students of science). This might seem to suggest that they do so without necessarily endorsing the existence of the entities described by those true sentences or the properties those true sentences ascribe to them. And that of course seems to incur a contradiction or at the very least a paradox. In other words, perhaps the acceptance of the basic statements of scientists as true requires already some commitment to a metaphysics of relations that favors realism. Musgrave certainly seems to think so, but the long tradition of deflationary accounts of truth reviewed in chapter 6 militates against this assumption. At any rate, this is not a debate that we need to resolve to understand representation in modeling practice. As I made clear in part 2, the relation between deflationism about truth and deflationism about representation is quite thin, amounting to nothing more than a suggestive and helpful analogy. We do not need to settle issues of scientific truth to settle issues of representation. It is possible that deflationary accounts of truth fail while their cousins regarding scientific representation (such as the inferential conception but also those reviewed in chap. 6) succeed.

My aim here is to dispute the thought that merely having science aim at successful representation already settles the epistemic debate in favor of scientific realism. On the contrary, I claim that a deflationary account of representation leaves completely open whether truth, empirical adequacy, or even something weaker, such as instrumental reliability, is the aim.[27] Thus, I claim that the deflationary accounts of representation have the great virtue of staying neutral in the debate regarding scientific realism. Simply asserting that science aims at representation via models does not settle this debate, and it certainly does not settle this debate automatically in favor of realism. On the contrary, when properly understood, the statement leaves open what are significant epistemic aims of science beyond mere representation. It is even consistent with pluralism regarding these aims. Truth may be an appropriate aim in some cases, as may be empirical adequacy, but ultimately instrumental reliability per se may be more than enough in most cases. And there are many other aims that scientific investigations have: to explain a theory or a phenomenon, to apply that theory, to generalize it, to probe into nature, to intervene and control, to predict, simply to understand at some basic level, to communicate, and so on. The fact that scientific activities have aims does not entail that there is one aim that all scientific activities share. Only a failure in quantifier logic would suggest such a fallacy.

So NOA and the inferential conception of representation see eye to eye. In addition, the modeling attitude and its cognate representational semantic conception of theories also share some elements with NOA. First, there is of course the commitment to a generic philosophical vision of science in terms of stances or attitudes, as opposed to philosophical theses, principles, or statements. Here, NOA has been undoubtedly leading the way. Second, there is also in NOA a commitment to modeling as a central feature or activity in science: "No theory can lay out all of its possible applications and none does. Indeed, the road from theory to applications generally involves a number of intermediate modeling stages" (Fine 2001, 114).[28] Finally, the deflationary conception of representation that I have defended in this book refrains from any metaphysical commitments and aims to bring philosophical reflection back to the level of practice, where it arguably stays at its best. There is still much philosophical work to do on representation, but, just as NOA advises us to carry out epistemological reflection in a close and evolving relation with the sciences themselves, so too does the inferential conception—any deflationary account for that matter—invite us to study scientific representation philosophically always at the level of practice. A similar deflationary attitude to grand philosophical narrative runs through both.

PHILIP KITCHER'S PRAGMATIC GALILEAN REALISM

Kitcher's *Science, Truth and Democracy* signaled a major call for philosophers to embrace the study of the social, ethical, and political dimensions of scientific research seriously (Kitcher 2001b).[29] It was a call to defend a view of science that would bring it in close alignment with our democratic ideals and the sort of accountability that they demand. Some of its influence is felt in the way much scientific epistemology is now more clearly committed to considering the establishment of scientific belief and evidence against a background of the institutions and practices that characterize science (as opposed to just offering a very abstract rendition of the logical structure of beliefs).[30] Yet Kitcher's own discussion runs against a background of mildly realistic assumptions regarding scientific knowledge that often serve to articulate the framework for many of his disquisitions regarding applied science and science policy. While I am wholeheartedly in agreement with the overall spirit of the work and I find many of the proposals for a well-ordered science apposite,[31] I also have misgivings about the picture that it presupposes regarding both scientific representation and truth's role in scientific discovery.

To put it bluntly, it strikes me that a conventional substantive account of representation lies at the heart of Kitcher's position. Yet, as I shall argue, modifying this account does not significantly alter the resulting view about the role of values in science and the need to develop a well-ordered science. On the contrary, the adoption of a more neutral deflationary account of representation, if anything, makes those concerns more relevant and pressing. My critique thus aims at strengthening Kitcher's call to organize our inquiry in ways that respond to our ethical values as well as our social and political goals and institutions as a democratic society.

A good way to visualize and make explicit the difference in the underlying accounts of representation begins by discussing Kitcher's epistemology for science, which he refers to as *real realism*. This is how Kitcher (2001a, 155) defines this view: "We thus envisage a world of entities independent not just of each but of all of us, a world that we represent more or less accurately, and we suppose that what we identify as our successes signal the approximate correctness of some of our representations." This is evidently some sort of realism that imposes no bounds on humans' knowledge of their surroundings. I shall first briefly describe this view's main ingredients and then focus on the ostensible use it makes of a substantive conception of representation.[32] Real realism is meant as a statement of a pretheoretical or even prereflective cognitive scientific attitude, which emerges as the conjunction of three different elements. First, there is Arthur Fine's NOA toward science's existential commitments. Next, there is an analogous assumption regarding our epistemic stance that Kitcher refers to as the *natural epistemic attitude* (NEA). Finally, there is a third assumption that Kitcher refers to as the *Galilean strategy* (so called because it was inaugurated by Galileo's use of the telescope). Kitcher contends that the former two assumptions are presuppositions in our everyday cognitive practice because they are built into even the most ordinary instances of cognition from infancy. By contrast, the third assumption (the Galilean strategy) is the bit added by modern science to allow us to move cognitively from the level of ordinary observable objects and their properties to the level of the unobservable objects postulated by science, whenever they can be detected through the aid of modern instrumentation. Thus, Kitcher claims that anyone committed to modern science must be committed to this assumption too. The conjunction of the three assumptions is intended to render real realism, a very homely and moderate form of realism available to and on display in anyone who is both knowledgeable of the scientific method and endowed with ordinary cognitive capacities, regardless of their further epistemological persuasions.

I have already discussed NOA, which, in this debate with Kitcher, is helpful common ground. The Galilean strategy seems acceptable also to anyone who rejects strict positivist or constructive empiricist strictures on science, and in the previous section of this chapter I did question van Fraassen's claim to bound empirical knowledge at the limits of what is observable. Furthermore, we have not encountered any obstacle to the representation of unobservable (even inexistent) entities in the scientific models we have studied in this book. Nor is there any boundary to what can be represented in any of the conceptions of representation we have studied, whether substantive or deflationary. As regards the inferential conception, there do not seem to be any grounds for skepticism about our inferential capacities reaching beyond merely the observable domain. The inferential capacities of our models often, if not always, do transcend the domain of the observable. In fact, getting us beyond the observed and the tested seems an essential part of the work that models do for us. Bringing in any such skepticism at this stage would appear strangely to clash with all our discussion so far. Therefore, I shall essentially go along with Kitcher at this point and accept the Galilean strategy too, as a natural consequence of a trusting attitude that relies on what appear on the face of it to be the best methods of modern science.

The sticking point is thus squarely NEA, which is very much the new ingredient that Kitcher brings to the cooking pot for real realism. Here is how Kitcher (2001a, 154) describes NEA: "We are animals that form representations of the things around us; that is, the world sometimes puts human beings into states that bear content. Those states, in turn, guide our behavior. In observing, or thinking about, other people, we take it for granted that their representational states sometimes adequately and accurately represent objects, facts, and events that we can also identify." This is then the assumption that adds representational content, and it provides for the moderate but firm realist commitment in the statement of real realism. Yet, as readers of this book are now easily able to judge, there are substantive reductionist and deflationary versions of this representational assumption, and not surprisingly they yield very different epistemological positions. To be precise, NEA is ambiguous between a realist-leaning version that employs a substantive notion of representation and more neutral deflationary account.

On a substantive reductionist reading, recall from part 2 that NEA could be paraphrased as follows: "We are animals that construct mirror images of the things around us. We take it for granted that representational states are sometimes adequate and accurate matches of objects, facts, and events that we can also identify." Since this is the realist version, or NEA-realist, let us refer to it as the NEAR. It is quite clear that, in conjunction with NOA and the Galilean strategy, NEAR yields precisely the form of real realism that Kitcher seeks. However, the problem is that there are other interpretations of NEA relying not on a substantive but on a deflationary account of representation. Consider, for instance, what NEA would look like under the inferential conception of representation defended in part 2. The closest paraphrase would then be as follows: "We are animals that draw inferences regarding the things around us; that is, the world sometimes puts human beings into states that bear content. Those states, in turn, guide our behavior. In observing or thinking about other people, we take it for granted that their inferences are sometimes adequate or accurate regarding objects, facts, and events that we can also identify." The resulting version of NEA is no longer so committed to any realism from the word go. It is not antirealist (the content in which humans are put could still be about reality out there), but it is markedly nonrealist. There is no need for all the inferences to be to true conclusions or always accurate. And, even if they were, this would not necessarily entail that they are about an independently existing reality. We can refer to this weaker version of NEA as NEA-nonrealist, or NEAN. Note that, unlike NEAR, NEAN entails no relation between the source and the target of a representation, other than the source's capacity in the appropriate context to yield inferences about the target.

We find ourselves at a quandary. The version of NEA that yields real realism is implicitly committed to a substantive account of representation. It is not surprising then that it does yield realism since (recall the discussions in pt. 2) substantive accounts already prejudge the issue of realism by building the standards of accuracy or correctness into the very definition of representation. What is more, the claim that NEAR is part of our everyday cognition of ordinary objects, acquired since infancy, seems far-fetched, to say the least. By contrast, NEAN does seem to be built into our ordinary cognitive practices from a very early age. It squares nicely with Kitcher's (2001a, 154) claim that NEA "already plays a large role in our everyday lives, for example, in our guidance of children's development." Indeed, it is much easier to see the drawing of inferences playing this role than the recognizing of either similarity or isomorphism relations. To anyone acquainted with some early childhood psychology, NEAN looks not just part of ordinary cognitive practice but very much its starting point. Toys are often used by children for the purpose of surrogate inference. These are among the first cognitive activities performed by children in their encounter with identifiable objects, as is the recognition that other children are also capable of similar inferences. Such inferential practices seem to precede even the development of language. They certainly precede any judgments of relevant relations between objects, never mind such complex judgments as those involving similarity or isomorphism, a subtle form of pattern recognition that comes much later.

Yet, contrary to Kitcher's goals, the conjunction of the more primitive NEAN with NOA and the Galilean strategy does not yield real realism, as any reasonable reconstruction of Kitcher's argument will show. We can, for example, summarize Kitcher's argument in four succinct premises and a conclusion:[33]

(1) NEA is undeniably the unreflective epistemic assumption of ordinary life. Since it is an uncontroversial part of ordinary life cognition, it must be accepted by realists and antirealists alike.
(2) NOA trusts in the existence of the entities postulated by science and is just the core of ontological assumptions shared by both realism and antirealism.
(3) The Galilean strategy is the paradigm form in modern science of those inferences from the observable to the unobservable brought about by the Scientific Revolution. It must be accepted by anyone committed to the rationality of the scientific enterprise as we know it.
(4) The conjunction of NEA, NOA, and the Galilean strategy yields a basic and moderate form of scientific realism, real realism, that must therefore be accepted by all parties in this debate.

The version of NEA that makes premise 1 true is NEAN, the deflationary or nonrealist version, which is indeed built into ordinary practice. By contrast, NEAR is needed in premise 4 to extract a form of realism from the conjunction of the three core assumptions in Kitcher's argument. But, for the argument to be valid, sound, and cogent, the same version of NEA would need to appear in all four premises. However, if we run the argument with NEAN, premise 4 is false, the argument is unsound, and the conclusion does not follow, whereas, if we run the argument with NEAR, premise 1 is false, so the argument is once more unsound, and the conclusion again fails to follow. There is no hope of deriving real realism on these grounds, and Kitcher may have to rest content with the version of real realism that is provided only by NEAN. That is, admittedly, a very deflated version and one acceptable to any antirealist too.[34]

How does all this affect Kitcher's proposals for a well-ordered science? I do not believe that it affects them very much. It is still the case that there is a strong normative component to scientific research—in the form of normative practices for settling issues of representational force and inferential capacities. And it is still the case that there is an objective difference between those representations that are accurate, complete, and empirically adequate and those that are not. Recall that a representation is *true* if it licenses no inferences to false conclusions about the target; *complete* if it is true and fully informative, licensing inferences to all truths about the target; and *empirically adequate* if it is complete with respect to all the observable or measurable aspects of the target, licensing inferences to all the truths about those aspects. These notions are still fully operative as guides or goals in a well-ordered science. Nothing in Kitcher's description needs to go when the conventional substantive account of representation (embraced in Kitcher [2001b, pt. 1]) is jettisoned in favor of a deflationary account. The roles for expertise in democratic society (so aptly described in Kitcher [2001b, pt. 2]) remain essentially unchanged, and our reasons to support a well-ordered science are, if anything, enhanced.

HELEN LONGINO'S SOCIAL VALUES

Helen Longino's (1990) was among the first book-length treatments of the implications for scientific epistemology of the ineliminable role of values in science and remains perhaps the most articulate and passionate exposition of the view. To be clear, it is of course not as if this view had not in some ways been voiced before. There are evident precedents in the postpositivistic philosophy of science (in Thomas Kuhn and Paul Feyerabend in particular), and the literature on the values of simplicity, coherence, and explanatory power stands out too. But Longino (1990) was perhaps the first extended treatment to put the issue of social values up front and discuss the relevant patterns of socialization that underpin some of the essential concepts in the demonstration of scientific evidence. The original argument was predicated on the ineliminable underdetermination of theories by data. Longino's main claim was that background beliefs, infused by a plethora of social values, are determinants of the acceptance of otherwise radically underdetermined theories (Longino 1990, 60ff.). This was inevitably the result of adopting an account of the relation of theory and evidence rather different from the received view of the logical positivists. According to Longino (1990, 48): "What would count [in the logical positivistic tradition] as evidence for a hypothesis is determined by the form of hypothesis sentences and evidence sentences, not by their content. This means that inference to a hypothesis is not mediated by . . . value-laden assumptions." By contrast, in Longino's persuasive analysis, values, judgments, and choices lie everywhere in the processes that take us from evidence and data gathering to the formulation of hypotheses and their confirmation.

The convincing arguments in Longino (1990) relied on reasoning and inferential practices, and Longino particularly saw values enter as background assumptions that made plausible or permissible diverse patterns of reasoning. This is much in the spirit of my own account of representation, which relies strongly on the normative inferential practices of scientific modelers. The inferential conception also sees the normative commitments coming at the level of reasoning practices—in the sorts of surrogative reasons characteristic of the inferences drawn from models. However, there are important differences in the way Longino and I treat the notion of representation, which is, after all, at the heart of many of her arguments. Longino (1990) took a more conventional approach to representation. While not being very explicit, she seemed committed to the sorts of semantic relations typical of propositional knowledge. Thus, while studiously avoiding a purely syntactic characterization, she seemed happy to consider "the relation between a state of affairs said to be evidence and a statement or proposition, the hypothesis, for which the former is said to be evidence" (Longino 1990, 39). True, in her account, unlike in the logical positivist account, values infuse the entire system of propositions since they provide warrant for the background conditions that are necessary at each point to firm up reasoning patterns and thus ultimately to overcome the problem of underdetermination. Yet the underlying notion of representation that supposedly holds between the models and the theories provided by the scientists and their targets is amenable to any of the substantive conceptions that we discarded in earlier parts of this book.

The impression is confirmed in the later, more explicit treatment of this issue in Longino (2002, esp. chap. 5), where Longino moves away from any propositional account and comes out squarely in favor of a version of the semantic conception. This is not, however, the structuralist version of the view (what I referred to as the SSV in chap. 6) but what Longino calls *the model theory,* which she admits can be quite plural: "The approach treats theories as models or families of models. There are, however, quite a few different conceptions of what a model is, ranging from a mathematical entity (i.e., a set of equations) to a two-, three-, or four-dimensional visual representation. What the conceptions have in common as treatments of scientific theories is a rejection of the standard view: that theories are sets of propositions, organized as axiomatic systems" (Longino 2002, 113). There is thus also apparently agreement on the diversity of representational means. Is Longino then a fellow deflationary inferentialist traveler?

There are some striking points of contact between the inferential conception of representation defended in this book and some of Longino's pronouncements in the fascinating text that is *The Fate of Knowledge* (2002).[35] Most striking perhaps is Longino's introduction of the term *conformation* (2002, 115), which resembles my own *conformity.* It does not appear that Longino was prompted to introduce her term in the wake of Hertz (whose work indeed suggested mine, as discussed in chap. 7).[36] But the striking coincidence does signal a similar attempt at abstracting away from the diversity of concrete relations. Yet, while Longino does not go deflationary as regards the constituent of representation, she does go along with the same pluralism of means that I have defended in this book. It is doubtful, considering the preceding discussions, that her pluralism can be made consistent with any substantive account of representation. Rather, the natural suggestion is that her aims would be better served by a deflationary account since that is the surest path to a strong defense of radical pluralism about the means of our representations.[37]

There is also much in common—barring some nuanced issues of balance—between the inferential conception and a central feature of Longino's work, namely, the strong emphasis on the social character of scientific knowledge. Longino's advice is to regard both scientific practice and scientific knowledge as social through and through. Yet we must be cautious when using the term: " 'Social,' here, does not mean 'common,' 'collective,' or 'shared,' but 'interactive' " (Longino 2002, 99). As should be clear by now, the inferential conception goes along with these pronouncements, when applied narrowly to the inferential practices that constitute representation. These practices, as I have argued, are thoroughly social. While always minimally answering to the general rules of deduction and logical consequence, they typically incorporate the norms of correct surrogative reasoning from models that are needed in addition to give models proper empirical content and that were discussed in chapters 6 and 7. Such norms in the application of representational force and inferential capacities are the result of productive agreement and interaction between individual modeling scientists. They are often embodied or institutionalized in textbooks and training courses on the use of models but originate in negotiation procedures between individual modelers.

Longino in addition inveighs against what she calls the *rational/social dichotomy*, and she attacks those philosophers who hanker after an individualistic view of knowledge, cognition, and scientific practice. I am in less sympathy with this more polemical part of her work, and I am rather skeptical of her dichotomy. It does not strike me as an accurate portrayal overall of what remains stubbornly an activity carried out by individual agents (for a convincing critique along these lines, see Kitcher [2002]). But this may be a generational effect, the result of scholars like Longino having changed the field to make more acceptable the fact that science is a collective enterprise in which individuals do agree to collaborate against a background of established practices and institutions. Certainly, my claim that representation is a social activity must be understood to involve not just social interactions between individual modelers but the institutions and practices that support, constrain, and enhance those interactions as well as those that ground and maintain the rules of use and application of those models.[38]

Scientific Understanding and the Inferential Conception

Throughout the chapter, I have been at pains to emphasize different ways in which the inferential conception turns out to be useful in scientific epistemology. It is often the case that a mere change in our background assumptions regarding representation suffices to articulate a different view in a debate or a different version of some established view. While the result rarely contributes an entirely new epistemology, it has the virtue of clarifying and making the source of the diverse sets of commitments more transparent. Often, the application of a deflationary conception of representation—such as the inferential conception—renders the resulting epistemology more palatable or viable. This is because it results in a view less loaded metaphysically, one that is therefore more acceptable to the different parties in any debate. The realist will naturally want to add to the aim or goal of representation, while the antirealist will refrain from adding much, but both can live with the underlying minimal account of representation. At the very least, this contributes useful sanitary work in cleansing extraneous background commitments out of the way. Yet, in fairness, it can be argued that the inferential conception does not change epistemology per se, only its parameters.

In this final section of the book, I would like briefly to consider an area of contemporary philosophical debate where the application of the inferential conception might make a more substantial difference. Recent years have seen intense debates surrounding the notion of understanding in science and how it might differ from that older acquaintance of philosophers of science, scientific explanation (see, e.g., de Regt 2018; Elgin 2017; Khalifa 2017; and Potochnik 2017). A large part of the discussion has focused on two central issues that might distinguish understanding from explanation: the veridicality or facticity thesis (i), which typically requires (the explanans in) an explanation to be true or to reflect the facts, and the explanatory thesis (ii) that scientific explanations rely on established theory or laws of nature as part of their explanans. Some of the most distinguished accounts of scientific explanation, such as Carl Hempel's (1965) covering law account, satisfy both requirements and thus lie in the background of some of these discussions. Since understanding is supposed to come short on either count, Hempel's account provides a contrasting benchmark for most theories of scientific understanding. (That is, scientific explanation seems to be more robust, or at least more demanding, than what has come to be known as scientific understanding, which often fails either or both commitments listed above.) The claim in this final section is that an inferential conception of representation allows us to make some inroads into the critical differences that do obtain in practice between scientific explanation and scientific understanding.

My discussion here will mainly employ Henk de Regt's (2018) account of understanding as a useful foil since I regard it as central, not least in carefully balancing aspects of the other extant accounts. Thus, that account rejects the facticity thesis (thesis i) for understanding but nonetheless has understanding critically depend on background theory (thesis ii). By contrast, most other accounts either adopt both theses jointly for understanding as well as explanation (Khalifa 2017) or reject both for understanding (Elgin 2017; Potochnik 2017). I side with Elgin's and Potochnik's views and arguments in this debate. But I believe that there is much to learn from the others too. Khalifa's is an extremely cogent defense of the received view, and de Regt's is perhaps the most articulate and long-standing in the field.[39]

De Regt's *criterion for understanding phenomena* states: "A phenomenon P is understood scientifically if and only if there is an explanation of P that is based on an intelligible theory T and conforms to the basic epistemic values of empirical adequacy and internal consistency" (de Regt 2018, 92). Several notions appear in this definition that require some unpacking, and we have already discussed how *empirical adequacy* in particular fares rather differently on substantive and deflationary conceptions of representation. By *internal consistency*, de Regt basically means logical consistency between all the theory's principles or axioms, a standard measure of self-consistency or coherence. Finally, a theory is *intelligible*, according to de Regt, when it fulfills the following *criterion for the intelligibility of theories*: "A scientific theory T (in one or more of its representations) is intelligible for scientists (in context C) if they can recognize qualitatively characteristic consequences of T without performing exact calculations" (de Regt 2018, 102).

The resulting account of understanding is heuristically powerful, particularly suited to physics, and guided by a sizable part of the same history (reviewed in chap. 2) that gives rise to the modeling attitude.[40] A lot of that historical scholarship in fact lies at the heart of de Regt's rejection of the facticity thesis. It is thus surprising to my mind that de Regt would retain a commitment to thesis ii, aptly expressed in his criterion for understanding phenomena. My sense is that this is partly due to a difference in our use of the term *theory*. In chapter 6, I defended the position that, on our best present-day accounts, a theory is either a set or a family of representations, as in the RSV, or, alternatively, a hodgepodge of practices, assumptions, instruments, models, principles, and exemplars, as in the pragmatic conception of theories. De Regt seems to rely on a more robust notion of theory as a set of organized explanatory principles, perhaps akin, or at any rate closer to, a robust syntactic or structuralist semantic conception (SSV). This would naturally also go hand in a hand with a commitment to a more substantive notion of representation reviewed early in the book. On such views, theories possess a stronger form of explanatory power, typically brought about by the substantive relations that they hold to their representational targets.[41] Yet such views are in tension with those that give rise to the modeling attitude, which instead are presided over by Maxwell's insights regarding models as conductors for appropriate reasoning. Hence, one might wonder what would result if we were to adopt one of the more deflationary views instead.

On an inferential conception of representation, recall, modeling a phenomenon is in essence nothing but the activity of informative inference from a source toward its established representational target. Yet this is already a considerable cognitive achievement all by itself, and it does provide a minimal sense of understanding of the phenomenon or target. The different models reviewed throughout this book provide a great and varied number of instances of such models for understanding, capable in all cases of providing some sort of enlightenment regarding the possible causes of the phenomenon and sometimes, in addition, a sense of how those causes interact with or depend (logically or conceptually) on each other. And all this is true regardless of how felicitous, accurate, or appropriate the model of a phenomenon may be—and even regardless of whether the target that the phenomenon is ascribed to is in fact real. Maxwell's model of the ether is a nice paradigmatic example of all this. The model serves as a conductor for our reasoning regarding certain properties of the ether, which in turn stand in correspondence with phenomena such as electromagnetic induction. The model thus provides understanding of the phenomenon in its own mechanical terms, a sort of understanding so sought after and cherished at the time that it turned Maxwell into a leading light of Victorian science. The Maxwellians were the followers of this mode of understanding who sought enlightenment on all sorts of practical issues related to magnetic and electric phenomena by relating them to the mechanical ether.[42]

Our previous discussions of the use of models thus suggest some tentative definitions and a contrast between stronger and weaker forms of explanation and understanding as follows. One gains minimal understanding of a phenomenon merely by modeling it in whatever terms are apposite for the purpose of informative surrogative inference from the model. In other words, representing B by means of a model source A provides *minimal understanding* of B (in terms of A). Note that, in agreement with our deflationary conception of representation, this in no way and at no stage (present or future) demands true understanding or even approximately true understanding. It should be clear that no notion of understanding along these lines will agree with the facticity thesis. But, in view of the preceding discussions, this seems as it should be. Most models here reviewed contain simplified, idealized, or fictional assumptions. They are operative assumptions within a representation whenever they facilitate surrogative inference, and in some cases they become indispensable for the representation to have any force or inferential capacity at all. Thus, they are not optional or merely convenient for the purpose of efficient or familiar inference. They are necessary assumptions. In the words of Catherine Elgin (2017), these are the "felicitous falsehoods" that make much of current science possible. Some of these idealizations are close enough to the truth in ways that we can investigate and determine, as Elgin herself points out. But not all are so. Some are wide of the mark, and it is their being wide of the mark that provides them with the power for understanding that makes them so useful for us.[43]

All the instances of models studied throughout this book satisfy this requirement on minimal understanding described above. Nevertheless, some of them undeniably go further and provide more than a merely minimal understanding of the target phenomena since they support their conclusions in some well-grounded and widely accepted scientific theory. While this is obviously not the case (or no longer the case) of Maxwell's ether model, models of phlogiston, and so on, it is certainly the case in the models of stellar structure that we studied closely in chapter 3. Thus, there must be an increased level of understanding—amounting to the full traditional concept of explanation—that does rely on theory in accordance with thesis ii above. I accept this, which brings my account of understanding closer to the received views of de Regt (2018) and Khalifa (2017). But, in view of the discussions in chapter 6, I also urge us to avoid falling back on any notion of scientific theory that will commit us to untenable substantive views on scientific representation. On the extant views of theory that remain viable (a representational semantic conception or a pragmatic conception of theories), it is possible to commit to the notion of a "model in a theory" as merely one of the inferential representations of targets in the domain of the theory that do indeed appear as part of said theory. Thus, a given model source A of a theory T represents some phenomenon B if A has force toward B and A can be used by informed and competent agents appropriately to draw surrogative inferences regarding B. In this case, we could indeed say that this model source A constitutes an explanation of the target B that grounds our understanding on some established theory T. A is a theoretical explanation, and it certainly goes beyond merely providing minimal understanding.[44]

There are examples of pure theoretical models of this sort, of course, and hence there are many examples of theoretically explanatory models that provide such robust theoretical explanations of diverse phenomena. Stellar structure models in astrophysics are certainly written in the language of theory even though radiation theory, fluid dynamics, and thermodynamics all come to bear on them simultaneously. And many of the examples canvassed in de Regt (2018) also naturally illustrate my category of theoretical explanatory models. Most models explored by de Regt are presented as complying with requirement ii regarding reliance on theory while failing the facticity thesis (thesis i).[45] But some of the most recent ones, particularly in quantum mechanics, prima facie satisfy both thesis i and thesis ii.

However, I am not sure that there are that many robust theoretical explanations of this sort in science—or at least not as many as would appear from the history textbooks and the introductory lessons in science that we give our youngest scholars. Newtonian mechanics and Einsteinian physics are obviously cases in point, as are parts of quantum mechanics. In evolutionary biology, there are all kinds of models that naturally rely on natural selection while often importing heavier assumptions from other disciplines or fields. But elsewhere most modeling is patchwork and often done in the absence of much theory. A lot of modeling takes place at the level of experimental activity and practice rather than that of theory. I have been arguing that the experimental attitude is also a modeling attitude and hence properly often representational too. In those areas of science where theory is lacking and experiment gains a life of its own, there is much successful minimal understanding being acquired, but it is rarely the more honorific and respected explanatory kind. And this can also happen with mathematical yet nontheoretical models, as was noted in chapter 3. In fact, by and large, the array of models that we have studied in this book and the case studies in chapter 3 suggest that most models are, in principle at least, merely aiming at minimal understanding.

History has, if anything, a way to show us that theory backs up only some of these pedestrian and experimental models much farther down the road and only then eventually turns some of them into more respectful or robust theoretical explanations. Maxwellian electromagnetism is a wonderful but perhaps exceptional example. Certainly, it is a contention of this book that, all by itself, successful scientific representation does not amount to successful robust explanation. It is not within representation's magic powers to turn a conventional experimental model into a robust theoretical explanation of anything. Representation is, after all, relatively easy to come by and not that hard to achieve, even in the absence of a body of theory. But that is precisely what makes it so pervasive throughout the experimental sciences too and beyond, and it is the reason why representations are so useful in all walks of our cognitive life. A model that provides us with understanding, without any recourse to theory, is valuable already. In as much as it does, for as long as it does provide that understanding, however tentative or provisional, we have reason to accept the model. And this is not a bad thing. Subsuming a model under a theory is a lofty goal, not often within the purview of any realistic science. In other areas of our cognitive lives where theories are rather absent (e.g., in much painting and art, as studied in chap. 8), it is not even an option, yet undeniably those cognitive representations certainly can be said to provide some or even a lot of understanding of their targets, sometimes more than could be achieved by any other means. Mere representational success, understood as minimally as it has been in this book, is already a tremendous achievement of its own.

Chapter Summary

This concluding chapter comes full circle to the arena of epistemology where the debate regarding the nature of scientific representation first raised its head.

I begin by raising doubts about some entrenched dichotomies in the field, such as Ian Hacking's famous one between representing and intervening. On a deflationary account, representing is also a kind of activity in the world, and it is ruled by the same sorts of community norms and standards that inform any other scientific activity, including experiment. This in turn vindicates a deflationary approach to the position known in the literature as *experimental realism*, which gives rise to a wider experimental modeling attitude. Bas van Fraassen's constructive empiricism has often been understood to imply a structuralist semantic conception of theories and thereby a substantive conception of representation. But alternative definitions of empirical adequacy are possible that preserve the epistemic advantages of constructive empiricism. Arthur Fine's NOA provides a useful template, while both Philip Kitcher's real realism and Helen Longino's social constructivism look more promising when cast in a deflationary light. The chapter closes with a brief disquisition on the notion of understanding, which, it argues, stands as the cornerstone and main goal of much scientific modeling activity.

Notes

Chapter One

1. I am myself guilty of helping establish this mantra in some of my earliest talks on the topic. Two examples of papers influenced in this way are Callender and Cohen (2006) and Frigg (2006).

2. And indeed a few papers concerned with representational matters from a structuralist point of view were published in the 1980s coinciding with the advent of the semantic conception. See, e.g., Mundy (1986) and Swoyer (1991).

3. And, anyway, chap. 6 argues that the inferential conception is compatible with any of the three extant accounts of theory—syntactic, semantic, and pragmatist—when these are all properly understood.

4. For some good examples of historians and sociologists engaging with the practices of representation in science, see the essays in de Chadarevian and Hopwood (2004) and Lynch and Woolgar (1990).

5. A movement that originates in the Research Group in Models in Physics and Economics, led by Mary Morgan and Margaret Morrison at the London School of Economics, Tinbergen Institute, and the Wissenschaftkolleg zu Berlin in the mid-1990s. Cambridge University Press published the results of this collective enterprise as Morgan and Morrison (1999).

6. The eventual result was Hughes (1997).

7. The quietism advocated by the inferential conception has a long history in analytic philosophy and enjoys related antecedents in other traditions, such as the originally Spanish heretical form of *quietismo* so prevalent during the religious disputes in sixteenth-century Europe.

8. It is questionable, besides, whether there exists any fully comprehensive account—one fair to the letter in every respect—for any truly complex historical intellectual development, taking place in many different locations, across different cultures and in several different languages. I for one am not a believer in *total history*, a judgment shared by most professional historians I know.

9. There is by now a large literature within the philosophy of science on the distinction between philosophical theses and methodological "attitudes" or "stances" (Fine 1987/1996; van Fraassen 2002). I do not enter the debate except insofar as I adopt the almost universally agreed-on position that there *is* a significant difference and that it is best understood along the theory/praxis axis.

10. I do not enter the ongoing normative debate regarding the status of models vis-à-vis their putative aims. Some critical views, particularly in the medical and social sciences, point out that they can sometimes misapply or shape policy in pernicious ways. See, e.g., Alexandrova (2008) and Northcott and Alexandrova (2015). The emergence of the modeling attitude is not a uniformly positive development in the history of science, but that sort of evaluative project deserves a separate study. In this book, I am concerned only with the historical emergence of the modeling attitude, the practice that it gives rise to, and its philosophical implications.

11. These terms were introduced in Suárez (2003, sec. 2); the terminology *source/target* seems, happily, to be catching on in the field.

12. I am not using the term *relation* in a metaphysically loaded sense in this paragraph or indeed throughout this book. In later sections of this chapter, and in chaps. 5 and 6, I draw on a distinction between two kinds of relations. One is a metaphysically loaded sort that carries an ontological commitment to the existence of the relata. Another, more mundane sort simply asserts something regarding the relational character of a function, concept, or proposition. The distinction is akin to the distinction broached in chap. 6 between *denotation*, which is a success term in the sense requiring actual reference, and *denotative function*, which is not or at least not in the same sense of requiring a referent. I here use the term *relation* in this latter, deflationary, nonsuccess term sense.

13. The distinction between successful (true, empirically adequate) representation and representation per se is now entrenched in the literature. See Bailer-Jones (2003), Callender and Cohen (2006), Contessa (2007), Frigg (2006), Giere (2006), and Teller (2001). But it is worth stressing that it did not seem so entrenched, say, twenty years ago. A striking example is Hughes (1997), in many ways an admirable pioneering effort, but the account there of representation makes no obvious room for misrepresentation, i.e., grossly incorrect but nonetheless genuine representation. I have been emphasizing the distinction between representation and accurate representation since 1998 in papers and talks (see, e.g., Suárez 1999b), and I take some pride in having helped establish it.

14. Most saliently, Paul Horwich's (1990) and Crispin Wright's (1992) deflationary and minimalist accounts of truth, respectively.

15. Some of Hacking's slogans were aptly transformed into substantial arguments for experimental realism by Nancy Cartwright, particularly in the renowned *How the Laws of Physics Lie* (1983).

16. The classic reference for constructive empiricism is van Fraassen (1980), while some of the most poignant criticisms are collected in Churchland and Hooker (1985).

17. For the sort of nonfactive account of understanding that I endorse, see Elgin (2017). For a more substantial account that looks up instead to theoretical explanation, see de Regt (2018). There are several other distinctions in the literature that the inferential conception may well shed light on, such as the distinction between explaining why and explaining how, that between explaining facts and explaining the possibility of facts, and that between explaining factually and explaining fictionally. See Bokulich (2009), Massimi (2022), and Verrault-Julien (2019). There is, however, no scope to develop them fully in one book.

18. A contention that there is happily not much need to justify since it is shared widely among writers on modeling and scientific representation, both within the semantic conception tradition (see, e.g., Giere 1988; and van Fraassen 1989) and within the mediating models tradition (see, e.g., Morrison and Morgan 1999). Yet it makes sense to broach the topic in this introduction since outsiders to the philosophy of science debate often feel that it needs addressing.

19. Brandom's inferentialism has been put to use in the context of models by De Donato and Zamora-Bonilla (2009) and Kuorikoski and Lehtinen (2009), but a study of science through its linguistic categories and syntax, which was valuable prior to the semantic conception and mediating models revolution in the late 1990s, looks too restrictive now. A more promising expressivist approach, in line with the inferential conception (Khalifa, Millson, and Risjord 2022), appeared as this book was going to press.

Chapter Two

1. The word *scientist* was coined in 1833 by William Whewell (see Snyder 2006, 2), and it gained widespread acceptance at the end of the nineteenth century. Whewell is also relevant to the establishment of a modeling attitude through his notion of *colligation*. See Cristalli and Sánchez-Dorado (2021).

2. For an account of the progressive institutionalization of science in research centers and universities, particularly in France and Germany, see Fox and Weisz (1980, esp. 1–28), McClelland (1980), Turner (1971), and Weisz (1983).

3. An illustrious example, often discussed in this context, is Galileo Galilei, whose discourses combine modeling physical systems, often by geometric means, and philosophical disquisition about modeling in almost equal measure. For an illuminating discussion of Galileo's modeling strategies, see Hughes (2010, chap. 5).

4. Warwick (2003) is a wonderful treatise on the pedagogy of tacit knowledge that is of a piece with my emphasis on schools. Besides, its account of the Cambridge modeling tradition informs my history of the English-speaking modeling school in obvious ways.

5. Similarly, when I focus on artistic representation later in the book, I also adopt the view that the best understanding comes from a combination of the philosopher's theoretical outlook and the artist's practical engagement.

6. The development of mathematical logic in the twentieth century has sometimes been thought to provide a key to our understanding of modeling empirical phenomena, particularly in connection with the semantic conception of theories. It is undeniable that some elements of mathematical logic can clarify empirical models (Suppes 2002). But the claim that they constitute a key to such an understanding is debatable to say the least. And, when it comes to understanding the activity of modeling in the empirical sciences, mathematical logic is positively an unhelpful benchmark. (For critical assessment, see Giere [1999b].) The outlook in this book does not align with the idea that mathematical logic is central to our understanding of scientific modeling, and the urge to revisit the vision of the nineteenth-century modelers is of a piece with this outlook.

7. A stance or an attitude is a loose set of normative commitments that informs a practice. The notion is common currency in today's philosophy of science in the wake of Arthur Fine's (1987/1996) NOA and Bas van Fraassen's (2002) "empirical stance."

8. The nineteenth-century modelers introduce at least three novel elements. There was the new division of labor already mentioned; there was a self-conscious emphasis on the hypothetical and even fictitious nature of the models; and, finally, modeling became very sophisticated mathematically, particularly in the hands of what I call the *German-speaking school*, spreading from three-dimensional geometry into complex versions of the calculus and the algebra of equations. None of these elements are present in pre-nineteenth-century modeling, yet they are all central to contemporary modeling practice.

9. As documented in, among others, Davie (1961), Harman (1991), Olson (1975), Siegel (1991), and Smith and Wise (1989).

10. The founder of Scottish commonsense philosophy, Thomas Reid (1710–96), had been regent of Aberdeen's King's College between 1751 and 1764 and from 1764 to 1780 professor at Glasgow University, where he introduced major innovations into the curriculum and founded several key associations.

11. Particularly as developed by Reid's leading disciple, Dugald Stewart (1753–1828), who, through his influential hold on the university and the city as well as his many pupils in Britain and abroad, was instrumental in the successful spreading of commonsense philosophy around the world, particularly in France and the United States.

12. Davie (1961, 191–200), Harman (1991, chap. 2), and Olson (1975, chap. 12) document Maxwell's philosophical upbringing and education.

13. Reid and Simson in fact interacted as colleagues for a few years at Glasgow University. This and other links are revealed in their entries in the *Oxford Dictionary of National Biography*.

14. For a thorough and compelling historical study, see Davie (1961). Harman's (1991) edition of James Clerk Maxwell's papers reveals that Maxwell's first scientific writings were in fact concerned with the geometry of curves. One of these early papers on curves caught the attention of James Forbes and led him to invite Maxwell to follow his courses at Edinburgh University, thus precipitating Maxwell's scientific career.

15. It is even arguable that they derived the latter from the former. This yields an interpretation of the modeling attitude as a direct consequence of the application of the Scottish metaphysical conception of mathematics to empirical phenomena. Such an interpretation clearly lurks in the background of the discussions of both Davie (1961) and Olson (1975). I do not, however, need to argue for such a definite genealogy of the modeling attitude. It is enough for my purposes that, as is well documented and uncontroversial, the modeling attitude is both consistent and coexistent with the Scottish metaphysical school of abstract mathematics.

16. The passage became a famous exemplification that characterizes the Scottish school. It was reproduced in all the textbooks with which Maxwell, Thomson, and their mentors were instructed throughout their Scottish schooling, such as, prominently, Playfair (1795). Davie (1961, 132ff.) provides a good account of this rich history as well as reproducing the notorious passage. See also Harman (1991, 21).

17. Some of the caveats relate to a proper interpretation of the terms employed in this passage since they have more precise meanings now.

18. I focus my discussion on the German- and English-speaking schools, but a full picture of the modeling attitude in the nineteenth century would need to bring in the crucial contribution of the French mathematical physicists of the early part of the century who played critical roles particularly in relation to thermometry.

19. Siegel (1991) is a very good and scholarly account of the transition, in the works of Maxwell, from an instrumentalist conception of analogy to what it aptly refers to as *deep-theory modeling*.

20. There seems to be a threefold scaffolding structure to Maxwell's and Thomson's use of analogy. First, there is the bare structure embodied in the method of abstraction and the thesis of the relativity of knowledge. Second, there is the later addition of a superstructure that adds a dose of realism. Finally, there is an independent requirement that the analogy be of a mechanical type. It is important to distinguish all these layers carefully. I argue in this book that the modeling attitude per se is committed only to the first layer while being able to accommodate the

second layer in a deflationary spirit. The third layer, by contrast, is a feature of fin de siècle physics entirely incidental to the modeling attitude that the nineteenth-century modelers promoted and one that can be safely ignored in contemporary analytic accounts of representation.

21. A paragraph that is worth quoting in full: "The substance here treated of must not be assumed to possess any of the properties of ordinary fluids except those of freedom of motion and resistance to compression. It is not even a hypothetical fluid which is introduced to explain actual phenomena. It is merely a collection of imaginary properties which may be employed for establishing certain theorems in pure mathematics in a way more intelligible to many minds and more applicable to physical problems than that in which algebraic symbols alone are used. The use of the word 'Fluid' will not lead us into error, if we remember that it denotes a purely imaginary substance" (Maxwell 1856–57/1890, 160). Maxwell here reveals himself to be aware of the fictitious nature of his assumptions in a way that is entirely apposite to contemporary accounts of fictions in science. For insightful discussions of the transitions in Maxwell's use of analogy, see Cat (2001), Harman (1991), and Hon and Goldstein (2012).

22. Helmholtz (1858), following on the original Thomson (1847/2011, 1848/2011) model. For discussion, see Silliman (1963).

23. For the relevant history of Maxwell's model and its impact on his contemporaries, see Hunt (1991). For the details and internal logic of the model, see Siegel (1991). Harman (1991) expounds on the natural philosophy behind Maxwell's model. Hon and Goldstein (2012) focus on methodology. Nersessian (2008) provides an account of the import of Maxwell's vortex model for scientific reasoning via models that is close to my views, although it does not draw the relevant philosophical lessons in terms of an inferential conception of representation.

24. The kind of understanding that such fictive models can provide is explored later in the book.

25. William John Macquorn Rankine (Edinburgh 1820–72) was also trained in natural philosophy under Forbes, then elected fellow of the Royal Society in 1850 and made professor of civil engineering and mechanics at Glasgow University in 1855.

26. The use of the model in both these hallmark discoveries is sometimes minimized or even dismissed in contemporary textbooks. This is part of the overall rejection currently of any reasoning that begins with the mechanical properties of the electromagnetic ether itself, an entity obviously discarded after the advent of Einstein's relativity theories. However, historians and philosophers alike are rather in agreement that the model did play a fundamental role in Maxwell's reasoning, which led to new discoveries such as the displacement current and eventually to the development of the full new theory in his celebrated *Treatise on Electricity and Magnetism* (Maxwell 1873b/1954). For various defenses of the critical role that the model played in Maxwell's reasoning, see Harman (1991), Nersessian (2008), and esp. Siegel (1991).

27. For the details of how the model led to the form of the equation for the displacement current adopted by Maxwell, see Siegel (1991, chap. 4).

28. The influence of the Cambridge tradition on Maxwell is carefully documented in Harman (1985a) and Harman (1991, chap. 2).

29. By contrast, the displacement current in Maxwell's theory is orthogonal to the direction of motion of the electric signal in a wire and thus cannot be interpreted as physical motion along the wire.

30. I return to the distinction between mere representational understanding and deeper classical mechanical explanation in the last chapter of the book.

31. As a matter of fact, the ingenuity of late Victorian British physicists in modeling the ether far surpassed anything Thomson ever achieved. For a full historical account of the development

of Maxwellian electrodynamics that appropriately emphasizes the continuity in British science of analogical thinking by models throughout the second half of the nineteenth century, see Hunt (1991, esp. chaps 4–5). I discuss some of these models in greater detail later in the book.

32. Boltzmann spent part of 1871 in Berlin working at Helmholtz's laboratory, and the interaction seems to have played a role in furthering his already very considerable interest in Maxwell's work. But the function of analogy was not the focus of discussion, which rather concerned the introduction by Maxwell of statistical mechanics and the assumption of a statistical distribution of molecular velocities in a gas. For some illuminating discussion, see Klein (1973).

33. For some examples of Boltzmann's praise for Hertz's experimental discoveries, see Boltzmann (1905/1974, 84–85).

34. Boltzmann (1905/1974, 90–91) explicitly records how impressed he was by the discussion of the picture theory in the introduction to Hertz's *Principles of Mechanics* (1894/1956). By contrast, Hertz (1894/1956) contains no references to Boltzmann at this point. But this in no way entails that Boltzmann developed his own *Bildtheorie* out of Hertz's. For instance, Wilson (1989) mounts a convincing historical argument that Boltzmann did not come to *Bildtheorie* via Hertz's introduction to the *Principles of Mechanics* at all, that he was fully committed to his own variant well in advance of Hertz's first reflections on the issue. (Wilson then uses this fact to argue against some entrenched genealogy regarding the sources of Wittgenstein's picture theory of meaning in Hertz's *Bildtheorie*, a priority issue in Wittgenstein scholarship that need not detain us here.)

35. Another significant date is 1902, the year in which Boltzmann's "Models" appeared in the *Encyclopaedia Britannica*. See Boltzmann (1902/1974). But this had less of an impact both among the public and among the modeling community in the German-speaking world.

36. The irony will not be lost on the reader that the leading proponents and main propagandists of the "English" (Pierre Duhem's [1905/1954] term) and "German" schools were based in Scotland and Austria, respectively.

37. Darrigol (2000, esp. chaps 5–6) is a wonderful and impressive account of the transition from action-at-a-distance theories to Maxwellian field theories. Moreover, it nicely addresses the evolution of the modeling attitude away from mechanical models, particularly in connection with Hertz and Boltzmann. Buchwald (1994) is another impressive account, centered on Hertz's own personal transition, and more focused on the experimental side. Most commentators seem to believe that Helmholtz's conversion to field theories was never complete, but Hertz's definitely was.

38. Helmholtz applied it even further—to all his scientific endeavors—since he seems to have assumed that the formulation of a problem in terms of potential functions is integral to any plausible scientific resolution. As Buchwald (1994, 15) puts it: "It could . . . be said of Helmholtz that after the early 1870s nothing was clear to him until it could be formulated in terms of interaction energies."

39. This is curiously in striking contrast with the pedagogical resistance that field theories initially encountered precisely in Cambridge. See Warwick (2003, 306ff.).

40. For a discussion of the reactions to and influence of Helmholtz's proof, see Buchwald (1994, 15–16).

41. On Hertz's views regarding the underdetermination of ontology, see the essays in Baird, Hughes, and Nordmann (1998). For similar claims on behalf of Boltzmann, see Blackmore (1995).

42. The Helmholtz scholarship is large by now, and I can provide only a summary of some important features of these key elements. A comprehensive collection of papers is Cahan (1993). On Helmholtz's methodological practice, Buchwald (1993, 1994) are unsurpassed. These works,

together with Darrigol (2000) and Eckert (2006), provide detailed background to Helmholtz's own work in electrodynamics. For Helmholtz's work on perception, see Hatfield (1991) and Patton (2009), which also contain discussions of Helmholtz's sign theory. Heidelberger (1993, 1998), Hoffman (1998), Patton (2010), and Schiemann (1998) provide enlightening commentaries on Helmholtz's philosophical and scientific upbringing.

43. Yet he was an empiricist about perception on pragmatic, not on epistemic, grounds (Turner 1993). Hence, it would be a mistake to identify his empiricism with traditional British-style empiricism regarding the objects of our sensations.

44. There are also obvious questions to ask about the relation of signs to the objects they stand for. Schiemann (1998) argues that Helmholtz adopted a one-to-one correspondence where perceptual signs' referents are unique and stay constant. But this seems an additional assumption to both the sign theory of perceptions and the *Bildtheorie* in general. And indeed, as we shall see later, both Hertz and Boltzmann go on to reject the analogous claim for *Bilder*.

45. In other words, I shall be supporting Ibarra and Mormann's (2000) rendition of this passage in Hertz—and *Bildtheorie* generally—as a statement of homology against the perhaps more common reading in terms of isomorphism, such as the one found in Leroux (2001). For the philosophical reception of Hertz's ideas, see Majer (1998) and Mulligan (1998).

46. At this point, I admittedly elaborate on Hertz's *Bildtheorie*, but I do so to provide the most coherent rendition. Hertz is uncharacteristically unclear about the relation between the requirement of conformity and some of these other additional requirements, in particular, correctness. Thus, at one point he confusingly claims that an incorrect image is one that does "not satisfy our first fundamental requirement," and it would seem from the context that he has in mind conformity (Hertz 1894/1956, 2). I argue later in pt. 2 of the book that correctness can be a requirement not of representation per se but only of accurate or faithful representation.

47. De Regt (2005) explores the reach of Boltzmann's realism.

48. Thus, Boltzmann (1902/1974, 213) writes: "The essence of the process is the attachment of one concept having a definite content to each thing but without implying complete similarity between thing and thought, for naturally we can know but little of the resemblance of our thoughts to the things to which we attach them. What resemblance there is lies principally in the nature of the connection, the correlation being analogous to that which obtains between thought and language, language and writing, the notes on the stave and musical sounds, etc."

Chapter Three

1. Arthur Fine's (1987/1996) NOA and Bas van Fraassen's (2002) empirical stance are well-known, much-discussed instances. For further discussion of the nature of stances and attitudes and how they critically differ from principles, theses, or dogmas, see Chakravartty (2004) and Rowbottom and Bueno (2011).

2. Black (1962, chap. 13) also introduced into the philosophy of science the useful notion of a *conceptual archetype*, which I employ below.

3. This illustrates the distinction between a model at a 1:1 scale and identity. The 1:1 scale model in a lab of a particular geological terrain is not identical to the terrain and will often differ markedly in numerous ways, including the materials that make up said terrain, the color or texture of surfaces, weight, and so on.

4. The weakness in Black's distinction may be thought to lie with his rather primitive conception of scale models. He seems to have aligned them with replicas and consequently found them

of limited interest in science. I already discarded replicas in the preceding paragraphs, where I also accepted Sterrett's apt claim that scale models can be both complex and sophisticated.

5. There are rare exceptions, such as, e.g., the development of the mathematics of the delta function by P. A. M. Dirac as part of his endeavors to represent quantum states mathematically (Dirac 1930). That is arguably a rare instance in which a physical target and its properties provided enhanced understanding of the mathematical model source employed to represent it.

6. "Mathematics can be expected to do no more than draw consequences from the original empirical assumptions. If the functions and equations have a familiar form, there may be a background of pure mathematical research readily applicable to the illustration at hand" (Black 1962, 223).

7. The literature is large. My main sources are the account by the leading British historian of art Michael Baxandall (Baxandall 1985), the acclaimed book on bridge engineering by Henry Petroski (Petroski 1995), the review by Shipway (Shipway 1990), and the original, painstakingly documented report on the Forth Rail Bridge by the engineer of its central section, Wilhelm Westhofen (Westhofen 1890).

8. This section expands on previous work of mine, such as Suárez (2010b).

9. Suárez (2011) discusses just this point in relation to van Fraassen's (2008) similar claims.

10. I develop the case study in Suárez (2016), and some of the technical details that are described later in the book appear in Suárez and Pero (2019).

11. Carnap's and Reichenbach's attempts are discussed at length in Friedman's works. See, e.g., Friedman (1999).

12. Balázs (2017) also finds that the focus ought to be on Maxwell's early work.

13. I have expounded on stellar astrophysics models and their idealizations before, e.g., in Suárez (2013), of which what follows is a version. My exposition is indebted to Collins (1989), Hansen, Kawaler, and Tremble (2004), Prialnik (2000), and Tayler (1970/1994).

14. The HR law was independently discovered by the Danish astronomer Ejnar Hertzsprung (1873–1967) and the American astronomer Henry Norris Russell (1877–1957).

15. In other words, the *fundamental empirical properties* of astrophysics are not observable in Bas van Fraassen's (1980) sense of the term. They are rather derived from other properties that can be said to be observable in van Fraassen's sense but not by a simple procedure devoid of further theoretical assumptions. I discuss constructive empiricism in the last chapter of the book.

16. New asteroseismological data from the CoRoT mission of the European Space Agency and successive Kepler missions of the National Aeronautics and Space Administration impose even stronger constraints, which I review in Suárez (2023).

Chapter Four

1. This part of the book takes further some ideas of mine that have appeared in previous publications (Suárez 2003, 2004, 2010b, 2014, 2015, 2016). This chapter and the next are expanded and updated versions of material already published in Suárez (2003, 2010b).

2. Other critical early contributions were Bailer-Jones (2003), Frigg (2006), Giere (2004), Suárez (1999b), and van Fraassen (1994).

3. The reader who is already acquainted with my work on scientific representation is invited to proceed to the development of the positive account in chaps. 6 and 7 as well as pt. 3 of the book, which deals with its implications.

4. Ambrosio (2014) is a contemporary account from the point of view of scientific representation.

5. Peirce is not always explicit. For exegesis along these lines, see Hookway (1985, chap. 4).

6. The following is just a sample. Morgan and Morrison (1999) brought to the fore the autonomy of model building from both theorizing and data collection. The series of books edited by Lorenzo Magnani and his coeditors arising out of the Pavia conferences, such as Magnani, Nersessian, and Thagard (1999), focused on how model-based reasoning differs from typical theorizing. Most recently, this has given rise to an encyclopedic handbook of model-based science (Magnani and Bertolotti 2017). De Chadarevian and Hopwood (2004) emphasized how important the historical context is for determining the purpose of models. Pincock and Vorms (2013) explored the role of formats in models. Suárez (2009a) is an edited collection on the use of fictions in models.

7. A celebrated, albeit controversial, example is Mirowski (1991), a study of transference between field theories in physics and value theory in econometrics at the turn of the century.

8. Contessa (2007) exemplifies the analytic inquiry into the constitution question. Knuuttila (2009) is an example of practical inquiry into pragmatic questions regarding means.

9. Frigg and Nguyen (2020, chap. 1) refer to the issue of the constituents as *the problem of epistemic representation*, and the related definition offered above corresponds to their *ER scheme*. More generally, my extended notion of cognitive representation (of which scientific representation is a type) corresponds to what other authors in the literature, following Contessa (2007), call *epistemic representation*.

10. In the most basic form of analytic inquiry, R is a binary relation, and x is the empty set.

11. Questions of representation can, of course, be settled in practice simultaneously with questions of accuracy, and in fact they often are, as when modelers adjust their target in line with the reach and propriety of their predictions. My point is that, as a matter of conceptual necessity, questions of representation can never be settled strictly *after* questions of accuracy since they are presupposed in the latter.

12. Compare primitivism and reductionism in the analysis of laws, causation, dispositions, and time. A primitivist thinks that these notions are not further analyzable. A reductionist will want to analyze them in terms of other properties or relations. For instance, a Humean (a follower of David Hume's skeptical philosophy) may want to reduce laws to regularities and causation to constant conjunction. A logical empiricist, such as Carnap (1934/1937) or Reichenbach (1956), will endeavor similarly to reduce dispositions to robust counterfactuals, causation to open conjunctive forks or "oriented" correlations, and the arrow of time to causal asymmetry.

13. This is not to say that primitivist substantivism does not find application elsewhere. For illustration, consider the case of laws, causation, or time. A primitivist about these concepts claims that they are explanatory primitives. For instance, David Armstrong (1983) is widely held to defend primitivism about laws, Tim Maudlin (2007) is a primitivist about the passage of time, Wesley Salmon (1998, chap. 16) appears to have been a primitivist about conserving-quantities causal processes in his later years, and so on. By contrast, a reductionist Humean would attempt to reduce away such problematic concepts to those that he or she considers epistemically unproblematic. Thus, laws and causation are to be reduced to regularities, as in Mackie's (1980) theory. Alternatively, causation is to be reduced to probability, as in Reichenbach (1956), or to counterfactuals, as in Lewis (1986); and time direction is to be reduced to open conjunctive forks or oriented correlations, as in Reichenbach (1956).

14. For extended presentations of the different options, see Blackburn and Simmons (1999), Horwich (1980) and Wright (1992).

15. I gloss over the distinction between sentences and propositions entirely since it does not affect the discussion. In Quine's deflationism, and arguably in Ayer's, truth is predicated of sentences. Other theories, such as Horwich's and arguably Ramsey's, apply deflationism to propositions instead.

16. Except for some pronouncements of James's to the effect, this theory was not actually defended by any of the traditional pragmatists, and it does not really deserve the name *pragmatist* (which is why I adopt the more appropriate label *utility*). For informed discussion of Peirce's rejection of such a crude pragmatist theory, see Hookway (2000). For a similar discussion regarding Dewey, see Burke (1994).

17. This opens the door to the charge of truth relativism, one of several objections to the coherence theory. For a contemporary defense, see Walker (1989).

18. For similar distinctions, see Horwich (1980) and Wright (1992). I simplify the debates involved. See Blackburn and Simmons (1999). My purpose here is just to introduce a suggestive analogy, not to come to grips with the nature of truth, reference, and so on, about which, recall, this book takes a quietist attitude.

19. And, indeed, the Humean offers substantive accounts of regularity and constant conjunction. And the logical empiricist defends a reduction of mysterious dispositions to the solid grounds of substantive semantics for counterfactuals, the apparently vague relation of causation to the more substantive grounds of conjunctive forks, and the always elusive arrow of time to the more secure and evident asymmetry of causation. In all these cases, Humeans and logical empiricists alike are guided by their belief that the epistemically accessible grounds are the only substantive grounds.

20. However, there may be undeniable gains in the economy of our knowledge or the architecture of science. Why maintain multiple nonsubstantive notions when fewer would do? Ockham's razor may apply in favor of applying reduction even to the thinnest, most deflationary entities. Such principles of parsimony may well provide the deflationist with some good reasons to move toward reductionism.

21. The terms *[iso]* and *[sim]* were first introduced in Suárez (2003), and the definitions I offer are updated versions. Bas van Fraassen (1980) has been thought to defend something close to [iso] and Ronald Giere (1988) something close to [sim]. However, these are not their considered views, as was made clear in van Fraassen (2008) and Giere (2004). French (2003) provides arguments in favor of isomorphism as a necessary and sufficient condition on representation and thus can be taken to defend [iso], while Aronson, Harré, and Way (1993) came close to defending [sim], albeit under a more sophisticated account of similarity.

22. The identity theory of similarity is not trivial, and it has its own complications. I deal with them, as well as alternatives, later in this book. The main disadvantage of the identity account of similarity is that, on purely formal grounds, any object can turn out to be similar to any other since it is possible formally to define *property* so no pair of objects will fail to share some property (e.g., "larger than an Armstrong," "not blue-and-nonblue," etc.). So it is important to supplement the theory with some criterion of relevance, as I do in the text.

23. The claim may be generic for any pair of types of such systems or particular for one system of billiard balls with respect to a particular set of gas molecules enclosed in a container.

24. For more on the critical role of dissimilarities in inquiry, see Boesch (2019b).

25. In very early work (Suárez 2003), I chose the phrase *exemplify structure*, which is less suggestive of the problems of underdetermination that arise, so I switched to *instantiate structure* in Suárez (2010b) and subsequent works. *Exemplify* is also used now in connection with very different approaches that identify representation with Goodman's work (Elgin 2009; Frigg and Nguyen 2018), so it is best avoided in this context.

26. If this seems like a very poor attempt to capture what is interesting and relevant about the billiard ball model, it is because it is. We shall see later how we might compensate for the impoverished rendition by means of weakened morphisms. Ultimately, however, conceptual models remain a challenge to all substantive theories, particularly versions of [iso].

Chapter Five

1. The logical argument, the nonsufficiency argument, and the nonnecessity argument were first advanced in Suárez (1999b). The complete set appears for the first time in Suárez (2003), of which this is an updated and refined version. Many authors have gone on to reiterate such arguments in the literature as well as providing their own versions or adding arguments of their own to the same effect.

2. In addition, questions relative to the instantiation of structure are tricky. What structure is instantiated by a concrete object is highly relative to context, purpose, and description. Consider, e.g., the many important structures that a bridge instantiates besides geometric shape: the structure of weights and forces, the distribution of colors of each of the parts, the relative resistances of each part to air and water friction, etc. This underdetermination of structure is a major objection to structuralism (Suárez 2010b, 2016). (For a related discussion, see Pincock [2012].) But the issue is tangential to my concerns here.

3. For a full account of Victorian mechanical models of the ether in their incredible ingenuity and brilliance, see Hunt (1991). The most prominent of Maxwell's own models (see Maxwell 1861–62/1890) has already been reviewed in chap. 2 (see fig. 2.1).

4. For discussions of similarity in this context, see Toon (2012a, 2012b).

5. The analogy will naturally lead us to a further consideration of the role of representation in art and aesthetics in chap. 8.

6. This explains why [iso] and [sim] have been particularly attractive to defenders of the semantic view of theories, theories not being linguistic entities on this view, as we shall see in the next chapter.

7. For the technical details, see Suárez (2003, 2004, 2009b).

8. For related claims, see Goodman (1972).

9. Blunt (1969), writing during the cold war, probably overemphasizes the political aspects of *Guernica*. Chipp (1989), writing during the controversy over *Guernica*'s return to the new Spanish democracy and involved in the international diplomatic efforts that ensued, underemphasizes them. The most balanced account remains Arnheim (1962).

10. I employ *Guernica* to the same effect in Suárez (1999b). French (2003, 5) misreports my argument as one of ambiguity between different targets, and then, confusingly, goes on to write in response that "it is not difficult to find other examples from the history of art which might be called non-representational." Ambiguity is no problem for [iso] since it is always possible for different objects to exemplify isomorphic structures. Nor am I claiming that *Guernica* is non-representational: that would be an absurd claim, and moreover one that would bypass what is

at stake, namely whether there can be representation without isomorphism. My claim is that it makes no sense, and it would be false, to assert that either of these two genuine targets of *Guernica* is isomorphic, or similar in the relevant respects, to the canvas.

11. This objection is independent of the concerns regarding underdetermination voiced in the first section of chapter 5.

12. There does not seem to be a theory of mental intentionality that is free of problems or that has not already been refuted. Here is a sample list. Similarity accounts of intentionality are refuted by, among others, Cummins, who also has strongly criticized covariance or causal accounts. See, respectively, Cummins (1989, chaps. 4–6) and Cummins (1998). Yet Egan (1998), MacDonald (1998), and Millikan (2000) offer strong and convincing arguments against Cummins's own isomorphism theory of mental representation.

13. Since I first wrote on this topic, there has been some exciting work done on cognitive representation, such as Shea (2018) or Martínez (2019), that may eventually find application in this area. Already over a decade ago, Daniela Bailer-Jones (2003, 2009) emphasized the role of mental representation in scientific modeling, although my sense is that efforts to reduce scientific representations to mental states overall have not prospered.

14. This is not to say that representational forces cannot be studied scientifically. A good deal of historical and sociological research, e.g., is in one way or another devoted to objectively settling issues of past representational forces, and historians and sociologists have developed tools to carry out these tasks. Baxandall (1985) and Lynch and Woolgar (1990) were influential milestones in the history of art and the sociology of science, respectively. But, although science can study practices and complex community commitments and norms and their values, it cannot reduce them to facts.

15. Pero and Suárez (2016), Suárez (1999a), and Suárez and Cartwright (2008) argue that it does not, despite protestations to the contrary in Bueno and French (2011) and Bueno, French, and Ladyman (2002).

16. This is in fact Swoyer's penultimate definition. His final proposal includes an additional refinement to account for the further distinction between cases in which the representation correlates elements of B uniquely to elements of A and those in which the representation correlates elements of A uniquely to elements of B. Since the distinction is required only to cover cases of linguistic, or word-to-object, representation, I ignore it here.

17. For a sample of works endorsing or broadly in agreement with these views, see Boesch (2017, 2019a), Downes (2009), Elgin (2009), Fang (2017, 2019), Frigg (2006), Frigg and Nguyen (2016), Gelfert (2011, 2016, 2017), Giere (2004), Jebeile and Graham Kennedy (2015), Knuuttila (2009, 2011), Levy (2015), Nguyen (2016a), Peschard (2017), Toon (2012b), van Fraassen (2008), and Vorms (2011, 2017).

18. Those who attempt to get around the arguments to build up substantive accounts include Ambrosio (2014), Bueno and French (2011), Callender and Cohen (2006), Contessa (2007), French (2003), Pincock (2012), Poznic (2016), Sánchez-Dorado (2019), Shea (2014), Shech (2015), and Weisberg (2013).

19. This is really Tversky's (1977) original account. Weisberg's (2013, 145ff.) application of this account to modeling is more complex and appeals to a distinction between attributes and mechanisms in both model sources and model targets. However, the complexity adds nothing significant to the original account of similarity, so my assessment in the text applies.

20. For the original objection, see Suárez (2016, 451–52).

21. For a similar point, see Boesch (2019b).

22. The full details are developed in Pero and Suárez (2016).

23. One might suppose that this problem could be overcome by incorporating a fourth factor in the measure of similarity successfully representing the weight of features of the source that are explicitly denied in the target. It is, however, still to be seen if this is possible, and at any rate the three other objections remain.

24. Sánchez-Dorado (2018, 2019) are excellent disquisitions on the nature of these judgments. Nersessian (2008) is an early discussion of creativity in model building.

25. Bartels does not appeal to a normative practice of model building, as I shall do later in the book, but this can be understood to be presupposed in his discussion.

26. Such as, e.g., the fact that what Bartels calls *homomorphism* turns out to be what is technically known instead as *epimorphism*. For the technical details, see Pero and Suárez (2016, 83).

27. Interestingly the inverse negative analogy was appreciated by Campbell, and in fact he found it relevant to his discussion of the kinetic theory. It somehow gets lost in Hesse's simplified model of analogy, but it plays a hugely important part in the assessment of the limitations of the model. Campbell critically did not think one could regard the inverse negative analogy as merely a part of a heuristics, to be abandoned once the theory of the domain is finally established. Rather, he thought that such analogies must be understood to be part of the theory itself: "It is often suggested that the analogy leads to the formulation of the theory, but that once the theory is formulated the analogy has served its purpose and may be removed and forgotten. Such a suggestion is absolutely false and perniciously misleading" (Campbell 1920/1957, 129).

Chapter Six

1. I outlined such an argument against the semantic view in (Suárez 1999b). Frigg (2006) built on that. For similar views, see also Contessa (2007). However, as we shall see, these criticisms are predicated on a particular version, the structural semantic view, outlined below.

2. The focus is thus not historical since—as I made clear in the introduction—I do not share the view that philosophical work on representation historically derives from the upsurge of the semantic view of theories in the 1970s. However, I do not believe either that the semantic view is irrelevant to representation or that it contradicts a pragmatist account of representation. Hence the attempt in this chapter to show that a representational version of the semantic view is compatible with an inferential conception of representation.

3. In doing so, as I hope will become clear, I do not belittle the alternative syntactic and pragmatic views of theory. It would be easier to show that they countenance and fit in well with the conception of representation here defended, but the exercise would not be so illuminating regarding the tight connection between representation and models.

4. The terms *syntactic* and *semantic* are misnomers, as anticipated in chap. 1. On the one hand, the syntactic conception of scientific theories is in fact also semantic in as much as it imports an interpretation of the terms that appear in scientific theories. On the other hand, the semantic conception is really neither properly syntactic nor properly semantic since it is not strictly speaking linguistic.

5. Carnap (1956) is perhaps the clearest repudiation of the operationalist analysis of theoretical terms. Nagel (1961, 97ff.) contains another illustrative set of arguments. Overall, the logical empiricists were never guilty of the failed reduction of theoretical concepts to empirical sense data that their critics often charge them with. The neo-Kantian constitutive tradition that

inspired most logical empiricists and informed their understanding of theoretical science was in fact contrary to any such reductionist ambitions (Friedman 1999; Ryckman 2007).

6. Nagel (1961, 96) nominally employs a different model from genetic biology to illustrate the difference between the correspondence rules and the model interpretation of the abstract calculus. However, he turns to Newton's second law to illustrate the diffuse and wide-ranging application of correspondence rules (1961, 105). The illustration from Newton's gravitational theory in the text provides a cruder but more intuitive summary of the same ideas.

7. Carnap (1939) attempted to reduce all the nondescriptive language of science to expressions in first-order logic, a project that he eventually abandoned as being too narrow a base. In Carnap (1956), he expanded the base to include mathematical relations and functions, thus linking the theoretical language L_T also to the extension of mathematical functions. Psillos (2000) chronicles the transition and reviews the status of the complex Ramsey sentences for theoretical terms that result in Carnap's most sophisticated account.

8. The earliest objections are those of Hilary Putnam (1962/1975)—here as elsewhere a pioneer—but the criticisms reached their peak in the writings of Frederick Suppe (1972, 1973/1977) and Bas van Fraassen (1980, 1989). Suppe's (1973/1977) edited collection is an excellent resource containing all the relevant objections. For a more recent and complete review, one from a pragmatist perspective not dissimilar to my own, see Winther (2020a).

9. For valiant attempts to retain core aspects of the received view, see Hudetz (2019) and Lutz (2012, 2017). For inspired criticisms of the syntactic/semantic distinction, see Glymour (2013) and Halvorson (2012).

10. For the pervasive character of inconsistency in scientific theories, see Frisch (2005). For the relevance to the issue of representation, see Frisch (2014). As for inconsistent fictional assumptions within scientific models, the literature is by now large, starting with Vaihinger (1924), and including the essays contained in Suárez (2009) and Woods (2011).

11. See esp. Lutz (2017, 327), which mainly refers to Carnap (1939). Carnap (1966, 226) acknowledges that the distinction is vague in scientific practice: "There is a continuum which starts with direct sensory observations and proceeds to enormously complex, indirect methods of observation. Obviously, no sharp line can be drawn across this continuum; it is a matter of degree." Hempel (1965, 111) instead refers to an antecedently understood vocabulary of "primitive or basic terms," which leaves it open whether this is in any way the observational vocabulary. Nevertheless, that there *must* be some sharp distinction in the vocabulary, however vague its application, is evident through their writings. For instance, Carnap (1956, 41) critically asserts: "The terms of V_O are predicates designating observable properties of events or things (e.g., 'blue,' 'hot,' 'large,' etc.) or observable relations between them (e.g., 'x is warmer than y,' 'x is contiguous to y,' etc.)." The bottom line is that, no matter how the distinction between V_T and V_O is drawn, the criticisms that arise out of part ii reemerge with full force.

12. The example is contested. Some argue that the formulations are not actually equivalent (e.g., Muller 1997a, 1997b), which would mean that it is just right that the received view should distinguish them as distinct theories and the example is thus in fact an argument in its favor. Others argue that the received view possesses a more sophisticated notion of theory equivalence (e.g., Glymour 2013, 292), and this might perhaps be employed to show the required equivalence of matrix and wave mechanics. At any rate, more work is evidently required on behalf of the received view to align its account of theories with scientific practice since the equivalence is widely assumed in practice.

13. This is so because state-space descriptions can be easily decoded as either linguistic predicates ascribing values to variables or representational models—of a mathematical kind—that stand in lieu of physical systems.

14. With minor—albeit not insignificant—differences, SSV is the account of scientific theories defended by Balzer, Moulines, and Sneed (1987), da Costa and French (2003), Stegmüller (1979), Suppe (1989), and Suppes (2002), among others. Van Fraassen seems to me to have significantly and reasonably moved from an SSV account (van Fraassen 1980, 1989) toward a representational semantic account (van Fraassen 2008). Something similar can be said for R. I. G. Hughes, whose influential textbook on quantum mechanics (Hughes 1989) often appeals to SSV, while his later writings explicitly reject it in favor of a representational semantic account (Hughes 2010). Ronald Giere (2004) was perhaps most consistent in the view that the word "model" in the semantic bare claim 1 should not be taken in the model-theoretical sense—and as it turns out this also renders impossible a set-theoretical reading (Hudetz 2019).

15. This is a simplified version of axiom 6 in the original presentation in McKinsey, Sugar, and Suppes (1953, 258). The original version is somewhat more complex and refers to component forces, but the proviso in the assumption as I have outlined it in the text takes care of any additional forces that might be exerted on particle *p*.

16. While both claim 2 and claim 2′ are "structural" claims, the primed claim goes beyond its unprimed counterpart since claim 2′ makes a statement regarding all representations, while the more restricted claim 2 makes that claim only about scientific models, a subset of all representations in the terminology employed here.

17. To be clear, my claim is that the RSV is the most plausible (or, perhaps even more conservatively put, the RSV is the only plausible) semantic account of scientific theories. Thus, if scientific theories are to be reified, they had better be reified as sets of representations, not as set-theoretical structures. Note, however, that this is intended as a conditional claim, and I do not necessarily subscribe to the antecedent. On the contrary, as will become clear later in this chapter, I am sympathetic to the view that theories cannot in fact be reified but that they are rather better conceived as heterogeneous combinations of models, modeling techniques, and practices jointly regarded (essentially the view defended in Suárez and Cartwright [2008] and referred to as *the pragmatic view of theories* by Winther [2020a]). This pragmatic view of theories also fits in well with the inferential account of representation defended in this book since, as we shall see, the emphasis is entirely on representing activities (modeling) rather than on representational sources (models). Thus, to summarize my claim most precisely: the only plausible semantic account of scientific theories as homogeneous entities is the RSV since it aligns theories and theorizing most closely with the practice-based account of representational activities defended in this book.

18. Some of the reasons have already emerged in the historical chap. 2. While Boltzmann and Hertz did not defend any conception of scientific theories in anything like the contemporary syntactic/semantic terms, there are ways in which one can project back some of the relevant insights of the RSV. It would be much harder to read the SSV into their pronouncements, not least because neither was acquainted with the sort of higher-order logical calculi and set-theory characteristics of contemporaneous mathematical logic. Nor did any of them, to my knowledge, ascribe a particularly important role to set theory in the understanding of scientific theoretical knowledge or activity.

19. The point has been often made that ultimately the views are not that different since the syntactic view is also semantic and carries a model interpretation of the abstract calculi as well

as the notorious correspondence rules. See Friedman (1982), Halvorson (2012, 2013), Lutz (2017), Nagel (1961), and Worrall (1984). However, as Halvorson accurately notes, the equivalence works only on the assumption that the models appealed to in the semantic bare claim (claim 1) are set-theoretical structures: "Many philosophers of science now disagree that models should be mathematical objects. Those views will not be subject to the critique. . . . My critique is aimed at views that try to explicate the concept of a scientific theory using the concepts of contemporary mathematics" (2012, 188). It should be evident by now that I am one of those many philosophers of science to whom Halvorson refers since I defend the view that models are in general representations and that representations are typically not mathematical structures.

20. There are additional cognitive and epistemic issues that can be raised as objections to the SSV. For instance, if theories are set-theoretical structures, one would expect most scientists, and certainly theoretical scientists, to make constant use of set-theoretical methods in constructing and applying their theories, but this is rarely the case. And, if the SSV is seriously meant to be close to practice, it should also be close to the cognitive skills acquired and deployed by scientists, yet heavy training in abstract set theory is uncommon in almost all branches of science.

21. These philosophers often claim to follow the lead of Suppes (1962, 2002), even though Suppes himself went on to develop a much more nuanced position. See, e.g., Suppes (1968) or the report on his more recent views in Sneed (1994), which recognizes the severe limitations of set theory as a framework for scientific theories.

22. For a persuasive argument for this conditional, Frigg (2006).

23. This is a line that I have defended since (Suárez 1999b) and, more recently and explicitly, in Suárez and Pero (2019).

24. It is not necessary to explore here any further features of theories on such an RSV, but it clearly cannot amount to merely the claim that theories are motley sets or arbitrarily put together collections of models. The theory must contain some organizing principles, and it must be sufficiently abstract to enjoy general application through a much larger domain than that of the specific models it contains. Moreover, these principles must act as rules for the application of the models that are contained in the theory as well as generating tools of the construction of further models within the theory's purview. Giere (1988, chap. 3), Odenbaugh (2019, 11–12), Suárez (1999b), and Suárez and Cartwright (2008) contain suggestions regarding these various effects, suggestions that bring the RSV closer to a heterogeneous account of theories as activities and their rule books. A full elaboration would demand another book on the topic of the nature of theory. My more modest claim in this section is that there is nothing in the semantic view per se that impugns the inferential conception of representation about to be defended.

25. An even earlier deflationary account specifically for models is Downes (1992), already containing an extensive discussion of the implications for the semantic view. Downes updates his deflationary views in Downes (2021).

26. In the terminology defined in chap. 4, Giere's view can be understood as a kind of use-based account, while Elgin's and Frigg and Nguyen's are a sort of abstract minimalism (or, alternatively, when suitably dropping the denotation requirement—as I shall go on to argue in the text that one must do—they come out as no-theory theory accounts).

27. The qualification *not necessarily* is important as Hughes is explicit that he means not to define representation but only to describe some typical features of its application to science.

28. Hence, I believe that Frigg and Nguyen (2020, chap. 7) err when they present Hughes's DDI account as an earlier version of their own DEKI account. The DEKI account is explicitly

meant to capture the concept of representation fully through a set of individually necessary and jointly sufficient conditions, something that Hughes thought implausible and advised against even attempting to do (Hughes 1997, 329; Hughes 2010, chap. 5; personal communications with the author). The original DDI account requires some modifications to turn it into a fully pragmatic activity-based account, as I am about to point out. See also Suárez (2015, 43–45). At any rate, Hughes (1997) was prescient in emphasizing the pragmatic and deflationary aspects of representation. Indeed, he went on to develop an idiosyncratic synthesis of hermeneutics and pragmatism in his wonderful and eclectic final work (Hughes 2010).

29. Russell (1905) is a classic. For discussion specifically in the context of representation, see Goodman (1968/1976) and Elgin (1983, 2009).

30. In Suárez (2015, 43), I even ascribed this tempting identification to Hughes himself. Yet close study of his writings does not bear out any endorsement on his part of the relevant role for the relation of denotation. I now think that Hughes took denotation from Goodman's account and followed commonly established use in the terms that were available to him but that he was never entirely comfortable with that use. His evolving writings, culminating in Hughes (2010), show that he would have preferred, in the full deflationary use-based spirit, to focus entirely on the activity and leave aside the relation.

31. See Chang and Keisler (1990), which informs my discussion. This is the same definition discussed in the first section of chap. 6 in relation to the semantic view, and it gets picked out by Lutz (2017) in his defense of that view. Inasmuch as the semantic view is understood as the claim that theories are suitably interpreted structures, and as long as *interpretation* is understood model theoretically, it retains exactly this character, remains yoked to a particular syntax, and thus becomes essentially indistinguishable from the syntactic view. This observation adds substance to the claim, defended in the text, that this model-theoretical sense of *interpretation* is ill-suited for the DDI account's third typical component of modeling since throughout his life Hughes remained a defender of a liberal semantic account of theories as sets or families of representational models. In other words, it seems reasonable to suppose that Hughes never had a model-theoretical interpretation in mind for the "I" component in his DDI account.

32. The current niche of academic interest in the application of mathematics (Bueno and Colyvan 2011; McCullough-Brenner 2020) concerns just a study of this application component in scientific representations by mathematical means—or, in the present terminology, those cases of scientific representation in which the representational sources are a piece of mathematics. But, as noted already in Pincock (2012) and Suárez (2010b), and as is becoming increasingly evident, the underlying issues are quite general and best regarded as instances in more general accounts of scientific representation. The phrase *inferential conception of the application of mathematics* is thus revealed to be a bit of a rhetorical device to re-create (and appropriate?) the inferential conception of representation here defended.

33. In the inferential conception that I shall be developing in the next section, the activities involved in demonstration and application essentially suffice to establish a representational model, while denotation is effectively redundant.

34. See the essays contained in Godfrey-Smith and Levy (2020), Suárez (2009), and Woods (2011).

35. For discussion, see Nersessian (2008).

36. Similar claims in the literature include Isabelle Peschard's (2017) that representational relations often are constituted within the practice of representation and Tarja Knuuttila's (2011) that targets are often constructed with the representation itself.

37. This is so unless, of course, *relation* is understood so minimally as to coincide with the ability to draw claims. That is, unless it is true that, *if* a rule of inference is set up that allows an agent to reason—however fallibly, and regardless of context—from claims about a system or object A to claims about a system or object B, *then* A and B stand in a relation. But how plausible is this extremely thin and unusual notion of relation? It is certainly devoid of any ontological purchase since it would, e.g., apply equally well to fictional or abstract characters: within a fictional narrative, we are often invited to infer claims about some fictional character from claims about another. It would then follow, e.g., that Don Quixote and chivalry stand in a relation since in Cervantes's fiction it is possible to proceed from claims about Don Quixote's mental illness to claims about the nature of chivalry, and vice versa. But just as "truth in a fiction" is not equivalent to "truth about a fiction," never mind truth per se (Lewis 1978), "relation in a fiction" is also not equivalent to our relation to the fiction or to any other relation for that matter.

38. There is some conflation of process and product in Winther's writings on this topic, something of which he is probably aware of and may want to defend. Even the distinction between theory and theorizing could be considered dubious. Winther (2020a, 39) aptly quotes Suárez and Cartwright (2008, 79): "What we know 'theoretically' is recorded in a vast number of places in a vast number of different ways—not just in words and formulae but in machines, techniques, experiments, and applications as well." In that paper, Cartwright and I argued that the process and the product of theorizing are not as cleanly distinct as standard views of theory would have it. Winther (2020a) makes a very strong case (see also Winther 2020b, chap. 5), and this is of a piece with my defense of a pragmatist and deflationary RSV in the main text.

Chapter Seven

1. For discussion of this point, which, as we shall see, is critical in accommodating instances of inaccurate, imperfect, or incomplete representation, see Boesch (2017), Contessa (2007), and Suárez (2004).

2. There is no modal magic at work here. Many dispositional properties of objects, such as their fragility, portability, and so on, are also relational in just this sense.

3. The inferential conception goes beyond the "Gricean stipulation" proposal advanced in e.g., Callender and Cohen (2006). While the inferential conception agrees with these authors that scientific representation is not a natural—or naturally occurring—representation, it does not go along with their view that a mere conventional act of stipulation suffices for cognitive representation.

4. This is also van Fraassen's (2008, 23–24) *hauptsatz* (main proposition) for representation, and it is adopted in one way or another, under one or another terminology, by virtually every author now working in the field.

5. This sheds some light on how representational sources can appear, on the one hand, vague and indeterminate and, on the other, ambiguous and multiply representational. Since the establishment of force is a contextual matter, reliant as it is on some modeling practice, the target of a representation can be vague or indeterminate when the practice itself is vague or indeterminate. Luckily, this is rare in science, where practices are tightly agreed on and enforced through the teaching of exemplars and so on. (It is a lot more common in the arts, humanities, narrative, fiction, and so on.) Cases of multiplicity are, however, not so rare. A source is multiply representational when different modeling practices employ the same object or system to represent their

different targets. Thus, as I put it in the original paper (Suárez 2004, 768): "A spiral staircase, for example, may be taken to represent the mechanics of a spring, or the structure of DNA: the source's force varies with intended use. In each case an informed and competent agent will be led, upon considering the source, towards the correct target; if an agent is simultaneously competent and informed about the use of both representations, or ambivalent, the force of the source will be double or ambiguous respectively."

6. *Cognitive representation* is the name I employ for the kind of representation used in inquiry. It is roughly equivalent to Contessa's (2007) *epistemic representation*, which has been adopted by other authors (e.g., Bolinska 2013; and Frigg and Nguyen 2020). I still prefer the term *cognitive*. It does not have the connotation that a valuable representation must add to our stock of knowledge at the same time that it contributes to our cognitive life.

7. As this book goes to press, another attempt to provide a fully expressivist account of representation appeared. Khalifa, Millson, and Risjord (2022, 50) claim to have developed a "thoroughgoing inferentialism much in the spirit of Suárez's inferentialism . . . that exhibits a more plausible form of deflationism." I am certainly not opposed to a fully expressivist account of scientific representations going beyond linguistic representations. But I would caution against denying the strength of Huw Price's (2011, 2013) distinction between internalist representations (*i-representations*) and externalist representations (*e-representations*) (see Price 2013, 36), which I believe the inferential conception [inf] accommodates with ease.

8. There is, of course, considerable debate over this. See, e.g., Floridi (2007, 2011) and Scarantino and Piccinini (2010).

9. The analogy between modeling and Shannon information is explored more fully in Suárez and Bolinska (2021), where the reader can find all the technical definitions of the terms involved, including *noise*, *equivocation*, and so on. It is important for the purposes of the analogy not to conflate the *representational* source referred to in this book with the *information* source designated by communication theory.

10. See also some of the essays in Cassini and Redmond (2021).

11. The list cannot be comprehensive since it is an open question whether new types of sources and targets and their relations will emerge. In addition to similarity and isomorphism, in all their varieties, there are already candidates in truth—of linguistic models or descriptions (see e.g., Bailer-Jones 2003)—and exemplification (Frigg and Nguyen 2020).

12. This is the case even if abstract minimalism allows a definition of the abstract notion of cognitive representation in terms of representational force and inferential capacities, by insisting that they be considered in the abstract as jointly sufficient as well as necessary. Recall that, even then, the definition does not serve by itself to pick out the extension of the concept. Additional conditions apply.

13. The situation depicted in fig. 3.3 can then be regarded as a nested model since the figure itself depicts a further depiction of yet something else that does not appear presently. As was noted in chap. 5, there are issues concerning transitivity in such cases, and fig. 3.3 is arguably a rare case that obeys transitivity for the licensed inferences. The cantilever principle can be inferred from the human model depicted, and this principle applies to the object represented in the diagram right above the three men. Alternatively, the anthropomorphic model and the diagram depicted can be taken to be independent representations of the same physical principle at work in the bridge. For further discussion of the model and its subsequent uses, see Petroski (2013).

14. The history is recounted in multiple sources, including Brush (2003), Everitt (1975), Garber, Brush, and Everitt (1986), Harman (1982, 1987, 1991), Klein (1970), and Porter (1981, 1986).

Maxwell himself tells some of the relevant history in his delightful entry on "Molecules" in the journal *Nature* (Maxwell 1873a).

15. Probability was in the air in the 1850s, and other influences on Maxwell probably included Mill's *System of Logic* (1843/2002), and Boole's *Investigation of the Laws of Thought* (1854/2003).

16. Rather than abandoning representation altogether as, e.g., Coopmans, Vertesi, Lynch, and Woolgar (2014) do in the latest rendition of the sociological approach to the topic. The modeling attitude is hence also an invitation to sociologists and historians to reengage with representation.

17. The part of the modeling activity unearthed by the mediating models movement (Morgan and Morrison 1999).

Chapter Eight

1. Some of these cases have been put to similar use in previous works of mine, starting with Suárez (1999b) and including Suárez (2003, 2004, 2010b). Van Fraassen (1994) was perhaps a pioneer in bringing examples from art to bear on issues of scientific modeling, and Hughes (1997) touches lightly on some cases too, following Goodman (1968/1976). But the finest treatment of the subtle interrelations of judgments within the arts and sciences probably remains Catherine Elgin's (1996, 2009, 2017, 2018).

2. "Le chef d'oeuvre entre touts les portraits" (Taine 1866/1990, 225). Sir Joshua Reynolds also supposedly referred to the painting as "the finest . . . in Rome" (see https://www.doriapamphilj.it/en/portfolio/velazquez), but the quote may be apocryphal.

3. For some discussion of the circumstances of the painting's making, its reception, and its enduring influence through the years, see Brown (1986), Carr (2006), Harris (1982, 1999), and Justi (1889).

4. Including in Ernst Gombrich's influential (also among philosophers of art) *The Story of Art* (1950).

5. For discussions of Holbein and his painting in the context of his time, see Foister, Roy, and Wyld (1977). Mary Hervey (1900) was largely responsible for unearthing the story behind the painting and its central characters, who for a long time were assumed to have been English courtesans at the court of Henry VIII. They turned out to be two Frenchmen. The one standing on the left in regal costume is Jean de Dinteville, the French ambassador to the court of Henry VIII during 1533, the year the painting was made, and the commissioner—and thus also first owner—of the painting. The portrait stayed in the Dinteville family house in Polisy for over a century before being acquired by the second Earl of Radnor, who took it to England in 1808. The character pictured on the right, in sober yet magnificent clothing, is Georges de Selve, the bishop of Lavour at the time, and a temporary visitor to Dinteville for a short period during that time.

6. This detail has in fact been revealed reasonably recently—in the last century—since that part of the painting had been covered in successive attempts at conservation through the centuries. Particularly, the Polish Baltic oak planks that support the imprimatur were damaged through torsion, and they had been filled in a variety of ways (Foister, Roy, and Wyld 1977, 88ff.). The first major restoration took place soon after *The Ambassadors* was bought by the National Gallery (in 1890), but it did incur distortion of its own, particularly as it affects the anamorphic skull, which was seriously misreconstructed. A thoroughly modern restoration took place in 1997 before a major exhibition and produced a full image of the anamorphic skull in the correct

trapezoidal projection that must have been used by Holbein (Foister, Roy, and Wyld 1977, 44–58).

7. The term *content* has been studiously avoided so far in this book, partly on account of how difficult it is to make any sense of the concept, and partly because I do not find it felicitous for the study of scientific representation. Yet it has currency in the context of representational theories of depiction, where it is often assumed that any depiction possesses some content, which is sometimes linked to a hypothetically experienced visual field. This in turn relates to an object or a set of objects in nature in roughly the way a two-dimensional projection of an object relates to its three-dimensional target.

8. This is an example that I have appealed to and used in similar ways in the past. See Suárez (1999b, 2003, 2004). Others have also discussed it in the same context (e.g., Ambrosio 2008, 2010; and Sánchez-Dorado 2018).

9. The bibliography on *Guernica* is very extensive (it is possibly the most documented painting in the history of art), and it cannot be exhaustively listed here. My account is particularly indebted to some of the classics, including Arnheim (1962), Blunt (1969), Calvo Serraller (1981), and Chipp (1989), as well as several press cuttings and reports in Spanish from the time when the painting was moved back to Spain in September 1981, most prominently Tusell (1981). Many of these accounts were themselves taking partisan sides in some of the ideological, political, and diplomatic disputes that surrounded *Guernica*. Thus, e.g., Blunt (1969) was written in the midst of the Cold War (and the author was notoriously retrospectively recognized to have been deeply steeped in it), Calvo Serraller (1981) was devoted to the cause of bringing the painting back to Spain after democracy had been reestablished there (in accordance with Picasso's own expressed wishes), and Chipp (1989) was partly devoted to a eulogy of the political significance of the painting after its long exile had ended and its safe return completed. Still, with all due caveats, they are all compelling and fact-based accounts that provide much insight into the meaning and representational force of the painting.

10. Oppler (1988) is an excellent collection of essays making just this point. (I thank a referee for the reference.)

11. Most prominently in Mondrian (1920/2017, 1926/2017, and 1937/1971).

12. It must be noted that not all abstract art is in any way like this. I already mentioned *Guernica*, which is not only blatantly and clearly a piece of representational art but also richly so, in the sense of having diverse and multiple simultaneous targets at different levels of abstraction. And, indeed, the targets of most cubist and surrealist art are similarly elusive, complex, and diverse. Yet there is always by and large an attempt to project beyond the canvas. Even very austere abstract art such as American abstract expressionism can be representational. Rothko famously did say that his art (e.g., the renowned Seagram murals at the Tate Gallery) does represent, although what it represents is intensely subjective feeling, including Rothko's own angst at the Holocaust (Rothko was of Latvian Jewish descent). Mondrian really stands very much on his own in his open and explicit avowal of the nonrepresentational nature of his art, although many have undoubtedly followed suit. For some influential relevant essays, see Martin, Nicholson, and Gabo (1937/1971).

13. Mondrian's view (1937/1971, 55) is, roughly, that, while figurative art is always subjectively responsive to some vision or other, nonfigurative art is by contrast capable of abstracting away from any subjectivity or at least that it certainly aims toward pure objective form. In a forthcoming paper, tentatively entitled "Maxwell/Mondrian: Abstraction as a Process," I explore different ways in which Mondrian's art can be taken to have representational force toward pure abstract form.

14. My critique dovetails with Dominic Lopes's (1996), and the rest of this discussion is particularly indebted to him. See also Lopes (2005, 25–28). Thus, the same spirit that leads me to distinguish theories of the constituent of representation from theories merely of its typical means also leads Lopes to distinguish resemblance theories that aim to explain or explicate depiction from those that merely assert that resemblance is a common or even universal relation between depicting and depicted things. As he writes: "Nobody denies that visual resemblance has something to do with pictures, nor does anybody advocate severing the intuitive link between pictures and resemblance. Nevertheless, it is by no means self-evident that resemblance explains depiction. According to the resemblance theory, identifying what a picture represents is a matter of recognizing a similarity between its design and its subject. But to deny that we understand pictures by noticing similarities is not to deny the intuitive significance of resemblance for depiction" (Lopes 1996, 16).

15. In Lopes's terms, these theories violate what he calls the *independence challenge*, namely, the requirement to specify "those representation-independent resemblances in virtue of which we identify pictures' subjects" (Lopes 1996, 17).

16. It is clear that the analogy between artistic and scientific representation loses some of its grip here since the visual experience of a scientific model typically adds nothing to it or to our use of it—and similarly for the visual experience of the target system of a scientific model, which is typically immaterial to our use and in fact might even be impossible (e.g., in those cases, far from uncommon in many scientific areas, of target systems that are by their nature invisible to the naked eye, something particularly relevant in astrophysics, atomic physics, genetics, economics and so on). Nevertheless, as I shall suggest in due course, there are relevant analogues to be found in our experience of the uses of models that can be exploited in a more general experiential account of representation.

17. Peacocke (1987, 385) takes them to be aspects of the *sensational* contents of our experience, which he distinguishes from what he calls the *representational* contents. The distinction is tangential to our present purposes and adds a complication that I would like to avoid as it emphasizes a contrast between depiction and representation more generally. My claim is precisely that whatever stands at the nexus of all representation is as thin and deflationary as the inferential conception takes it to be but that—in line with the abstract minimalism explored in pt. 2—there can be additional norms that govern the application of representation in particular domains, such as, e.g., in the case of depiction. If so, as it pertains to relations between the sensational contents of our experience, Peacocke's proposal would also implicitly be advancing a norm for representation within the confines of specifically depictive representations.

18. For further critical discussion of this point in relation to Peacocke's proposal, see Lopes (1996, 24ff.).

19. For a similar proposal making explicit use of the notion of isomorphism between the expected visual field when observing a skull directly and the experienced visual field when looking at the relevant part of *The Ambassadors* from an appropriately oblique angle, see Budd (1993). Hyman (2006, 93) also uses the painting to justify a modification of his "occlusion shape" principle.

20. The application of Alberti's rule to *The Ambassadors* is thus no threat to its representational character, but it does question its integrity as a "whole" painting.

21. There are currently many extensive arguments against both objective and experienced or subjective resemblance theories. See Hyman (2006, chaps. 4–5), Kulvicki (2006, chap. 4), Kulvicki (2014, chap. 3), Lopes (1996, chap. 1), Lopes (2005, 25–45), and Newall (2011, chap. 4).

22. In Lopes's (1996, 16) terms, many, if not all, depictive resemblances are *representation dependent*. This point is widely recognized, going back arguably at least as far as Descartes. Most eloquently, perhaps, Goodman (1968/1976, 39) wrote: "Resemblance and deceptiveness, far from being constant and independent sources and criteria of representational practice are in some degree products of it." We also saw in chap. 3 how Max Black (1962, 1979), following I. A. Richards (1936), emphasized a similar claim regarding scientific metaphor.

23. This nominalism notably gives rise to several objections to Goodman's theory as applied specifically to depiction. See, e.g., Hyman and Bantinaki (2017, 13ff.) and Lopes (1996, chap. 3). While the holders of a name may share nothing other than the name, it is unclear, to say the least, how the objects or scenes depicted by a particular painting might share nothing other than being so denoted. While I accept the objections, my concern is rather with another set of more general worries regarding denotation that are revealed in cases of fictional representation of nonexistent entities.

24. Even Kulvicki (2006, 2014), who champions a version of Goodman's conventionalist theory, accepts the need to accommodate the phenomenology of twofoldness in order to address depiction at least. Most of these philosophers locate these differences in either the phenomenology of the experience of art (mainly but not only its perceptual experience) or our recognition capacities (which enable our perceptual experience).

25. See also Kulvicki (2014, 92ff.), which adds "incidental" factors.

26. For more on the fascinating issues raised by the application of Goodman's symbolic systems to scientific representations, see esp. Kulvicki (2014, chap. 7).

27. For attempts to overcome the objection, see Kulvicki (2014) and Lopes (1996).

28. As already mentioned, there have been attempts in the literature to get around this problem while retaining Goodman's central commitment to denotation. Elgin (2009) remains the outstanding reference. Recall from chap. 6 that the critical notion is denotative function, which can be successful even if denotation is empty.

29. The focus on perceptual experience goes back at least to Descartes, who in the *Optics* wrote: "The problem is to know simply how they [pictures or engravings] can enable the soul to have sensory perceptions of all the various qualities of the objects to which they correspond—not to know how they can resemble these objects" (Descartes 1630/1985, 165–66). In one fell swoop, Descartes both disposes of the resemblance theory and advances a perceptual/phenomenological theory instead.

30. The insight is critical to Gombrich (1959/2002), who made it mainstream again through his "illusion theory of depiction." In different guises, the same thought—that representation essentially calls for a phenomenology or study of perception—is also key to Hopkins (1998), Lopes (1996, 2005), Nanay (2004, 2005), Newall (2011), Polanyi (1970), and Wollheim (1968/1980, 1987). And it plays a prominent, if not critical, role in Hyman (2006), Kulvicki (2006, 2014), Voltolini (2015), and Wollheim and Hopkins (2003, 149–67 [Hopkins's contribution]).

31. Here, I mainly follow Hyman (2006, chap. 7) and Kulvicki (2014, chap. 1).

32. At least as understood by Gombrich's critics. See, e.g., Wollheim's (1963) review of *Art and Illusion*.

33. In other words, seeing-in does not just narrowly characterize depiction but is a more general phenomenon that appears to play some role in much artistic representation. Yet all depiction appears to involve seeing-in. While Wollheim initially harbored some doubts about the types of work known as *trompes l'oeil*, whose visual perception seems at first not to involve any experience of the surface, it becomes clear that there are ways of experiencing the surfaces of

trompes l'oeil too, just by changing relative viewing positions or the circumstances surrounding them. See Hyman (2006, 131ff.).

34. Hopkins (2010, 167ff.), another champion of a phenomenological approach, usefully distinguishes between divisive and unitary accounts of pictorial experience.

35. An excellent example is provided by the birth of quantum mechanics, which imported a large number of important notational innovations, from Max Born's famous probability rule (inviting us to identify the square modulus of the amplitude of an elementary particle's wave function with the probability that it is located some place or other), to Paul Dirac's notational innovations, including the delta function and the "bracket" notation for states and self-adjoint operators in order to calculate the ensuing values of the particle's dynamic quantities.

36. In Suárez (1999b), I made the further claim that nonrepresentational models so devoid of representational force become abstract formal theories, and I used the well-known slogan "Maxwell's theory is Maxwell's equations" as an instance. I now think this was a mistake (for the recognition of which I owe thanks to conversations with the late Margaret Morrison). For Maxwell's equations to be appropriately interpreted as such, their terms (***E***, ***B***, ***H***, ***i***, etc.) must receive a physical interpretation (as electric field, magnetic field and flux, displacement current, etc.). And this is already enough to provide the set of equations with representational force, just as, when appropriately interpreted, the stellar structure equations provide a model of stellar interiors or the Lotka-Volterra model is already a representation of an ecological population, however abstract the target might be conceived to be. For the Maxwell equations to be mere flat surface, we would have to interpret the symbols not physically but in purely formal terms, to the point that the equation would become an empty piece of formalism. Now I instead conceive of a general theory such as Maxwell's as an abstract representation of great generality, in other words, as a very abstract and general model.

37. This is Knuuttila's way of providing a detailed narrative for the models as mediators' movement articulated in Morgan and Morrison (1999).

38. There is correspondingly a sense in which deflationary approaches to representation are in the business of not explaining but rather merely "showing how it works," to put it in Wittgensteinian language. Thus, Lopes (1996, 49–50) complains that Wollheim's seeing-in account cannot really explain depiction, and Kulvicki (2014, 24) demands "to hear more." But, on my understanding of the nature of representation, that is exactly as it ought to be. There really is nothing more to say about the essence of representation other than representational force and inferential capacities, which in the case of pictorial art comes out as the phenomenology of seeing-in. What is left for us to carry out is a detailed philosophical study of the practice. This kind of work, as regards scientific models, began properly only in the twenty-first century.

39. Historians of art have, however, considered such questions in depth, and the seminal works of Baxandall (1985) and Gombrich (1950) have played a critical role in the argument at several points in this book. One could even harbor a hope that the application of the inferential conception to art might bring a welcome rapprochement of historical and philosophical approaches to the study of art.

Chapter Nine

1. A movement perhaps best exemplified by the "Society for Philosophy of Science in Practice" and "Models and Simulations" conference series.

2. For some informed and interesting leads, see Shea (2018).

3. I describe the epistemological implications of Hacking's rightly influential thought in an unavoidably cursory manner here and very much in my own terms. While I do hope to do it justice, it is also necessary to frame it in terms amenable to the issues discussed in this book.

4. This reference to Dewey in relation to Hacking's work is in no way accidental or fortuitous. On the contrary, Hacking's own criterion of reality, in his experimental realism, echoes Dewey's (1920, 144–45) own, namely: "Conceptions, theories and systems of thought . . . are tools. As in the case of all tools, their value resides not in themselves but in the capacity to work, as shown in the consequences of their use." On an inferential conception, representations too are tools for inference.

5. An autonomous life perhaps best chronicled in the mediating models literature, including Suárez (1999a) and Morgan and Morrison (1999) generally.

6. A more apt name for Hacking's book would have been "Theorizing and Experimenting," and it is best in my view to see the book as a close cousin of Peter Galison's (1997) *Image and Logic*, in its probing of the epistemological differences between those activities. (The lineage between both authors within the so-called Stanford school is, of course, well-known.)

7. And indeed van Fraassen (2008) is a first attempt to move in a different direction by appealing to a deflationary conception of representation not unlike that defended in this book.

8. I do not belittle understanding. On the contrary, it is the only sort of enlightenment that most sciences aim for. The ways, forms, and methods of refinement of the understanding provided by our models are critical to much of the explanatory work that most sciences achieve.

9. Cartwright first proposed a causal argument for experimental realism in her celebrated *How the Laws of Physics Lie* (1983); although the critical chapter was prepublished as Cartwright (1979). She moved on to what I would consider a metaphysical version of experimental realism in the subsequent *Nature's Capacities and Their Measurement* (1989) and *The Dappled World* (1999a). I take up some of these later works critically, while endorsing the original position, in (Suárez 2008).

10. For an informed critique, see Glymour (2010).

11. Although the claims have been contested on roughly the grounds discussed in the text.

12. I am not suggesting that these definitions exhaust Hacking's and Cartwright's statements on the topic, but they do provide good starting renditions of their core commitments. Please note that the definitions provided improve on those in my previous work on the topic (Suárez 2008) and do not always coincide.

13. For a sample, see Hitchcock (1992), Massimi (2004), Morrison (1990), Reiner and Pierson (1995), and Resnik (1994).

14. The disjunction is inclusive, so the experimental realist can fail both ways. And Resnik (1994, 410) seems to think so when he notes: "Experimenters do not operate without genuine scientific theories and laws about the phenomena they investigate: the gulf between experiment and theory is not nearly as large as Hacking supposes."

15. In line with the theory of causal inference and explanation espoused in, e.g., Woodward (2004).

16. I am not dogmatic about this. It may turn out in the end that most or perhaps all bona fide experimental warrant also grounds some causal inference or other, according to some account of causation, including Woodward's (2004) manipulability account. But it strikes me that the reduction of all experimental warrant to causal warrant is still to be shown.

17. Musgrave (1976) is another good introduction to the history of the overturning of phlogiston by Lavoisier's oxygen theory. Warwick (1995) is a wonderful account of Joseph Trouton's failed efforts to build a perpetuum mobile machine out of the earth's interaction with the ether.

18. Or of any other sensorial mode, whether any of the established senses or more complex cross-modal forms (Spence and Driver 2004). There is debate in the literature about whether *observable* should be taken to be a term covering a wider spectrum of senses than just narrowly vision, but it is not relevant to my purposes here.

19. Constructive empiricism has been evolving. The earliest formulation (van Fraassen 1980) is perhaps less explicit as regards the voluntarism; later forms in response to critics embrace it more explicitly. See particularly van Fraassen's responses to his critics in Churchland and Hooker (1985) and Monton (2007) as well as the most sophisticated rendition of an empiricist stance in van Fraassen (2002).

20. By and large because, admittedly, were such commitments entirely missing, the theory could hardly be said to be accepted. But the critical inference here goes the other way round, from acceptance to pragmatic commitment.

21. Van Fraassen (1980, 44ff.) pursues the example of Newton's theory (both mechanics and gravitation), but there are other empirically adequate, even explanatory, yet false theories, including the caloric, phlogiston, and ether theories in physics (Laudan 1981).

22. These are informal definitions. For more explicitly formal accounts, see Chang and Keisler (1990) and Lutz (2014).

23. This raises the thorny issue of how the appearances can be said to have a structure in the first place and what it could mean to say that the theory saves the appearances when what it saves as a matter of fact is instead yet another representation of the appearances. In later work, van Fraassen (2008) lays down a pragmatic equivalence between these two claims. For further discussion, see Nguyen (2016b) and Suárez (2010b).

24. This definition is in line with Percival's (2007) notion of a fully empirically informative theory. For a similar proposal, see Lutz (2014).

25. Arthur Fine (2001) associates the view with John Dewey, who first coined the term *instrumentalism* (see Dewey 1943, 463) and already anticipated this use. Fine goes on to suggest—as I do earlier in the text—that Deweyan instrumentalism provides the most promising model for constructive empiricism.

26. NOA is advertised in these papers as jointly developed with Micky Forbes. In the new afterword to the second edition of *The Shaky Game*, Fine reassesses NOA, considering different critical arguments and objections. See Fine (1987/1996).

27. *Instrumental reliability* is a term employed in Fine (2001, 2018), Suárez (1999a), and Winsberg (2010).

28. This is perhaps most evident in Fine's (1993/2009) writings on fictionalism, which fully embrace the modeling attitude.

29. This has been followed by subsequent work on similar themes, esp. Kitcher (2009).

30. The literature on the role of values in science is now large, including Douglas (2009), Kincaid, Dupré, and Wylie (2007), and Wilholt (2009) and, more recently, Irzik and Kurtulmus (2019) and Matzovinos (2020).

31. See esp. Kitcher (2001a, chap. 10), which lays down the desiderata. They are all apposite proposals *for an ideal world*, and ours is far from ideal of course. Kitcher is quite clear on the difficult issues of application of such an abstract account, although, in addition, there are elements to the account that I find a little Anglocentric, particularly as to how democratic institutions tend to work procedurally. Still, as an ideal, being well ordered remains in my view a sound proposal for almost any science.

32. Kitcher (2001b) develops real realism in the first instance, but the first few chapters of Kitcher (2001a) are also perfectly in line and fully devoted to his epistemological views. Suárez (2012b) criticizes real realism, and Kitcher (2012b) responds to the critique. This section summarizes the exchange.

33. This is a reconstruction of the argument offered in Suárez (2012b, 285ff.).

34. Even van Fraassen could accept it were the Galilean strategy suitably modified to allow only inferences from the domain of the observed to the domain of the observable.

35. This is a book that I came to read only in the last stages of writing my own. I thank Natalia Carrillo and Tarja Knuuttila for calling it to my attention during a meeting of a reading group in Vienna on a draft of the present book.

36. Clearly, Longino does not intend *conformation* as a rendition of the "relation" between a *Bild* and its target that informs the inferential conception. I am also obviously not suggesting that her views are similarly rooted in nineteenth-century science, nor is she committed to the natural modeling attitude that is cognate to the inferential conception. Still, the similarities are remarkable.

37. Longino's pluralism at this point echoes Sandra Mitchell's (2003, 2009) and, more recently, Michela Massimi's (2022). Massimi's book adopts an inferential outlook yet came out too late for me to do it justice here, but at any rate her account of modeling is not deflationary. Mitchell (2020), however, does develop a view of modeling quite close to a deflationary account. Her "pragmatic pluralism" is a good rendition of the inferential conception's pluralism regarding means, although she undoubtedly goes much further in proclaiming that most, if not all, scientific representations are patchy representations of only certain aspects of reality. This is an ontologically more committed pluralism than anything that follows from a deflationary account of representation. It is in fact quite close to the "dappled world" metaphysics that was discarded earlier in this chapter as unnecessary for a deflationary account of representation.

38. The institutional dimension of science is curiously missing in a lot of the recent philosophical literature on values in science, as I already pointed out in discussing Kitcher's work. I put this down to an Anglocentric bias that takes for granted the sort of liberal institutions that operate in the Anglosphere. It is striking that scholars writing on the institutional constraints on science (e.g., Irzik and Kurtulmus 2019; Matzovinos 2020; and Wilholt 2009) are often located outside that sphere.

39. Together with Lipton (2009), de Regt was instrumental in drawing philosophers' attention to the inherent contextuality of understanding.

40. De Regt (2018, chap. 6), e.g., reviews mechanical modeling in nineteenth-century physics as well as Boltzmann's *Bildtheorie*.

41. Hence, in tacking understanding to intelligible explanatory theories in this way, de Regt brings his position closer to the received view defended by Khalifa (2017) and others. My point above is that this creates tensions with other, more pragmatic features of his account.

42. Hunt (1991) is a particularly perspicuous exposition of this sort of mode of explanation as well as of the historical development of this movement. See further references in chap. 2 above.

43. Potochnik (2017, esp. chap. 4) is particularly good on this, and I agree with her that many idealizations are conducive for understanding precisely in as much as they are far from the truth. It is of a piece with the earlier discussion in this chapter—and it became salient in discussing the natural modeling attitude—that science serves many different aims, including several epistemic aims, and that truth is only one among many often-conflicting aims.

44. My notions of minimal understanding and theoretical explanation overlap somewhat but are ultimately distinct from Khalifa's. Thus, Khalifa (2017, 14) defines *minimal understanding* in personal terms as "S has minimal understanding of why p if and only if, for some q, S believes that *q explains why p*, and *q explains why p* is approximately true." This evidently inverts the order of my definitions in the main text, where it is theoretical explanation that piggybacks on the more minimal understanding by models, not the other way round.

45. De Regt presents Maxwell's model of the ether as a model of classical mechanics, something I would not do since the properties ascribed to the ether are contradictory in mechanical terms. But undoubtedly—and on this we would agree—it was exploring such contradictions that gave impetus to Maxwell's model.

References

Abell, Catharine, and Katerina Bantinaki, eds. 2010. *Philosophical Perspectives on Depiction*. Oxford: Oxford University Press.

Alberti, Leon Battista. 1450/1966. *De pictura / On Painting*. Translated by John R. Spencer. New Haven, CT: Yale University Press.

Alexandrova, Anna. 2008. "Making Models Count." *Philosophy of Science* 75:383–404.

———. 2010. "Adequacy-for-Purpose: The Best Deal a Model Can Get." *Modern Schoolman: A Quarterly Journal of Philosophy* 87, nos. 3–4:295–301.

Ambrosio, Chiara. 2008. "Iconicity and Network Thinking in Picasso's *Guernica*: A Study of Creativity across the Boundaries." PhD diss., University College London.

———. 2010. "Iconicity as Homomorphism: The Case of Picasso's *Guernica*." In *Ideas in Action: Proceedings of the Applying Peirce Conference* (Nordic Studies in Pragmatism 1), ed. M. Bergman, S. Paavola, A.-V. Pietarinen, and H. Rydenfelt, 151–62. Helsinki: Nordic Pragmatism Network.

———. 2014. "Iconic Representations and Representative Practices." *International Studies in the Philosophy of Science* 28, no. 3:255–75.

Ankeny, Rachel. 2001. "Model Organisms as Models: Understanding the Lingua Franca of the Human Genome Project." *Philosophy of Science* 68:251–61.

———. 2009. "Model Organisms as Fictions." In *Fictions in Science: Philosophical Essays on Modeling and Idealization*, ed. Mauricio Suárez, 193–204. New York: Routledge.

Ankeny, Rachel, and Sabina Leonelli. 2021. *Model Organisms*. Cambridge Elements in the Philosophy of Biology. Cambridge: Cambridge University Press.

Armstrong, David. 1983. *What Is a Law of Nature?* Cambridge: Cambridge University Press.

Arnheim, Rudolf. 1962. *Picasso's Guernica: The Genesis of a Painting*. Berkeley: University of California Press.

Aronson, Jerry, Rom Harré, and Eileen Way. 1993. *Realism Rescued*. London: Duckworth.

Ayer, Alfred J. 1936. *Language, Truth and Logic*. London: Gollanz.

Bailer-Jones, Daniela. 2003. "When Scientific Models Represent." *International Studies in the Philosophy of Science* 17, no. 1:59–74.

———. 2009. *Scientific Models in Philosophy of Science*. Pittsburgh, PA: University of Pittsburgh Press.

Baird, Davis, R. I. G. Hughes, and Alfred Nordmann, eds. 1998. *Heinrich Hertz: Classical Physicist, Modern Philosopher*. Dordrecht: Kluwer Academic.

Balázs, Gyenis. 2017. "Maxwell and the Normal Distribution: A Coloured Story of Probability, Independence and Tendency towards Equilibrium." *Studies in History and Philosophy of Modern Physics* 57:53–65.

Balzer, Wolfgang, Ulises Moulines, and Joseph D. Sneed. 1987. *An Architectonic for Science: The Structuralist Program*. Dordrecht: Reidel.

Barr, Nicholas. 2000. "The History of the Phillips Machine." In *A. W. Phillips: Collected Works in Contemporary Perspective*, ed. Robert Leeson, 89–114. Cambridge: Cambridge University Press.

Bartels, Andreas. 2006. "Defending the Structural Concept of Representation." *Theoria* 21, no. 1:7–21.

Bartha, Paul. 2010. *By Parallel Reasoning: The Construction and Evaluation of Analogical Arguments*. Oxford: Oxford University Press.

———. 2019. "Analogy and Analogical Reasoning." In *Stanford Encyclopedia of Philosophy*, ed. Edward N. Zalta and Uri Nodelman. https://plato.stanford.edu/entries/reasoning-analogy.

Baxandall, Michael. 1985. *Patterns of Intention: On the Historical Explanation of Pictures*. New Haven, CT: Yale University Press.

Bird, Alexander. 2007. "Underdetermination and Evidence." In *Images of Empiricism: Essays on Science and Stances with a Reply from Bas C. van Fraassen*, ed. Bradley Monton, 62–82. Oxford: Oxford University Press.

Black, Max. 1962. *Models and Metaphors: Studies in Language and Philosophy*. Ithaca, NY: Cornell University Press.

———. 1979. "How Metaphors Work: A Reply to Donald Davidson." *Critical Inquiry* 6, no. 1:131–43.

Blackburn, Simon, and Keith Simmons, eds. 1999. *Truth*. Oxford University Press.

Blackmore, John, ed. 1995. *Ludwig Boltzmann: His Later Life and Philosophy, 1900–1906*. Dordrecht: Kluwer Academic.

Blunt, Anthony. 1969. *Picasso's Guernica*. Oxford: Oxford University Press.

Boesch, Brandon. 2017. "There Is a Special Problem of Scientific Representation." *Philosophy of Science* 84, no. 5:970–81.

———. 2019a. "The Means-End Account of Scientific, Representational Actions." *Synthese* 196:2305–22.

———. 2019b. "Scientific Representation and Dissimilarity." *Synthese* 198:5495–5513.

Bokulich, Alisa. 2009. "Explanatory Fictions." In *Fictions in Science: Philosophical Essays on Modeling and Idealization*, ed. Mauricio Suárez, 91–109. New York: Routledge.

———. 2015. "Maxwell, Helmholtz, and the Unreasonable Effectiveness of the Method of Physical Analogy." *Studies in History and Philosophy of Science* 50:28–37.

Bokulich, Alisa, and Wendy Parker. 2021. "Data Models, Representation and Adequacy-for-Purpose." *European Journal for Philosophy of Science* 11, no. 1:1–26.

Bolinska, Agnes. 2013. "Epistemic Representation, Informativeness and the Aim of Faithful Representation." *Synthese* 190, no. 2:219–34.

Boltzmann, Ludwig. 1877. "Über die Nature der Gasmoleküle." *Annalen der Physik* 236, no. 1:175–76.

———. 1902/1974. "Models." In *Encyclopaedia Britannica* (10th ed.), 788–91. Edinburgh: Adam & Charles Black. Reprinted in *Theoretical Physics and Philosophical Problems: Selected Writings*, ed. Brian McGuiness, 213–20. Dordrecht: Reidel.

———. 1905/1974. *Populäre Schriften*. Leipzig: Johan Ambrosius Barth. Reprinted in *Theoretical Physics and Philosophical Problems: Selected Writings*, ed. Brian McGuinness, 1–199. Dordrecht: Reidel.

Boole, George. 1854/2003. *An Investigation of the Laws of Thought*. With an introduction by John Corcoran. Buffalo: Prometheus.

Brandom, Robert. 1994. *Making It Explicit: Reasoning, Representing and Discursive Commitment*. Cambridge, MA: Harvard University Press.

———. 2000. *Articulating Reasons: An Introduction to Inferentialism*. Cambridge, MA: Harvard University Press.

Bridgman, Percy William. 1936. *The Nature of Physical Theory*. Princeton, NJ: Princeton University Press.

Brown, Jonathan. 1986. *Velázquez: Painter and Courtier*. New Haven, CT: Yale University Press.

Brush, Stephen G., ed. 2003. *The Kinetic Theory of Gases: An Anthology of Classical Papers with Commentary*. London: Imperial College Press.

Buchdahl, Gerd. 1964. "Theory Construction: The Work of Norman Robert Campbell." *Isis* 55, no. 2:151–62.

Buchwald, Jed Z. 1985. *From Maxwell to Microphysics: Aspects of Electromagnetic Theory in the Last Quarter of the Nineteenth Century*. Chicago: University of Chicago Press.

———. 1993. "Electrodynamics in Context: Object States, Practice and Anti-Romanticism." In *Hermann von Helmholtz and the Foundations of Nineteenth Century Science*, ed. David Cahan, 334–73. Berkeley: University of California Press.

———. 1994. *The Creation of Scientific Effects: Heinrich Hertz and Electric Waves*. Chicago: University of Chicago Press.

Budd, Malcolm. 1993. "How Pictures Look." In *Virtue and Taste: Essays on Politics, Ethics and Aesthetics*, ed. Dudley Knowles and John Skorupski, 154–75. Oxford: Wiley-Blackwell.

Bueno, Otávio. 1997. "Empirical Adequacy: A Partial Structures Approach." *Studies in History and Philosophy of Science* 28, no. 4:585–610.

Bueno, Otávio, and Mark Colyvan. 2011. "An Inferential Conception of the Application of Mathematics." *Noûs* 45, no. 2:345–74.

Bueno, Otávio, and Steven French. 2011. "How Theories Represent." *British Journal for the Philosophy of Science* 62, no. 4:857–94.

Bueno, Otávio, Steven French, and James Ladyman. 2002. "On Representing the Relationship between the Mathematical and the Empirical." *Philosophy of Science* 69:497–518.

Burke, Tom. 1994. *Dewey's New Logic*. Chicago: University of Chicago Press.

Cahan, David, ed. 1993. *Hermann von Helmholtz and the Foundations of Nineteenth Century Science*. Berkeley: University of California Press.

Callender, Craig, and Jonathan Cohen. 2006. "There Is No Special Problem about Scientific Representation." *Theoria* 21, no. 1:67–85.

Calvo Serraller, Francisco. 1981. *El Guernica de Picasso*. Madrid: Alianza Editorial.

Campbell, Norman. 1920/1957. *Physics: The Elements*. Cambridge: Cambridge University Press. Reprinted as *Foundations of Science: The Philosophy of Theory and Experiment*. New York: Dover.

Carnap, Rudolf. 1934/1937. *Logische Syntax der Sprache / The Logical Syntax of Language*. Translated by Amethe Smeaton. London: Routledge.

———. 1939. *Foundations of Logic and Mathematics*. Vol. 1, no. 3, of *International Encyclopedia of Unified Science*. Chicago: University of Chicago Press.

———. 1956. "The Methodological Character of Theoretical Concepts." In *The Foundations of Science and the Concepts of Psychology and Psychoanalysis* (Minnesota Studies in the Philosophy of Science, vol. 1), ed. Herbert Feigl and Michael Scriven, 38–76. Minneapolis: University of Minnesota Press.

———. 1966. *Philosophical Foundations of Physics*. New York: Basic.

Carr, Dawson W. 2006. *Velázquez*. London: National Gallery.

Cartwright, Nancy. 1979. "Causal Laws and Effective Strategies." *Noûs* 13, no. 4:419–37.

———. 1983. *How the Laws of Physics Lie*. Oxford: Oxford University Press.

———. 1989. *Nature's Capacities and Their Measurement*. Oxford: Oxford University Press.

———. 1999a. *The Dappled World: A Study of the Boundaries of Science*. Cambridge: Cambridge University Press.

———. 1999b. "Models and the Limits of Theory: Quantum Hamiltonians and the BCS Model of Superconductivity." In *Models as Mediators: Perspectives on Natural and Social Science*, ed. Mary Morgan and Margaret Morrison, 241–81. Cambridge: Cambridge University Press.

———. 2007. *Hunting Causes and Using Them: Approaches to Philosophy of Economics*. Cambridge: Cambridge University Press.

Cassini, Alejandro, and Juan Redmond, eds. 2021. *Models and Idealisations in Science: Artifactual and Fictional Approaches*. Cham: Springer.

Cat, Jordi. 2001. "On Understanding: Maxwell and the Methods of Illustration and Scientific Metaphor." *Studies in History and Philosophy of Modern Physics* 32, no. 3:395–441.

Chakravartty, Anjan. 2004. "Stance Relativism: Empiricism versus Metaphysics." *Studies in History and Philosophy of Science* 35, no. 1:173–84.

Chang, Chen C., and Jerome H. Keisler. 1990. *Model Theory*. Amsterdam: North-Holland.

Chang, Hasok. 2010. "The Hidden History of Phlogiston: How Philosophical Failure Can Generate Historiographical Refinement." *HYLE: International Journal for Philosophy of Chemistry* 16, no. 2:47–79.

Chipp, Herschel B. 1989. *Picasso's Guernica: History, Transformations, Meanings*. Berkeley: University of California Press.

Churchland, Paul M. 1985. "The Ontological Status of Observables: In Praise of the Superempirical Virtues." In *Images of Science: Essays on Realism and Empiricism with a Reply from Bas C. van Fraassen*, ed. Paul M. Churchland and Clifford A. Hooker, 35–47. Chicago: University of Chicago Press.

Churchland, Paul M., and Clifford A. Hooker, eds. 1985. *Images of Science: Essays on Realism and Empiricism, with a Reply from Bas C. van Fraassen*. Chicago: University of Chicago Press.

Collins, George W. 1989. *The Fundamentals of Stellar Astrophysics*. New York: W. H. Freeman.

Contessa, Gabriele. 2007. "Scientific Representation, Interpretation and Surrogative Reasoning." *Philosophy of Science* 74, no. 1:48–68.

Coopmans, Catelijne, Janet Vertesi, Michael Lynch, and Steve Woolgar, eds. 2014. *Representation in Scientific Practice Revisited*. Cambridge, MA: MIT Press.

Cristalli, Claudia, and Julia Sánchez-Dorado. 2021. "Colligation in Modelling Practices: From Whewell's Tides to the San Francisco Bay Model." *Studies in History and Philosophy of Science* 85:1–15.

Cummins, Robert. 1989. *Meaning and Mental Representation*. Cambridge, MA: MIT Press.

———. 1998. *Representations, Targets, and Attitudes*. Cambridge, MA: MIT Press.

da Costa, Newton, and Steven French. 2003. *Science and Partial Truth*. Oxford: Oxford University Press.

D'Agostino, Salvo. 1990. "Boltzmann and Hertz on the Bild-Conception of Physical Theory." *History of Science* 28, no. 4:380–98.

Darrigol, Olivier. 2000. *Electrodynamics from Ampère to Einstein*. Oxford: Oxford University Press.

———. 2018. *Atoms, Mechanics, and Probability: Ludwig Boltzmann Statistico-Mechanical Writings—an Exegesis*. Oxford: Oxford University Press.

Darwin, Charles. 1859. *On the Origin of Species*. London: John Murray.

Davidson, Donald. 1978. "What Metaphors Mean." *Critical Inquiry* 5, no. 1:31–47.

Davie, George E. 1961. *The Democratic Intellect: Scotland and Her Universities in the Nineteenth Century*. Edinburgh: Edinburgh University Press.

da Vinci, Leonardo. 1989. *On Painting*. Translated by Martin Kemp and Margaret Walker. New Haven, CT: Yale University Press.

de Chadarevian, Soraya, and Nick Hopwood, eds. 2004. *Models: The Third Dimension of Science*. Stanford, CA: Stanford University Press.

De Donato, Xavier, and Jesús Zamora-Bonilla. 2009. "Credibility, Idealization, and Model-Building: An Inferential Approach." *Erkenntnis* 70, no. 1:101–18.

de Regt, Henk. 1999. "Ludwig Boltzmann's 'Bildtheorie' and Scientific Understanding." *Synthese* 119:113–34.

———. 2005. "Scientific Realism in Action: Molecular Models and Boltzmann's Bildtheorie." *Erkenntnis* 63:205–30.

———. 2018. *Understanding Scientific Understanding*. Oxford: Oxford University Press.

de Regt, Henk, and Dennis Dieks. 2005. "A Contextual Approach to Scientific Understanding." *Synthese* 144:137–70.

Descartes, René. 1630/1985. *Philosophical Writings*. Translated by John Cottingham. Cambridge: Cambridge University Press.

Dewey, John. 1920. *Reconstruction in Philosophy*. Vol. 12 of *The Middle Works of John Dewey, 1899–1924*. Edited by Jo Ann Boydston. Carbondale: Southern Illinois University Press.

———. 1943. "The Development of American Pragmatism." In *Twentieth Century Philosophy*, ed. Dagobert D. Runes, 449–68. New York: Philosophical Library.

Díez, José. 1997. "Hacia una teoría general de la representación científica." *Theoria* 13, no. 1:157–78.

Dirac, Paul Adrien Maurice. 1930. *The Principles of Quantum Mechanics*. Oxford: Clarendon.

Douglas, Heather. 2009. *Science, Policy and the Value-Free Ideal*. Pittsburgh, PA: University of Pittsburgh Press.

Downes, Steven M. 1992. "The Importance of Models in Theorising: A Deflationary Semantic Account." *Proceedings of the Biennial Meeting of the Philosophy of Science Association* 1:142–53.

———. 2009. "Models, Pictures and Unified Accounts of Representation: Lessons from Aesthetics for Philosophy of Science." *Perspectives on Science* 17, no. 4:417–28.

———. 2021. *Models and Modeling in the Sciences: A Philosophical Introduction*. London: Routledge.

Duhem, Pierre. 1905/1954. *The Aim and Structure of Physical Theory*. Princeton, NJ: Princeton University Press.

Eckert, M. 2006. *The Dawn of Fluid Dynamics: A Discipline between Science and Technology*. New York: Wiley.

Egan, Frances. 1998. Review of Robert Cummins, *Representations, Targets, and Attitudes* (1998). *Philosophical Review* 107:118–20.

Elgin, Catherine. 1983. *With Reference to Reference*. London: Hackett.

———. 1996. *Considered Judgment*. Princeton, NJ: Princeton University Press.

———. 1997. *Between the Absolute and the Arbitrary*. Ithaca, NY: Cornell University Press.

———. 2009. "Exemplification, Idealization, and Scientific Understanding." In *Fictions in Science: Philosophical Essays on Modeling and Idealization*, ed. Mauricio Suárez, 77–90. New York: Routledge.

———. 2017. *True Enough*. Cambridge, MA: MIT Press.

———. 2018. "Nature's Handmaid, Art." In *Thinking about Science, Reflecting on Art*, ed. Otávio Bueno, George Darby, Steven French, and Dean Rickles, 27–40. London: Routledge.

Everitt, C. W. Francis. 1975. *James Clerk Maxwell: Physicist and Natural Philosopher*. New York: Scribner's.

Fang, Wei. 2017. "Holistic Modelling: An Objection to Weisberg's Weighted Feature-Matching Account." *Synthese* 194, no. 5:1743–64.

———. 2019. "An Inferential Account of Model Explanation." *Philosophia* 47, no. 1:99–106.

Feyerabend, P. K. 1958. "An Attempt at a Realistic Interpretation of Experience." *Proceedings of the Aristotelian Society* 58:143–70.

Fine, Arthur. 1984a. "And Not Anti-Realism Either." *Noûs* 18:51–65.

———. 1984b. "The Natural Ontological Attitude." In *Scientific Realism*, ed. Jarrett Leplin, 83–107. Berkeley: University of California Press.

———. 1987/1996. *The Shaky Game: Einstein, Realism and the Quantum Theory*. 2nd ed., with a new afterword. Chicago: University of Chicago Press.

———. 1993/2009. "Fictionalism." *Midwest Studies of Philosophy* 18:1–18. Reprinted in *Fictions in Science: Philosophical Essays on Modeling and Idealization*, ed. Mauricio Suárez, 19–36. New York: Routledge.

———. 2001. "The Scientific Image Twenty Years Later." *Philosophical Studies* 106, nos. 1–2: 107–22.

———. 2018. "Motives for Research." *Spontaneous Generations: A Journal for the History and Philosophy of Science* 9, no. 1:42–45.

Floridi, Luciano. 2007. "In Defence of the Veridical Nature of Semantic Information." *European Journal of Analytic Philosophy* 3, no. 1:1–18.

———. 2011. *The Philosophy of Information*. Oxford: Oxford University Press.

Foister, Susan, Ashok Roy, and Martin Wyld. 1977. *Holbein's Ambassadors*. London: National Gallery.

Fourier, Jean-Baptiste Joseph. 1822. *Théorie analytique de la chaleur*. Paris: Didot.

Fox, Robert, and George Weisz, eds. 1980. *The Organization of Science and Technology in France, 1808–1914*. Cambridge: Cambridge University Press.

French, Steven. 2003. "A Model Theoretic Account of Representation (or, I Don't Know Much about Art, but I Know It Involves Isomorphism)." *Philosophy of Science* 70:1472–83.

French, Steven, and James Ladyman. 1999. "Reinflating the Semantic Approach." *International Studies in the Philosophy of Science* 13, no. 2:103–21.

Friedman, Michael. 1982. Review of Bas van Fraassen, *The Scientific Image* (1980). *Journal of Philosophy* 79:274–83.

———. 1999. *Reconsidering Logical Positivism*. Cambridge: Cambridge University Press.

Frigg, Roman. 2006. "Scientific Representation and the Semantic View of Theories." *Theoria* 55:49–65.

Frigg, Roman, and James Nguyen. 2016. "Scientific Representation." In *Stanford Encyclopedia of Philosophy*, ed. Edward N. Zalta and Uri Nodelman. https://plato.stanford.edu/entries/scientific-representation.

———. 2018. "The Turn of the Valve: Representing with Material Models." *European Journal for Philosophy of Science* 8:205–24.

———. 2020. *Modelling Nature: An Opinionated Introduction to Scientific Representation*. Synthese Library, vol. 427. Dordrecht: Springer.

Frisch, Mathias. 2005. *Inconsistency, Asymmetry, and Non-Locality: A Philosophical Investigation of Classical Electrodynamics*. Oxford: Oxford University Press.

———. 2014. "Models and Scientific Representations; or, Who Is Afraid of Inconsistency?" *Synthese* 191, no. 13:3027–40.

Galileo, Galilei. 1638/1974. *Discourses and Mathematical Demonstrations Relating to Two New Sciences*. Translated by Stillman Drake. Madison: University of Wisconsin Press.

Galison, Peter. 1997. *Image and Logic*. Chicago: University of Chicago Press.

Garber, Elizabeth, Stephen G. Brush, and C. W. Francis Everitt. 1986. *Maxwell on Molecules and Gases*. Cambridge, MA: MIT Press.

Gelfert, Axel. 2011. "Mathematical Formalisms in Scientific Practice: From Denotation to Model-Based Representation." *Studies in History and Philosophy of Science* 42:272–86.

———. 2016. *How to Do Science with Models*. Dordrecht: Springer.

———. 2017. "The Ontology of Models." In *Handbook of Model-Based Science*, ed. Lorenzo Magnani and Thomasso Bertolotti, 5–23. Dordrecht: Springer.

Giere, Ronald. 1988. *Explaining Science: A Cognitive Approach*. Chicago: University of Chicago Press.

———. 1999a. *Science without Laws*. Chicago: University of Chicago Press.

———. 1999b. "Using Models to Represent Reality." In *Model-Based Reasoning in Scientific Discovery*, ed. Lorenzo Magnani, Nancy Nersessian, and Paul Thagard, 41–57. New York: Kluwer Academic.

———. 2004. "How Models Are Used to Represent Reality." *Philosophy of Science* 71, no. 5:742–52.

———. 2006. *Scientific Perspectivism*. Chicago: University of Chicago Press.

———. 2009. "An Agent-Based Conception of Models and Scientific Representation." *Synthese* 172:269–81.

Gingerich, Philip D. 2012. "Evolution of Whales from Land to Sea." *Proceedings of the American Philosophical Society* 156, no. 3:309–23.

Glymour, Clark. 2010. "What Is Right with Bayes' Nets Methods and What Is Wrong with 'Hunting Causes and Using Them.'" *British Journal for the Philosophy of Science* 61, no. 1:161–211.

———. 2013. "Theoretical Equivalence and the Semantic View of Theories." *Philosophy of Science* 80, no. 2:286–97.

Godfrey-Smith, Peter, and Arnon Levy, eds. 2020. *The Scientific Imagination*. Oxford: Oxford University Press.

Gombrich, Ernst H. 1950. *The Story of Art*. London: Phaidon.

———. 1959/2002. *Art and Illusion: A Study in the Psychology of Pictorial Representation*. 6th ed. London: Phaidon.

Goodman, Nelson. 1968/1976. *Languages of Art: An Approach to a Theory of Symbols*. Rev. ed. Indianapolis: Bobbs-Merrill.

———. 1972. "Seven Strictures on Similarity." In *Problems and Projects*, 437–47. New York: Bobbs-Merrill.

Gutting, Gary. 1985. "Scientific Realism versus Constructive Empiricism: A Dialogue." In *Images of Science: Essays on Realism and Empiricism with a Reply from Bas C. van Fraassen*, ed. Paul M. Churchland and Clifford A. Hooker, 118–31. Chicago: University of Chicago Press.

Gyenis, Balázs. 2017. "Maxwell and the Normal Distribution: A Coloured Story of Probability, Independence, and Tendency toward Equilibrium." *Studies in History and Philosophy of Modern Physics* 57:53–65.

Hacking, Ian. 1982. "Experimentation and Scientific Realism." *Philosophical Topics* 13:154–72.

———. 1983. *Representing and Intervening*. Cambridge: Cambridge University Press.

———. 1985. "Do We See through a Microscope?" In *Images of Science: Essays on Realism and Empiricism with a Reply from Bas C. van Fraassen*, ed. Paul M. Churchland and Clifford A. Hooker, 132–52. Chicago: University of Chicago Press.

———. 1989. "Extragalactic Reality: The Case of Gravitational Lensing." *Philosophy of Science* 56, no. 4:555–81.

Halvorson, Hans. 2012. "What Scientific Theories Could Not Be." *Philosophy of Science* 79, no. 2:183–206.

———. 2013. "The Semantic View, If Plausible, Is Syntactic." *Philosophy of Science* 80, no. 3:475–78.

Hansen, Carl J., Stephen D. Kawaler, and Virginia Tremble. 2004. *Stellar Interiors: Physical Principles, Structure, and Evolution*. New York: Springer.

Hanson, Norwood R. 1958. *Patterns of Discovery*. Cambridge: Cambridge University Press.

Harman, P. M. 1982. *Energy, Force and Matter: The Conceptual Development of Nineteenth Century Physics*. Cambridge: Cambridge University Press.

———. 1985a. "Edinburgh Philosophy and Cambridge Physics: The Natural Philosophy of James Clerk Maxwell." In *Wranglers and Physicists: Studies on Cambridge Physics in the Nineteenth Century*, ed. Peter M. Harman, 202–24. Manchester: Manchester University Press.

———. ed. 1985b. *Wranglers and Physicists: Studies on Cambridge Physics in the Nineteenth Century*. Manchester: Manchester University Press.

———. 1987. "Mathematical Reality in Maxwell's Dynamical Physics." In *Kelvin's Baltimore Lectures and Modern Theoretical Physics: Historical and Philosophical Perspectives*, ed. Robert Kargon and Peter Achinstein, 267–97. Cambridge, MA: MIT Press.

———. 1991. *The Natural Philosophy of James Clerk Maxwell*. Cambridge: Cambridge University Press.

Harré, Rom H. 2004. *Modelling: Gateway to the Unknown*. Edited by D. Rothbart. Amsterdam: Elsevier.

Harris, Enriqueta. 1982. *Velázquez*. London: Phaidon.

———. 1999. "Inocencio X." In *Velázquez* (Amigos del Museo del Prado), 169–96. Barcelona: Galaxia Gutenberg.

Harris, Todd. 2003. "Data Models and the Acquisition and Manipulation of Data." *Philosophy of Science* 70, no. 5:1508–17.

Hatfield, Gary. 1991. *The Natural and the Normative: Theories of Spatial Perception from Kant to Helmholtz*. Cambridge, MA: MIT Press.

Heidelberger, Michael. 1993. "Force, Law and Experiment: The Evolution of Helmholtz's Philosophy of Science." In *Hermann von Helmholtz and the Foundations of Nineteenth Century Science*, ed. David Cahan, 461–97. Berkeley: University of California Press.

———. 1998. "From Helmholtz's Philosophy of Science to Hertz's Picture-Theory." In *Heinrich Hertz: Classical Physicist, Modern Philosopher*, ed. Davis Baird, R. I. G. Hughes, and Alfred Nordmann, 9–24. Dordrecht: Kluwer Academic.

Helmholtz, Hermann von. 1858. "Über Integrale der Hydrodynamischen Gleichungen, welche den Wirbelbewegungen Entsprechen." *Journal für die reine und angewandte Mathematik* 55:25–55.

———. 1870. "Über die Bewegungsgleichungen der Elektrizität für ruhende leitende Körper." *Journal für die reine und angewandte Mathematik* 72:57–129.

Hempel, Carl. 1965. *Aspects of Scientific Explanation*. New York: Free Press.

Herschel, John. 1850/2014. "Quetelet on Probabilities." In *John Herschel: Essays from the Edinburgh and Quarterly Reviews with Addresses and Other Pieces*, 365–465. Cambridge: Cambridge University Press.

Hertz, Heinrich. 1888. "Über Strahlen Elektrisker Kraft." *Sitzungsberichte de Koniglich Preussischen Akademie der Wissenschaften su Berlin*, no. 50:1297–1307.

———. 1893/1962. *Electric Waves: Being Researches on the Propagation of Electric Action with Finite Velocity through Space*. Translated by D. E. Jones. New York: Dover.

———. 1894/1956. *The Principles of Mechanics Presented in a New Form*. Translated by D. E. Jones and J. T. Walley. New York: Dover.

Hervey, Mary. 1900. *Holbein's Ambassadors: The Picture and the Men*. London: George Bell & Sons.

Hesse, Mary. 1963/1966. *Models and Analogies in Science*. 2nd ed. Notre Dame, IN: Notre Dame University Press.

Hitchcock, Christopher. 1992. "Causal Explanation and Scientific Realism." *Erkenntnis* 37:11–178.

Hoffman, D. 1998. "Heinrich Hertz and the Berlin School of Physics." In *Heinrich Hertz: Classical Physicist, Modern Philosopher*, ed. Davis Baird, R. I. G. Hughes, and Alfred Nordmann, 1–8. Dordrecht: Kluwer Academic.

Hon, Giora, and Bernard R. Goldstein. 2012. "Maxwell's Contrived Analogy: An Early Version of the Methodology of Modelling." *Studies in History and Philosophy of Modern Physics* 43:236–57.

Hookway, Christopher. 1985. *Peirce*. London: Routledge.

———. 2000. *Truth, Rationality and Pragmatism: Themes from Peirce*. Oxford: Oxford University Press.

Hopkins, Robert. 1998. *Picture, Image and Experience*. Cambridge: Cambridge University Press.

———. 2010. "Inflected Pictorial Experience." In *Philosophical Perspectives on Depiction*, ed. Catharine Abell and Katerina Bantinaki, 151–80. Oxford: Oxford University Press.

———. 2012. "Seeing-in and Seeming to See." *Analysis* 72, no. 4:650–59.

Horwich, Paul. 1980. *Truth*. Oxford: Oxford University Press.

Hudetz, Laurenz. 2019. "The Semantic View of Theories and Higher-Order Languages." *Synthese* 196, no. 3:1131–49.

Hughes, R. I. G. 1989. *The Structure and Interpretation of Quantum Mechanics*. Cambridge, MA: Harvard University Press.

———. 1997. "Models and Representation." *Philosophy of Science* 64:S325–S336.

———. 2010. *The Theoretical Practices of Physics*. Oxford: Oxford University Press.

Humphreys, Paul. 2002. "Computational Models." *Philosophy of Science* 69:1–11.

———. 2004. *Extending Ourselves: Computational Science, Empiricism, and Scientific Method*. Oxford: Oxford University Press.

Hunt, Bruce J. 1991. *The Maxwellians*. Ithaca, NY: Cornell University Press.

Hyman, John. 2006. *The Objective Eye: Color, Form, and Reality in the Theory of Art*. Chicago: University of Chicago Press.

Hyman, John, and Katarina Bantinaki. 2017. "Depiction." In *Stanford Encyclopedia of Philosophy*, ed. Edward N. Zalta and Uri Nodelman. https://plato.stanford.edu/entries/depiction.

Ibarra, Andoni, and Thomas Mormann. 2000. "Una teoría combinatoria de las teorías científicas." *Crítica* 22, no. 95:3–46.

Irzik, Gürol, and Faik Kurtulmus. 2019. "What Is Epistemic Public Trust in Science?" *British Journal for the Philosophy of Science* 70:1145–66.

Jeans, James. 1940. *Kinetic Theory of Gases*. Cambridge: Cambridge University Press.

Jebeile, Julie, and Ashley Graham Kennedy. 2015. "Explaining with Models: The Role of Idealizations." *International Studies in History and Philosophy of Science* 29, no. 4:383–92.

Justi, Carl. 1889. *Velázquez and His Times*. Translated by H. A. Keane. London: Grevel.

Kargon, Robert, and Peter Achinstein, eds. 1987. *Kelvin's Baltimore Lectures and Modern Theoretical Physics: Historical and Philosophical Perspectives*. Cambridge, MA: MIT Press.

Kemp, Gary, and Gabriele M. Mras, eds. 2016. *Wollheim, Wittgenstein, and Pictorial Representation*. New York: Routledge.

Khalifa, Kareem. 2017. *Understanding, Explanation, and Scientific Knowledge*. Cambridge: Cambridge University Press.

Khalifa, Kareem, Jared Millson, and Mark Risjord. 2022. "Scientific Representation: An Inferentialist-Expressivist Manifesto." *Philosophical Topics* 50, no. 1:263–92.

Kincaid, Harold, John Dupré, and Alison Wylie, eds. 2007. *Value-Free Science: Ideals and Illusions*. Oxford: Oxford University Press.

Kitcher, Philip. 2001a. "Real Realism: The Galilean Strategy." *Philosophical Review* 110, no. 2:151–97.

———. 2001b. *Science, Truth and Democracy*. Oxford: Oxford University Press.

———. 2002. "The Third Way: Reflections on Helen Longino's *The Fate of Knowledge*." *Philosophy of Science* 69, no. 4:549–59.

———. 2009. *Science in a Democratic Society*. New York: Prometheus.

———. 2012a. *Preludes to Pragmatism: Towards a Reconstruction of Philosophy*. Oxford: Oxford University Press.

———. 2012b. "Second Thoughts." In *Scientific Realism and Democratic Society: The Philosophy of Philip Kitcher* (Poznan Studies in the Philosophy of the Sciences and the Humanities, vol. 101), ed. Wenceslao González, 353–89. Amsterdam: Rodopi.

Klein, Martin. 1970. "Maxwell, His Demon, and the Second Law of Thermodynamics." *American Scientist* 58, no. 1:84–97.

———. 1972. "Mechanical Explanation at the End of the Nineteenth Century." *Centaurus* 17:58–82.

———. 1973. "The Maxwell-Boltzmann Relationship." In *Transport Phenomena: Second Annual International Centennial Boltzmann Seminar*, 297–308. College Park, MD: American Institute of Physics.

Knuuttila, Tarja. 2009. "Some Consequences of the Pragmatist Approach to Representation: Decoupling the Model-Target Dyad and Indirect Reasoning." In *EPSA Epistemology and Methodology: Launch of the European Philosophy of Science Association*, ed. Mauricio Suárez, Mauro Dorato, and Miklos Rédei, 139–48. New York: Springer.

———. 2011. "Modelling and Representing: An Artefactual Approach to Model-Based Representation." *Studies in History and Philosophy of Science* 42:262–71.

———. 2017. "Imagination Extended and Embedded: Artifactual versus Fictional Accounts of Models." *Synthese* 198:5077–97.

———. 2021. "Epistemic Artifacts and the Modal Dimension of Modelling." *European Journal for Philosophy of Science*, vol. 11. https://link.springer.com/article/10.1007/s13194-021-00374-5.

Knuuttila, Tarja, and Andrea Loettgers. 2017. "Modelling as Indirect Representation? The Lotka-Volterra Model Revisited." *British Journal for the Philosophy of Science* 68, no. 4:1007–36.

Kot, Mark. 2001. *Elements of Mathematical Ecology*. Cambridge: Cambridge University Press.

Krantz, David, Duncan Luce, Patrick Suppes, and Amos Tversky. 1971. *Foundations of Measurement 1*. New York: Academic.

Kuhn, Thomas. 1962/1996. *The Structure of Scientific Revolutions*. 3rd ed. Chicago: University of Chicago Press.

Kulvicki, John V. 2006. *On Images*. Oxford: Oxford University Press.

———. 2014. *Images*. London: Routledge.

Kuorikoski, Jaakko, and Aki Lehtinen. 2009. "Incredible Worlds, Credible Results." *Erkenntnis* 70, no. 1:119–31.

Laudan, Larry. 1981. "A Confutation of Convergent Realism." *Philosophy of Science* 48, no. 1:19–49.

Leonelli, Sabina. 2016. *Data-Centric Biology: A Philosophical Study*. Chicago: University of Chicago Press.

———. 2019. "What Distinguishes Data from Models?" *European Journal for Philosophy of Science*, vol. 9, no. 2. https://link.springer.com/article/10.1007/s13194-018-0246-0.

Leroux, Jean. 2001. "'Picture Theories' as Forerunners of the Semantic Approach to Scientific Theories." *International Studies in the Philosophy of Science* 15, no. 2:189–97.

Levy, Arnon. 2015. "Modeling without Models." *Philosophical Studies* 172:781–98.

Lewis, David. 1978. "Truth in Fiction." *American Philosophical Quarterly* 15, no. 1:37–46.

———. 1986. *Philosophical Papers*. Vol. 2. Oxford: Oxford University Press.

Lipton, Peter. 2009. "Understanding without Explanation." In *Scientific Understanding: Philosophical Perspectives*, ed. Henk de Regt, Sabina Leonelli, and Kai Eigner, 43–63. Pittsburgh, PA: Pittsburgh University Press.

Lloyd, Elizabeth. 1988/1994. *The Structure and Confirmation of Evolutionary Theory*. Reprint, Princeton, NJ: Princeton University Press.

Longino, Helen. 1990. *Science as Social Knowledge: Values and Objectivity in Scientific Inquiry*. Princeton, NJ: Princeton University Press.

———. 2002. *The Fate of Knowledge*. Princeton, NJ: Princeton University Press.

Lopes, Dominic. 1996. *Understanding Pictures*. Oxford: Oxford University Press.

———. 2005. *Sight and Sensibility: Evaluating Pictures*. Oxford: Oxford University Press.

Lotka, Alfred. 1934/1998. *Analytic Theory of Biological Populations*. Translated and with an introduction by David P. Smith and Hélène Rossert. New York: Springer. Originally published as *Théorie analytique des associations biologiques*.

Lutz, Sebastian. 2012. "A Straw Man in the Philosophy of Science: A Defense of the Received View." *Journal of the International Society for the History of Philosophy of Science* 2, no. 1:77–120.

———. 2014. "Empirical Adequacy in the Received View." *Philosophy of Science* 81, no. 5:1171–83.

———. 2017. "What Was the Syntax-Semantics Debate in the Philosophy of Science About?" *Philosophy and Phenomenological Research* 95, no. 2:319–52.

Lynch, Michael, and Steve Woolgar, eds. 1990. *Representation in Scientific Practice*. Cambridge, MA: MIT Press.

MacDonald, Graham. 1998. Review of Robert Cummins, *Representations, Targets, and Attitudes* (1998). *British Journal for the Philosophy of Science* 49:175–80.

Mackie, John L. 1980. *The Cement of the Universe: A Study of Causation*. Oxford: Oxford University Press.

Magnani, Lorenzo, and Thomasso Bertolotti, eds. 2017. *Handbook of Model-Based Science*. Dordrecht: Springer.

Magnani, Lorenzo, Nancy Nersessian, and Paul Thagard, eds. 1999. *Model-Based Reasoning in Scientific Discovery*. New York: Kluwer Academic.

Majer, U. 1998. "Heinrich Hertz's Picture-Conception of Theories: Its Elaboration by Hilbert, Weyl, and Ramsey." In *Heinrich Hertz: Classical Physicist, Modern Philosopher*, ed. Davis Baird, R. I. G. Hughes, and Alfred Nordmann, 225–42. Dordrecht: Kluwer Academic.

Martin, Leslie, Ben Nicholson, and Naum Gabo, eds. 1937/1971. *Circle: International Survey of Constructive Art*. London: Faber & Faber.

Martínez, Manolo. 2019. "Representation as Rate-Distortion Sweet Spots." *Philosophy of Science* 86, no. 12:1214–26.

Massimi, Michela. 2004. "Non-Defensible Middle Ground for Experimental Realism: Why We Are Justified to Believe in Colored Quarks." *Philosophy of Science* 71, no. 1:36–60.

———. 2022. *Perspectival Realism*. Oxford: Oxford University Press.

Matzovinos, Crysostomos. 2020. "Institutions and Scientific Progress." *Philosophy of the Social Sciences* 51, no. 3:243–65.

Maudlin, Tim. 2007. *The Metaphysics within Physics*. Oxford: Oxford University Press.

Maxwell, James Clerk. 1856/1990. "Are There Real Analogies in Nature?" In *The Scientific Letters and Papers of J. C. Maxwell*, ed. P. M. Harman, 1:376–83. Cambridge: Cambridge University Press. The essay originally appeared as "Analogies in Nature: Essay for the Apostles."

———. 1856–57/1890. "On Faraday's Lines of Force, Parts 1 and 2." In *The Scientific Papers of James Clerk Maxwell*, ed. William Davidson Niven, 1:155–229. Cambridge: Cambridge University Press.

———. 1860/1890. "Illustrations of the Dynamical Theory of Gases, Parts 1 and 2." In *The Scientific Papers of James Clerk Maxwell*, ed. William Davidson Niven, 1:377–409.

———. 1861–62/1890. "On Physical Lines of Force, Parts 1–4." In *The Scientific Papers of James Clerk Maxwell*, ed. William Davidson Niven, 1:451–513. Cambridge: Cambridge University Press.

———. 1866–67/1890. "On the Dynamical Theory of Gases, Parts 1 and 2." In *The Scientific Papers of James Clerk Maxwell*, ed. William Davidson Niven, 2:26–78. Cambridge: Cambridge University Press.

———. 1873a. "Molecules." *Nature*, September 25, 437–41.

———. 1873b/1954. *A Treatise on Electricity and Magnetism*. 3rd. ed. New York: Dover.

———. 1877. "A Treatise on the Kinetic Theory of Gases: Review." *Nature*, July 26, 242–46.

May, Robert. 2001. *Stability and Complexity in Model Ecosystems*. Princeton, NJ: Princeton University Press.

McClelland, C. E. 1980. *State, Society and University in Germany, 1700–1914*. Cambridge: Cambridge University Press.

McCullough-Brenner, Colin. 2020. "Representing the World with Inconsistent Mathematics." *British Journal for the Philosophy of Science* 71:1331–58.

McKinsey, John Charles C., Alvin C. Sugar, and Patrick Suppes. 1953. "Axiomatic Foundations of Classical Particle Mechanics." *Journal of Rational Mechanics and Analysis* 2, no. 2:253–72.

Mikenberg, Irene, Newton C. A. da Costa, and Rolando Chuaqui. 1986. "Pragmatic Truth and Approximation to Truth." *Journal of Symbolic Logic* 51, no. 1:201–21.

Mill, John Stuart. 1843/2002. *A System of Logic: Ratiocinative and Inductive*. Reprint, Honolulu: University Press of the Pacific.

Millikan, Ruth. 2000. *Language, Thought and Other Biological Categories*. Cambridge, MA: MIT Press.

Mirowski, Philip. 1991. *More Heat Than Light*. Cambridge: Cambridge University Press.

Mitchell, Sandra D. 2003. *Biological Complexity and Integrative Pluralism*. Cambridge: Cambridge University Press.

———, ed. 2004. "The Pragmatics of Scientific Representation." Special section, *Philosophy of Science* 71, no. 5:742–804.

———. 2009. *Unsimple Truths: Science, Complexity and Policy*. Chicago: University of Chicago Press.

———. 2020. "Through the Fractured Looking Glass." *Philosophy of Science* 87:771–92.

Mondrian, Piet. 1920/2017. "Le néo-plasticisme." In *Piet Mondrian: The Complete Writings*, ed. Louis Veen, 186–98. Leiden: Primavera Pers.

———. 1926/2017. "L'art purement abstrait." In *Piet Mondrian: The Complete Writings*, ed. Louis Veen, 258–60. Leiden: Primavera Pers.

———. 1937/1971. "Plastic Art and Pure Plastic Art (Figurative and Non-Figurative Art)." In *Circle: International Survey of Constructive Art*, ed. Leslie Martin, Ben Nicholson, and Naum Gabo, 41–56. London: Faber & Faber.

Monton, Bradley, ed. 2007. *Images of Empiricism: Essays on Science and Stances with a Reply from Bas C. van Fraassen*. Oxford: Oxford University Press.

Monton, Bradley, and Chad Mohler. 2021. "Constructive Empiricism." In *Stanford Encyclopedia of Philosophy*, ed. Edward N. Zalta and Uri Nodelman. https://plato.stanford.edu/entries/constructive-empiricism.

Morgan, Mary. 2012. *The World in the Model*. Cambridge: Cambridge University Press.

Morgan, Mary, and Marcel Boumans. 2004. "Secrets Hidden by Two-Dimensionality: The Economy as a Hydraulic Machine." In *Models: The Third Dimension of Science*, ed. Soraya de Chadarevian and Nick Hopwood, 369–401. Stanford, CA: Stanford University Press.

Morgan, Mary, and Margaret Morrison, eds. 1999. *Models as Mediators: Perspectives on Natural and Social Science*. Cambridge: Cambridge University Press.

Morrison, Margaret. 1990. "Theory, Intervention, and Realism." *Synthese* 82:1–22.

Morrison, Margaret, and Mary Morgan. 1999. "Models as Mediating Instruments." In *Models as Mediators: Perspectives on Natural and Social Science*, ed. Mary Morgan and Margaret Morrison, 10–37. Cambridge: Cambridge University Press.

Muller, Fred. 1997a. "The Equivalence Myth of Quantum Mechanics, Part 1." *Studies in History and Philosophy of Modern Physics* 28, no. 1:35–61.

———. 1997b. "The Equivalence Myth of Quantum Mechanics, Part 2." *Studies in History and Philosophy of Modern Physics* 28, no. 2:219–47.

Mulligan, Joseph F. 1998. "The Reception of Heinrich Hertz's Principles of Mechanics by His Contemporaries." In *Heinrich Hertz: Classical Physicist, Modern Philosopher*, ed. Davis Baird, R. I. G. Hughes, and Alfred Nordmann, 173–81. Dordrecht: Kluwer Academic.

Mundy, Brent. 1986. "On the General Theory of Meaningful Representation." *Synthese* 67:391–437.

Musgrave, Alan. 1976. "Why Did Oxygen Supplant Phlogiston: Research Programmes in the Chemical Revolution." In *Method and Appraisal in the Physical Sciences*, ed. Colin Howson, 181–210. Cambridge: Cambridge University Press.

———. 1989. "Noa's Ark: Fine for Realism." *Philosophical Quarterly* 39, no. 157:383–98.

Nagel, Ernst. 1961. *The Structure of Science*. New York: Harcourt.

Nanay, Bence. 2004. "Taking Twofoldness Seriously: Walton on Imagination and Depiction." *Journal of Aesthetics and Art Criticism* 62, no. 3:285–89.

———. 2005. "Is Twofoldness Necessary for Representational Seeing?" *British Journal of Aesthetics* 45, no. 3:248–57.

Nersessian, Nancy. 2008. *Creating Scientific Concepts*. Cambridge, MA: MIT Press.

Newall, Michael. 2011. *What Is a Picture? Depiction, Realism, Abstraction*. New York: Palgrave Macmillan.

———. 2015. "Is Seeing-in a Transparency Effect?" *British Journal of Aesthetics* 55, no. 2:131–56.

Nguyen, James. 2016a. "How Models Represent." PhD diss., London School of Economics.

———. 2016b. "On the Pragmatic Equivalence between Representing Data and Phenomena." *Philosophy of Science* 83, no. 2:71–91.

Northcott, Robert, and Anna Alexandrova. 2015. "Prisoner's Dilemma Doesn't Explain Much." In *The Prisoner's Dilemma: Classic Philosophic Arguments*, ed. Martin Peterson, 64–84. Cambridge: Cambridge University Press.

Odenbaugh, Jay. 2019. *Ecological Models*. Elements in the Philosophy of Biology. Cambridge: Cambridge University Press.

———. 2021. "Models, Models, Models: A Deflationary View." *Synthese* 198:1–16.

Olson, Richard S. 1975. *Scottish Philosophy and British Physics, 1750–1880: A Study in the Foundations of the Victorian Scientific Style*. Princeton, NJ: Princeton University Press.

Oppler, Ellen C. 1988. *Picasso's Guernica*. Critical Studies in Art History. New York: Norton.

Palmieri, Paolo. 2018. "Galileo's Thought Experiments: Projective Participation and the Integration of Paradoxes." In *The Routledge Companion to Thought Experiments*, ed. Michael T. Stuart, Yiftach Fehige, and James Robert Brown, 111–27. London: Routledge.

Parker, Wendy. 2010. "Scientific Models and Adequacy-for-Purpose." *Modern Schoolman: A Quarterly Journal of Philosophy* 87, nos. 3–4:285–93.

———. 2020. "Model Evaluation: An Adequacy-for-Purpose View." *Philosophy of Science* 87, no. 3:457–77.

Patton, Lydia. 2009. "Signs, Toy-Models, and the a Priori: From Helmholtz to Wittgenstein." *Studies in History and Philosophy of Science* 40:281–89.

———. 2010. "Hermann von Helmholtz." In *Stanford Encyclopedia of Philosophy*, ed. Edward N. Zalta and Uri Nodelman. https://plato.stanford.edu/entries/hermann-helmholtz.

Peacocke, Christopher. 1987. "Depiction." *Philosophical Review* 96, no. 3:383–410.

Peirce, Charles S. 1931. *Collected Papers*. Vol. 2, *Elements of Logic*. Edited by Charles Hartshorne and Paul Weiss. Cambridge, MA: Harvard University Press.

Percival, Philip. 2007. "An Empiricist Critique of Constructive Empiricism: The Aim of Science." In *Images of Empiricism: Essays on Science and Stances with a Reply from Bas C. van Fraassen*, ed. Bradley Monton, 83–116. Oxford: Oxford University Press.

Pero, Francesca, and Mauricio Suárez. 2016. "Varieties of Misrepresentation and Homomorphism." *European Journal for Philosophy of Science* 6, no. 1:71–90.

Peschard, Isabelle. 2017. "Making Sense of Modelling: Beyond Representation." *European Journal for Philosophy of Science* 1, no. 3:335–52.

Petroski, Henry. 1995. *Engineers of Dreams: Great Bridge Builders and the Spanning of America.* New York: Vintage/Random House.

———. 2013. "Engineering: An Anthropomorphic Model." *American Scientist* 101, no. 2:103–7.

Pincock, Christopher. 2012. *Mathematics and Scientific Representation.* New York: Oxford University Press.

Pincock, Christopher, and Marion Vorms, eds. 2013. "Models and Simulations 4," special issue, *Synthese*, vol. 190, no. 2.

Playfair, John. 1795. *Elements of Geometry: Containing the First Six Books of Euclid.* Edinburgh: Bell & Bradfute.

Polanyi, Michael. 1970. "What Is a Painting?" *British Journal of Aesthetics* 10:225–36.

Porter, Theodore. 1981. "A Statistical Survey of Gases: Maxwell's Social Physics." *Historical Studies in the Physical Sciences* 12, no. 1:77–116.

———. 1986. *The Rise of Statistical Thinking, 1820–1900.* Princeton, NJ: Princeton University Press.

Potochnik, Angela. 2017. *Idealization and the Aims of Science.* Chicago: University of Chicago Press.

Poznic, Michael. 2016. "Representation and Similarity: Suárez on Necessary and Sufficient Conditions of Scientific Representation." *Journal for General Philosophy of Science* 47:331–47.

Preston, Paul. 1993. *Franco.* London: Fontana.

Prialnik, D. 2000. *An Introduction to the Theory of Stellar Structure and Evolution.* Cambridge: Cambridge University Press.

Price, Huw. 2011. "Expressivism for Two Voices." In *Pragmatism, Science and Naturalism*, ed. Jonathan Knowles and Henrik Rydenfelt, 87–113. Zurich: Peter Lang.

———. 2013. *Expressivism, Pragmatism, and Representationalism.* Cambridge: Cambridge University Press.

Psillos, Stathis. 1999. *Scientific Realism: How Science Tracks Truth.* London: Routledge.

———. 2000. "Carnap, the Ramsey-Sentence, and Realistic Empiricism." *Erkenntnis* 52, no. 2:253–79.

Putnam, Hilary. 1962/1975. "What Theories Are Not." In *Mathematics, Matter and Method: Philosophical Papers*, 1:215–27. Cambridge: Cambridge University Press.

———. 1981. *Reason, Truth and History.* Cambridge: Cambridge University Press.

———. 2002. *The Collapse of the Fact-Value Dichotomy.* Cambridge, MA: Harvard University Press.

Ramsey, F. P., and G. E. Moore. 1927. "Symposium: 'Facts and Propositions.'" *Proceedings of the Aristotelian Society* 7:153–206.

Rankine, William J. M. 1855. "On the Hypothesis of Molecular Vortices; or, Centrifugal Theory of Elasticity, and Its Connection with the Theory of Heat." *London, Edinburgh and Dublin Philosophical Magazine*, ser. 4, 10, nos. 67–68:840ff.

Reichenbach, Hans. 1956. *The Direction of Time.* Berkeley: University of California Press.

Reiner, R., and R. Pierson. 1995. "Hacking's Experimental Realism: An Untenable Middle Ground." *Philosophy of Science* 62:60–69.

Resnik, Nicholas. 1994. "Hacking's Experimental Realism." *Canadian Journal of Philosophy* 24:395–412.

Richards, I. A. 1936. *The Philosophy of Rhetoric*. New York: Oxford University Press.

Rorty, Richard. 1980. *Philosophy and the Mirror of Nature*. Oxford: Blackwell.

Rosen, Gideon. 1994. "What Is Constructive Empiricism?" *Philosophical Studies* 74:143–78.

Rouse, Joseph. 1987. *Knowledge and Power: Toward a Political Philosophy of Science*. Ithaca, NY: Cornell University Press.

Rowbottom, Darrell P., and Otávio Bueno. 2011. "How to Change It: Modes of Engagement, Rationality, and Stance Voluntarism." *Synthese* 178:7–17.

Russell, Bertrand. 1905. "On Denoting." *Mind* 14, no. 56:479–93.

Ryckman, Thomas. 2007. *The Reign of Relativity: Philosophy in Physics, 1915–1925*. Oxford: Oxford University Press.

Salmon, Wesley. 1998. *Causality and Explanation*. Oxford: Oxford University Press.

Sanches de Oliveira, Guilherme. 2022. "Radical Artefactualism." *European Journal for Philosophy of Science* 12, no. 2:1–33.

Sánchez-Dorado, Julia. 2018. "Methodological Lessons for the Integration of Philosophy of Science and Aesthetics: The Case of Representation." In *Thinking about Science, Reflecting on Art: Bringing Aesthetics and Philosophy of Science Together*, ed. Otávio Bueno, George Darby, Steven French, and Dean Rickles, 10–26. London: Routledge.

———. 2019. "Scientific Representation in Practice: Models and Creative Similarity." PhD diss., University College London.

Scarantino, Andrea, and Gualtiero Piccinini. 2010. "Information without Truth." *Metaphilosophy* 41, no. 3:313–30.

Schiemann, Gregor. 1998. "The Loss of World in the Image: Origin and Development of the Concept of Image in the Thought of Hermann von Helmholtz and Heinrich Hertz." In *Heinrich Hertz: Classical Physicist, Modern Philosopher*, ed. Davis Baird, R. I. G. Hughes, and Alfred Nordmann, 25–38. Dordrecht: Kluwer Academic.

Shannon, Claude E. 1948/1963. "The Mathematical Theory of Communication." *M.D. Computing: Computers in Medical Practice* 14, no. 4:306–17.

Shea, Nicholas. 2014. "Exploitable Isomorphism and Structural Representation." *Proceedings of the Aristotelian Society* 114:123–44.

———. 2018. *Representation in Cognitive Science*. Oxford: Oxford University Press.

Shech, Elay. 2015. "Scientific Misrepresentation and Guides to Ontology: The Need for Representational Codes and Contents." *Synthese* 192, no. 11:3463–85.

Shipway, J. S. 1990. "The Forth Railway Bridge Centenary, 1890–1990: Some Notes on Its Design." *Proceedings of the Institution of Civil Engineers* 88:1079–1107.

Siegel, Donald. 1991. *Innovation in Maxwell's Electromagnetic Theory: Molecular Vortices, Displacement Current, and Light*. Cambridge: Cambridge University Press.

Silliman, Robert H. 1963. "William Thomson: Smoke Rings and 19th Century Atomism." *Isis* 54, no. 4:461–74.

Simson, Robert, trans. 1756/1762. *Euclid: The Elements*. Glasgow: R. & A. Foulis.

Smith, Crosby, and Norton Wise. 1989. *Energy and Empire: A Biographical Study of Lord Kelvin*. Cambridge: Cambridge University Press.

Sneed, Joseph. 1994. "Structural Explanation." In *Patrick Suppes: Scientific Philosopher*, vol. 2, *Philosophy of Physics, Theory Structure and Measurement Theory*, ed. Paul Humphreys, 195–216. Dordrecht: Kluwer.

Snyder, Laura J. 2006. *Reforming Philosophy: A Victorian Debate on Science and Society*. Chicago: University of Chicago Press.

Spence, Charles, and Jon Driver. 2004. *Cross Modal Space and Cross Modal Attention*. Oxford: Oxford University Press.

Steer, George. 1937a. "Historic Basque Town Wiped Out." *New York Times*, April 28.

———. 1937b. "The Tragedy of Guernica, Town Destroyed in Air Attack, Eye-Witness Account." *The Times*, April 27.

Stegmüller, Wolfgang. 1979. *The Structuralist View of Theories*. Dordrecht: Springer.

Sterrett, Susan. 2017. "Physically Similar Systems—a History of the Concept." In *Handbook of Model-Based Science*, ed. Lorenzo Magnani and Thomasso Bertolotti, 377–410. Dordrecht: Springer.

———. 2020. "Scale Modeling." In *Handbook in the Philosophy of Engineering*, ed. Diane P. Mitchenfelder and Neelke Doorn, 394–407. London: Routledge.

Stuart, Michael T., Yiftach Fehige, and James Robert Brown, eds. 2018. *The Routledge Companion to Thought Experiments*. London: Routledge.

Suárez, Mauricio. 1997. "Models of the World, Data Models, and the Practice of Science: The Semantics of Quantum Theory." PhD diss., London School of Economics.

———. 1999a. "The Role of Models in the Application of Scientific Theories: Epistemological Implications." In *Models as Mediators: Perspectives on Natural and Social Science*, ed. Mary Morgan and Margaret Morrison, 168–96. Cambridge: Cambridge University Press.

———. 1999b. "Theories, Models and Representation." In *Model-Based Reasoning in Scientific Discovery*, ed. Lorenzo Magnani, Nancy Nersessian, and Paul Thagard, 75–83. New York: Kluwer Academic.

———. 2003. "Scientific Representation: Against Similarity and Isomorphism." *International Studies in the Philosophy of Science* 17:225–44.

———. 2004. "An Inferential Conception of Scientific Representation." *Philosophy of Science* 71:767–79.

———. 2005. "The Semantic Conception, Empirical Adequacy and Application." *Crítica* 37, no. 109:29–63.

———. 2008. "Experimental Realism Reconsidered: How Inference to the Most Likely Cause May Be Sound." In *Nancy Cartwright's Philosophy of Science*, ed. Luc Boven, Carl Hoefer, and Stephan Hartmann, 137–63. New York: Routledge.

———, ed. 2009a. *Fictions in Science: Philosophical Essays on Modeling and Idealization*. New York: Routledge.

———. 2009b. "Quantum State Diffusion Theory." In *Compendium of Quantum Physics*, ed. D. Greenberger, K. Hentschel, and F. Weinert, 608–9. Heidelberg: Springer.

———. 2010a. Review of R. I. G. Hughes, *The Theoretical Practices of Physics* (2010). *Notre Dame Philosophical Reviews*. https://ndpr.nd.edu/news/the-theoretical-practices-of-physics-philosophical-essays.

———. 2010b. "Scientific Representation." *Philosophy Compass* 5, no. 1:91–101.

———. 2011. "Van Fraassen's Long Journey from Isomorphism to Use." *Metascience* 20:417–49.

———. 2012a. "The Ample Modelling Mind." Review of Daniela Bailer-Jones, *Scientific Models in Philosophy of Science* (2009). *Studies in History and Philosophy of Science* 43, no. 1:213–17.

———. 2012b. "Scientific Realism, the Galilean Strategy, and Representation." In *Scientific Realism and Democratic Society: The Philosophy of Philip Kitcher* (Poznan Studies in the Philosophy of the Sciences and the Humanities, vol. 101), ed. W. González, 269–92. Amsterdam: Rodopi.

———. 2013. "Fictions, Conditionals, and Stellar Astrophysics." *International Studies in the Philosophy of Science* 27, no. 3:235–52.

———. 2014. "Scientific Representation." In *Oxford Bibliographies Online*. https://www.oxfordbibliographies.com/view/document/obo-9780195396577/obo-9780195396577-0219.xml.

———. 2015. "Deflationary Representation, Inference, and Practice." *Studies in History and Philosophy of Science* 49:36–47.

———. 2016. "Representation in Science." In *The Oxford Handbook of Philosophy of Science*, ed. Paul Humphreys, 440–59. Oxford University Press.

———. 2023. "Stellar Structure Models Revisited: Evidence and Data in Asteroseismology." In *Philosophy of Astrophysics: Stars, Simulations, and the Struggle to Determine What Is Out There*, ed. Nora Boyd, Siska de Baerdemaeker, Kevin Heng, and Vera Matarese, 111–29. Synthese Library, vol. 472. Cham: Springer.

Suárez, Mauricio, and Agnes Bolinska. 2021. "Informative Models: Idealization and Abstraction." In *Models and Idealisations in Science: Artifactual and Fictional Approaches*, ed. Alejandro Cassini and Juan Redmond, 71–85. Cham: Springer.

Suárez, Mauricio, and Nancy Cartwright. 2008. "Theories: Tools versus Models." *Studies in History and Philosophy of Science Part B: Studies in History and Philosophy of Modern Physics* 39, no. 1:62–81.

Suárez, Mauricio, and Francesca Pero. 2019. "The Representational Semantic Conception." *Philosophy of Science* 86, no. 2:344–65.

Suárez, Mauricio, and Albert Solé. 2006. "On the Analogy between Cognitive Representation and Truth." *Theoria* 55:27–36.

Suppe, Frederick. 1972. "What's Wrong with the Received View on the Structure of Scientific Theories?" *Philosophy of Science* 39, no. 1:1–19.

———, ed. 1973/1977. *The Structure of Scientific Theories*. 2nd ed. Urbana: University of Illinois Press.

———. 1989. *The Semantic Conception of Theories and Scientific Realism*. Urbana: University of Illinois Press.

Suppes, Patrick. 1957. *An Introduction to Logic*. Princeton, NJ: D. Van Nostrand.

———. 1962. "Models of Data." In *Logic, Methodology and Philosophy of Science*, ed. Ernst Nagel, Patrick Suppes, and Alfred Tarski, 252–61. Stanford, CA: Stanford University Press.

———. 1968. "The Desirability of Formalization in Science." *Journal of Philosophy* 65:651–64.

———. 2002. *Representation and Invariance of Scientific Structures*. Stanford, CA: CSLI Publications. Until it appeared in print, *Representation and Invariance of Scientific Structures* circulated as an unpublished manuscript completed in 1962 and titled "Set-Theoretical Structures in Science."

Swoyer, Christopher. 1991. "Structural Representation and Surrogative Reasoning." *Synthese* 87:449–508.

Taine, Hippolyte. 1866/1990. *A Rome: Voyage en Italie I*. With a prologue by Émile Zola. Brussels: Complexe.

Tan, Peter. 2021. "Inconsistent Idealizations and Inferentialism about Scientific Representation." *Studies in History and Philosophy of Science* 87:11–18.

Tayler, Roger. 1970/1994. *The Stars: Their Structure and Evolution*. 2nd ed. Cambridge: Cambridge University Press.

Teller, Paul. 2001. "Twilight of the Perfect Model Model." *Erkenntnis* 55:393–415.

Thompson, D'Arcy. 1917/1992. *On Growth and Form*. Reprint, New York: Dover.

Thomson, William. 1847/2011. "On a Mechanical Representation of Electric, Magnetic, and Galvanic Forces." In *Mathematical and Physical Papers*, 1:76–80. Cambridge: Cambridge University Press.

———. 1848/2011. "Note on the Integration of the Equations of Equilibrium of an Elastic Solid." In *Mathematical and Physical Papers*, 1:97–98. Cambridge: Cambridge University Press.

Toon, Adam. 2012a. *Models as Make Believe: Imagination, Fiction and Scientific Representation*. London: Palgrave Macmillan.

———. 2012b. "Similarity and Scientific Representation." *International Studies in the Philosophy of Science* 26, no. 3:241–57.

Toulmin, Stephen. 1963. *The Philosophy of Science: An Introduction*. London: Hutchinson.

Turner, R. Steven. 1971. "The Growth of Professional Research in Prussia: Causes and Context." In *Historical Studies in the Physical Sciences*, ed. Russell McCormmach, 137–82. Philadelphia: University of Pennsylvania Press.

———. 1993. "Consensus and Controversy: Helmholtz on the Visual Perception of Space." In *Hermann von Helmholtz and the Foundations of Nineteenth Century Science*, ed. David Cahan, 154–204. Berkeley: University of California Press.

Tusell, Javier. 1981. "La larga aventura del *Guernica*: Parts 1–6." *El país*, September 16–23.

Tversky, Amos. 1977. "Features of Similarity." *Psychological Review* 84:327–52.

Tversky, Amos, and Itamar Gati. 1978. "Studies of Similarity." In *Cognition and Categorization*, ed. Eleanor Rosch and Barbara Lloyd, 79–98. Hillsdale, NJ: Erlbaum.

———. 1982. "Similarity, Separability, and the Triangle Inequality." *Psychological Review* 89, no. 2:123–54.

Vaihinger, Hans. 1924. *The Philosophy of "as If": A System of the Theoretical, Practical and Religious Fictions of Mankind*. Translated by C. K. Ogden. London: Kegan Paul.

Van Dyck, Maarten. 2007. "Constructive Empiricism and the Argument from Underdetermination." In *Images of Empiricism: Essays on Science and Stances with a Reply from Bas C. van Fraassen*, ed. Bradley Monton, 11–31. Oxford: Oxford University Press.

van Fraassen, Bas C. 1980. *The Scientific Image*. Oxford: Oxford University Press.

———. 1987. "The Semantic Approach to Scientific Theories." In *The Process of Science*, ed. N. Nersessian, 105–24. Dordrecht: Kluwer.

———. 1989. *Laws and Symmetry*. Oxford: Oxford University Press.

———. 1994. "Interpretation of Science: Science as Interpretation." In *Physics and Our View of the World*, ed. Jan Hilgevoord, 169–87. Cambridge: Cambridge University Press.

———. 2002. *The Empirical Stance*. New Haven, CT: Yale University Press.

———. 2008. *Scientific Representation: Paradoxes of Perspective*. Oxford: Oxford University Press.

Verrault-Julien, Philippe. 2019. "How Could Models Possibly Provide How-Possibly Explanations?" *Studies in History and Philosophy of Science* 73:22–33.

Volterra, Vito. 1926. "Fluctuations in the Abundance of a Species Considered Mathematically." *Nature* 2972, no. 118:558–60.

Voltolini, Alberto. 2015. *A Syncretistic Theory of Depiction*. New York: Palgrave Macmillan.

Vorms, Marion. 2011. "Representing with Imaginary Models: Formats Matter." *Studies in History and Philosophy of Science* 42:287–95.

———. 2017. "Formats of Representation in Scientific Theorizing." In *Models, Simulations, and Representations*, ed. Paul Humphreys and Cyrille Imbert, 250–73. London: Routledge.

Walker, Ralph S. 1989. *The Coherence Theory of Truth: Realism, Anti-Realism, Idealism*. London: Routledge.

Warwick, Andrew. 1995. "The Sturdy Protestants of Science: Larmor, Trouton and the Motion of the Earth through the Ether." In *Scientific Practice: Theories and Stories of Doing Physics*, ed. Jed Z. Buchwald, 300–343. Chicago: University of Chicago Press.

———. 2003. *Masters of Theory: Cambridge and the Rise of Mathematical Physics*. Cambridge: Cambridge University Press.

Watson, Henry William. 1876. *A Treatise on the Kinetic Theory of Gases*. Oxford: Clarendon.

Weisberg, Michael. 2012. "Getting Serious about Similarity." *Philosophy of Science* 79, no. 5:785–94.

———. 2013. *Simulation and Similarity: Using Models to Understand the World*. Oxford: Oxford University Press.

Weisz, George. 1983. *The Emergence of Modern Universities in France, 1863–1914*. Princeton, NJ: Princeton University Press.

Westhofen, Wilhelm. 1890. "The Forth Bridge." *Engineering*, February 28.

Wilholt, Torsten. 2009. "Bias and Values in Scientific Research." *Studies in History and Philosophy of Science* 40, no. 1:92–101.

Williams, Bernard. 1985. *Ethics and the Limits of Philosophy*. Cambridge, MA: Harvard University Press.

Wilson, Andrew. 1989. "Hertz, Boltzmann and Wittgenstein Reconsidered." *Studies in History and Philosophy of Science* 20, no. 2:245–63.

Wilson, David B. 1985. "The Educational Matrix: Physics Education at Early Victorian Cambridge, Edinburgh and Glasgow Universities." In *Wranglers and Physicists: Studies on Cambridge Physics in the Nineteenth Century*, ed. Peter M. Harman, 12–48. Manchester: Manchester University Press.

Wilson, Mark. 1985. "What Can Theory Tell Us about Observation?" In *Images of Science: Essays on Realism and Empiricism with a Reply from Bas C. van Fraassen*, ed. Paul M. Churchland and Clifford A. Hooker, 222–24. Chicago: University of Chicago Press.

Winsberg, Eric. 2010. *Science in the Age of Computer Simulation*. Chicago: University of Chicago Press.

Winther, Rasmus. 2020a. "The Structure of Scientific Theories." In *Stanford Encyclopedia of Philosophy*, ed. Edward N. Zalta and Uri Nodelman. https://plato.stanford.edu/entries/structure-scientific-theories.

———. 2020b. *When Maps Become the World*. Chicago: University of Chicago Press.

Wise, Norton, and Crosbie Smith. 1987. "The Practical Imperative: Kelvin Challenges the Maxwellians." In *Kelvin's Baltimore Lectures and Modern Theoretical Physics: Historical and Philosophical Perspectives*, ed. Robert Kargon and Peter Achinstein, 323–48. Cambridge, MA: MIT Press.

Wollheim, Richard. 1963. Review of Ernst H. Gombrich's *Art and Illusion* (1960). *British Journal of Aesthetics* 31, no. 3:15–37.

———. 1968/1980. *Art and Its Objects*. 2nd ed. Cambridge: Cambridge University Press.

———. 1987. *Painting as an Art*. London: Thames & Hudson.

———. 1998. "Pictorial Representation." *Journal of Aesthetics and Art Criticism* 56:217–26.

Wollheim, Richard, and Robert Hopkins. 2003. "What Makes Representational Painting Truly Visual?" *Proceedings of the Aristotelian Society* 77:131–67.

Woods, John, ed. 2011. *Fictions and Models: New Essays*. Munich: Philosophia.

Woodward, Jim. 2004. *Making Things Happen: A Theory of Causal Explanation*. Oxford: Oxford University Press.

Worrall, John. 1984. "An Unreal Image." *British Journal for the Philosophy of Science* 35, no. 1:65–80.

Wright, Crispin. 1992. *Truth and Objectivity*. Cambridge, MA: Harvard University Press

Index

Page numbers in italics refer to figures.

abstraction, 22–24, 29–31, 35–36, 53–55, 65, 130; in art, 192–93, 200–202, 206
accuracy, 7–8, 87–89, 163, 171, 190–91, 202–3
action-at-a-distance, 33–35, 266n37
adequacy, empirical, 13, 190–91, 238–44, 246, 251, 256
adequacy conditions, 190–91
Alberti's rule, 205, 282n20
Alexandrova, Anna, 190
Ambassadors, The (Holbein), 194–96, *195*, 199, 202, 205–7, 212–13, 222, 280n5
Ambrosio, Chiara, 127, 269n4, 272n18
Ampère's law, 28–29
analogue models, 28, 49–53, 65–70, 95–96, 105–6, 167
analogy: analogical reasoning, 22–26, 29–31; inverse negative, 127, 130; Maxwell's use of, 24–31, 42, 265n26; negative, 66–67, 70, 96, 125, 127, 130; neutral, 66–68, 70, 96, 130; positive, 66–67, 70, 96, 125, 127, 130
analytic philosophy, 4, 7, 84–87. *See also* metaphysics
antirealism, 12, 13, 237–38, 244, 249–51, 254–55
appropriateness, 39–41, 162, 164, 190
art, 10, 11, 189–223; figurative to abstract, 192–202; representational vs. nonrepresentational, 192–93, 198, 200–221
ascription, 128, 146–47, 150–51, 156–57, 161–64
astrophysics, 70–77, *72*, 101, 108–9, 181–84
atomic hypothesis, 21, 35, 40–41, 238

Bacon, Francis, 111–12, *112*, 168–69, 203
Baker, Benjamin, 59, 60, *61*, 77, 88, 173–75, 216
Bartels, Andreas, 128–30, 141, 161, 273nn25–26
Baxandall, Michael, 64, 268n7, 272n14, 284n39
Berlin, 31, 33, 34, 266n32
Bildtheorie (theory of images), 21, 31–32, 35–42, 162–63, 165, 266n34, 267nn44–46
billiard ball model, 65–69, 96, 100–101, 106, 127–30, 167, 175–77. *See also* kinetic theory of gases
Black, Max, 3, 45–46, 49–51, 56, 58, 83, 128
Black-Gelfert classification, 45–46
Boesch, Brandon, 141, 173, 270n24, 278n1
Boltzmann, Ludwig, 2, 6, 20, 31–32, 34–36, 38, 69, 175; information-gain requirement, 40–42, 155, 164, 184
Bonn, 33, 40
Bouch, Sir Thomas, 59
Brandom, Robert, 14–15, 263n19
British school of modeling. *See* English-speaking school of modeling
Buchwald, Jed Z., 33, 266nn37–38, 266n40, 266n42

Cambridge, 24–26, 30, 263n4, 265n28, 266n39
Campbell, Norman, 2, 66–68, 70, 83, 273n27
cantilever principle, 59–60, *60–63*, 88–89, 95, 173–75
Carnap, Rudolf, 2, 68, 133–35, 140, 269n12, 273n5, 274n7, 274n11
Cartwright, Nancy, 140, 227, 229–30, 233–35, 278n38, 285n9, 285n12
causal realism, 35, 36, 235
causal warrant, 230, 231, 233–35
Chang, Hasok, 236, 277n31, 286n22
coherence theory, 91–92, 244, 270n17
commonsense philosophy (Scottish), 21, 22, 26, 29, 185, 264nn10–11

completeness, 40, 251
conceptual archetypes, 47, 50, 53, 79
conformity, 39–42, 155, 161–64, 253, 267n46
Constable, John, 205
constituent of representation, 86–89, 90–91, 118, 123–24, 127, 166–68, 172; in art, 191, 193, 204; definitions, 6–7, 87; in epistemology, 253
constructive empiricism, 12, 13, 228, 237–44, 260
Contessa, Gabriele, 141, 269n8, 279n6
conventionalist approaches to depiction, 207–10
correctness, 39–40, 162–65, 182–83, 234, 241, 250, 267n46
correspondence theory, 91, 92, 134, 245

Darwin, Charles, 57, 139
data models, 53–55, 103, 107. *See also* mathematical models
Davidson, Donald, 50–51
DDI (denotation-demonstration-interpretation), 132, 141–47, *142*, 156
deflationary accounts of representation, 4, 131–32, 140–41, 151–53, 221, 228, 232–33, 241–53, 260. *See also* DDI (denotation-demonstration-interpretation); DFDIF (denotative function–demonstration–inferential function); inferential conception
deflationism, 8, 141, 143–45, 155, 161, 166, 245, 270n15
demonstration, 141, 143–47, 150, 156, 161–62
denotation, 1, 4, 9, 15, 117, 141–50, 167, 208–10
denotation-demonstration-interpretation (DDI), 132, 141–47, *142*, 156
denotative function, 147–50, 159–60, 162, 167–70, 182, 262n12, 283n28
denotative function–demonstration–inferential function (DFDIF), 147–52, 154, 155–56, 159–60
density (syntactic/semantic), 209–10
depiction, 28, 193, 202–13; conventionalist/linguistic approaches to, 207–10; phenomenological/perceptual approaches to, 207, 210–13
de Regt, Henk, 255–56, 258, 279n47, 287nn39–41, 288n45
Dewey, John, 12, 226, 237, 285n4, 286n25
DFDIF (denotative function–demonstration–inferential function), 147–52, 155–56, 159–60
directionality, 104, 113, 117–20, 129, 168–70
displacement current, 28–30, 167, 265nn26–27, 265n29, 284n36
disquotational schema, 91–92
distinctness, 39, 40, 162–65, 171, 251

EA (thermal equilibrium assumption), 73–77, 183
Edinburgh, physics in, 21, 25, 178, 264n14. *See also* Forth Rail Bridge
EER (epistemic experimental realism) thesis, 234
Einstein, Albert, 30, 114, 258, 265n26
electrodynamics, 20–30, 32, 37, 66, 150, 259, 265n31; action-at-a-distance, 33–35, 266n37; field theory, 33, 34, 42
electromagnetism. *See* electrodynamics
elements of representation, 7–10
Elgin, Catherine, 141, 148, 255, 257, 276n26, 280n1, 283n28
EMA (experimental modeling attitude) (epistemic), 235–36
embedding, 12, 140, 239–41
empirical adequacy, 13, 190–91, 238–44, 246, 251, 256
empiricism, constructive, 12, 13, 228, 237–44, 260
empiricism, logical, 2, 3, 68, 83, 133–34, 237, 252
English-speaking school of modeling, 20, 21–31, 32, 38
entity realism, 226, 227, 229–36, 260
epistemic experimental realism (EER) thesis, 234
epistemology, 10–13, 19, 161, 165, 224–60
ether, models of, 20, 24, 27, 105, 149, 167, 257
exemplars, 9, 199, 217, 256
exemplification, 11, 117–18
experimental modeling attitude (epistemic) (EMA), 235–36
experimental realism (experimentalism), 226, 227, 229–36, 260
experimental warrant, 234–37, 285n16

facticity thesis, 164–65, 255–58
Fang, Wei, 126
Faraday, Michael, 26, 28, 29, 33, 34, 36
Fechner, Gustav Theodor, 33, 34
fictional models, 45–47, 148, 163, 167
field theory, 33, 34, 42
Fine, Arthur, 3, 13, 228, 243–48
FitzGerald, George Francis, 20–21, 30
fluid dynamics, 35, 37, 258
Forbes, James David, 21, 22, 25, 26, 178, 264n14, 265n25
force, representational, 9, 10, 15–16, 159–61, 166–73, 181–82; in art, 189, 192, 197–200, 207, 215–21; definition, 119; phenomenology of, 221–23
Forth Rail Bridge, 58–65, 88–89, 95, 166–67; anthropomorphic model of, 60, *61*, 77, 173–74, 216, 279n13; Benjamin Baker and, 59, 60, *61*, 77, 88, 173–75, 216; Sir Thomas Bouch and, 59; cantilever principle and, 59–60, *60–63*, 88–89, 95, 173–75; diagrams, *62*, *63*, *88*, *89*, 105, 173–75, 214; over Firth of Forth, 58, 59, 61, 89, 175; Sir John Fowler and, 59; Inch-Garvie pier and, 61, *62*; Queensferry pier and, *63*; side winds and, 59–61, 175; Tay Bridge and, 59–61; Wilhelm Westhofen and, *60*, 61, *61*, *63*, 64, 88, *89*, 173, 268n7

Fourier, Jean-Baptiste Joseph, 25
Fowler, Sir John, 59
French, Steven, 161, 270n21, 271n10
Frigg, Roman, 9, 141, 261n1, 269n9, 276n22, 276n26, 276n28

Galilean strategy, 13, 248–50, 287n34
Galilei, Galileo, 58, 108, 263n3; kinematic theory, 141–44, *142*, 146, 150–51, 156–58, 171
Gelfert, Axel, 45, 46
geometry, 22–24, 89, 100, 141–44, 146, 205
German-speaking school of modeling, 21, 25, 30–42, 263n8, 266n35
Giere, Ronald, 121, 133, 140, 141, 270n21, 275n14, 276n26
Glasgow, 21, 264n10, 264n13, 265n25
Gombrich, Ernst, 203, 211, 280n4, 283n30, 283n32, 284n39
Goodman, Nelson, 3, 9, 110, 117, 203, 207–10, 212, 215
Guernica (Picasso), 116, 196–200, *197*, 213, 281n9

Hacking, Ian, 12, 226–37, 285nn3–4; *Representing and Intervening*, 226, 229, 285n6
Hamilton, Sir William, 21, 22, 24, 26
Heaviside, Oliver, 21, 29
Helmholtz, Hermann von, 2, 27, 31–42; on perception, 21, 35–37, 38
Helmholtzianism, 33–38
Hempel, Carl, 133, 134, 255, 274n11
Herschel, John, 178–79
Hertz, Heinrich, 2, 6, 20–21, 29–36, 38–42, 155, 162–64, 171, 253, 266nn33–34, 267nn44–46; conformity requirement of, 39–42, 155, 161–64, 267n46; *Principles of Mechanics*, 33, 35, 36, 38–40, 266n34
Hertzsprung-Russell (HR): diagram, 71–73, *72*, 75–77, 108–9, 181–83; law, 71, 108, 268n14
Hesse, Mary, 3, 66–69, 96, 125, 127, 162
Holbein, Hans, 194–96, *195*, 199, 202, 205–7, 212–13, 222, 280n5
homomorphism, 98, 121–23, 128–30, 141, 161, 273n26
Horwich, Paul, 93, 143, 262n14, 270n15
HR (Hertzsprung-Russell): diagram, 71–73, *72*, 75–77, 108–9, 181–83; law, 71, 108, 268n14
Hughes, R. I. G., 3, 9, 84, 132–33, 141–56, 275n14, 276nn27–28, 277nn30–31

IA (isolation assumption), 74–77, 183
illusion theory, 203, 211, 283n30
inaccuracy, 88, 113, 115, 117, 169–71. *See also* accuracy
inertness, 159, 160
inference: rules of (*see* rules of inference); specific, 159
inferential capacity, 9–10, 155–61, 164, 166, 167, 169, 171, 257
inferential conception, 155–86; and art, 189–92, 199–200, 204, 211, 217, 221–22; case studies, 173–84; definition, 166; and epistemology, 224, 227–28, 233, 241–49, 252–53; and objections to substantive conceptions, 166–72; and scientific understanding, 254–59
inferential function, 147, 151
inflection, 215–17
information-gain requirement, 40–42, 155, 164, 184
informativeness, 10, 164–65, 173–84
instrumentalism, 3, 11–13, 220, 228, 243, 264n19
intervention, 12, 33, 46, 226–27, 229, 231, 234
isolation assumption (IA), 74–77, 183
isomorphism, 7, 12, 94, 97–102, 103–31, 140–41, 166–72; amending, 118–20; in epistemology, 239, 242, 244, 250; partial, 98, 122–23, 141

Kelland, Philip, 25
Kelvin, Lord (William Thomson), 2, 6, 20–32, 34, 265n31
Khalifa, Kareem, 255, 258, 279n7, 287n41, 288n44
kinematics, 28, 121, 214; Galileo's theory of, 141–44, *142*, 146, 150–51, 156–58, 171
kinetic theory of gases, 65–70, 97, 130, 175–81, 184. *See also* billiard ball model
Kitcher, Philip, 13, 228, 247–51, 286n31, 287n32
Knuuttila, Tarja, 220, 269n8, 277n36, 284n37
Kuhn, Thomas, 9, 217, 252

Las meninas (Velázquez), 110, 112
Leonelli, Sabina, 54, 55, 107
licensing, 150–51, 156–57, 161–65, 173–77, 185, 217. *See also* rules of inference
linguistic view of theory. *See* syntactic view of theory
Lloyd, Elizabeth, 121, 133
logical argument (against substantive accounts), 104, 109–13, 120–26, 129, 168–69
logical positivism (logical empiricism), 2, 3, 68, 83, 133–34, 237, 252
Longino, Helen, 13–14, 228, 251–54, 287nn36–37
Lopes, Dominic, 216, 282nn14–15, 283n22, 284n38
Lotka-Volterra model, 10, 55–57, *101*, 107, 114, 167, 172, 215
Lozenge Composition with Yellow, Black, Blue, Red, and Gray (Mondrian), 200–202, *201*, 203, 219–21
Lutz, Sebastian, 240, 274n11, 277n31

MacLaurin, Colin, 22, 23
manipulation/manipulability, 201, 220, 229–37, 285n16
mapping, 98–100, 145, 147, 150–51, 156, 161, 167, 239

mathematical models, 10, 45–46, 53–56, 77–78, 98, 106–8, 209; data models, 53–55, 103, 107
Maxwell, James Clerk, 2, 6, 20, 149–50; analogy, use of, 24–31, 42, 265n26; field theory, 33, 34, 42; ideal spheres model, 67, 69, 70, 175–81; influence on German-speaking school, 32, 40, 41; Maxwellians, 27, 30–31, 105, 257; and models as conductors of reasoning, 25, 29, 70, 109, 166, 168, 184, 242, 256, 257
MCER (metaphysical condition on experimental realism), 230–33, 236
means of representation, 6–8, 86–89, 103–5, 107–8, 120, 124–25, 166–68; definitions, 6–7, 87
mechanics (classical), 25–40, 117, 134, 137, 163, 176, 258
MER (metaphysical experimental realism) thesis, 230–33, 236; flectron objection to, 232–33
metaphor, 50–51, 53, 66, 128, 152, 194; Davidson's view of, 50–51; double-meaning view of, 50; interactive account of, 50–51, 53, 66, 79, 128, 283n22
metaphysical condition on experimental realism (MCER), 230–33, 236
metaphysical experimental realism (MER) thesis, 230–33, 236; flectron objection to, 232–33
metaphysical Scottish mathematics, 22–23
metaphysics, 4, 16, 78, 85, 148, 186; epistemology and, 224, 229–30, 233–34, 242, 245, 287n37
method of exhaustions (Archimedes), 23–24
Millikan, Robert A., 240
Millikan, Ruth, 272n12
minimalism, 8–9, 14, 45, 143, 189, 244; abstract, 91, 143, 161, 276n26, 279n12, 282n17
misrepresentation, 88, 104, 113–15, 117, 120–24, 127–29, 169–70. *See also* inaccuracy; mistargeting
mistargeting, 113–14, 121–22, 129, 169–70
modeling, surface features in. *See* force, representational; inferential capacity
modeling attitude, 5–6, 44–45, 83–84, 161–62, 184–86, 256; natural, 13, 228, 287n36; origins of, 19–43
modeling practice, 3–4, 44, 84–85, 106–7, 147–48, 160–63, 184–86, 245; dynamic aspects of, 175, 177, 179–81, 215–17
models, 3; as artifacts, 165, 201, 219–21; classification of, 45–46; definitions of, 137; mediating movement of, 3, 262n18, 263n19, 280n17; uses of, 5–7, 64–79, 86, 95, 177, 185
model theory, 253
molecular vortices model. *See* vortices and idle wheels/vortex model
Mondrian, Piet, 192, 200–202, *201*, 203, 215, 219–21, 281nn11–13
Morrison, Margaret, 3–4, 229, 231, 261n5
Müller, Johannes, 37
Mundy, Brent, 121
Musgrave, Alan, 245, 285n17

Nagel, Ernest, 133–35, 273n5, 274n6
natural epistemic attitude (NEA), 248–51
natural generativity, 210
naturalism, reductive, 7–8, 11–14, 91, 94
natural ontological attitude (NOA), 13, 228, 243–50. *See also* Fine, Arthur
NEA (natural epistemic attitude), 248–51
NEAN (NEA-nonrealist), 249–51
NEAR (NEA-realist), 249–51
neo-Kantianism, 32, 35–38, 273n5
Neumann, Franz Ernst, 33–34
Newton, Sir Isaac, 108, 137, 274n6, 286n21. *See also* mechanics (classical)
Nguyen, James, 9, 141, 269n9
Nichol, John Pringle, 21, 22
NOA (natural ontological attitude), 13, 228, 243–50. *See also* Fine, Arthur
nonnecessity argument, 8, 10, 104, 115–17, 120–24, 129, 171–72
nonsufficiency argument, 8, 104, 115, 117–18, 120–24, 172
normativity of representation, 14, 217–19
no-theory (redundancy theory), 91, 93, 143–44, 147, 276n26

observation/observability, 12, 71, 108–9, 134–36, 231, 234, 237–43, 248, 250
Ostwald, Wilhelm, 35

painting(s), 110, 192–213, 216–18; portraits, 85–86, 88, 110–12, 168–70, 193–95, 202–3, 212–13
Parker, Wendy, 54, 190
partition, 128, 146–47, 150–51, 156–57, 161–64
Peacocke, Christopher, 205, 282n17
Peirce, Charles Sanders, 9, 85–86, 144, 270n16
PER (psychological version of experimental realism) thesis, 231
permissibility, 39, 40, 163
phase space, 101, *101*, 106–9, 117–18
phenomenology: and depiction, 207, 210–13; of force, 221–23
Phillips-Newlyn machine, 51–53, *52*, 66, 95–97, 100, 105, 167
philosophy of art, 10–11, 114, 191–92, 202–23. *See also* art
philosophy of language, 14, 16, 224, 225
philosophy of mind, 119, 224, 225–26
Picasso, Pablo, 116, 196–200, *197*, 213, 281n9
Pincock, Christopher, 141, 269n6
Pope Innocent X (Velázquez), 110–13, *111*, 168–70, 193–94, 202–3, 205–6
Potochnik, Angela, 255, 287n43
pragmatic Galilean realism, 247–51

pragmatic view of theory, 3, 87, 132–33, 140, 152–53, 256, 258, 273n3, 275n17
pragmatism, 4, 12–14, 41, 126, 237
pragmatist theory of truth, 92, 270n16
Price, Huw, 279n7
primitivist accounts of representation, 90, 91, 94, 269nn12–13
psychological version of experimental realism (PER) thesis, 231
Putnam, Hilary, 114, 136, 191, 274n8

quantum mechanics, 10, 113, 136, 170, 171–72; quantum state diffusion model, 6, 113, 118, 170

Rankine, William John Macquorn, 28, 29, 265n25
realism, 3, 11, 13, 226, 237–38, 240, 243–51, 254. *See also* antirealism; causal realism; experimental realism (experimentalism); pragmatic Galilean realism; real realism
real realism, 13, 247–51, 287n32
received view of theory. *See* syntactic view of theory
reductive accounts of representation, 90, 91, 94, 249
reductive naturalism, 7–8, 11–14, 91, 94
redundancy theory (no-theory), 91, 93, 143–44, 147, 276n26
reflexivity, 9, 100, 110, 112, 121–24, 168, 204
Reid, Thomas, 24, 26, 264nn10–11, 264n13
relativity of knowledge, 5, 6, 22, 26, 29
repleteness, 209
representation: conceptual history of, 1–4, 84; conceptual integrity of, 11; elements of, 7–10; philosophical accounts of, 90–91 (*see also* deflationary accounts of representation; primitivist accounts of representation; reductive accounts of representation; substantive accounts of representation); successful, 7, 14, 104, 190, 245–46; and truth, 91–94
representational force. *See* force, representational
representational mechanism, 128–30
representational semantic view (RSV), 137–40, 152, 241, 256, 275nn17–18, 276n24
resemblance, 11, 95, 97, 102, 110, 192, 202–8, 210–11; depiction as, 202–3; objective/subjective, 204–5
Resnik, Nicholas, 231–32, 285n14
Richards, I. A., 51, 79, 128, 283n22
Riemann, Bernhard, 114
Rorty, Richard, 12
Rosen, Gideon, 240–42
RSV (representational semantic view), 137–40, 152, 241, 256, 275nn17–18, 276n24
rules of inference, 37–39, 42, 157–85, 217–21; horizontal, 145–46, 163, 173, 176, 182, 219–21, 242; vertical, 144–45, 162–63, 173–74, 176, 182, 184, 217–21

Salisbury Cathedral from the Meadows (Constable), 205
Sánchez-Dorado, Julia, 263n1, 273n24
San Francisco Bay model, 48–49, 166–67, 214–15
scale models, 47–49, 64–65, 97, 104–5, 166–67, 214–15, 267nn3–4
Scottish Enlightenment, 21–22, 24
Scottish school of physics, 21–26, 29–30, 264nn15–16. *See also* Edinburgh, physics in; Glasgow
seeing-as, 212–13, 218
seeing-in, 114, 191, 212–23, 283n33, 284n38
semantic view of theory, 1–4, 121, 132–40, 151–53, 273nn1–2, 276nn24–25, 277n31; representational claim, 137–38; semantic bare claim, 136–40; structural claim, 137–39. *See also* RSV (representational semantic view); SSV (structural semantic view)
Shannon's mathematical theory of communication, 164, *165*, 279n9
sign theory, 37–39, 41, 85, 266n42, 267n44
similarity, 7, 94–98, 102, 103–31, 166–72; amending, 118–20; in art, 192, 202, 204, 207; creative approach to, 127–28; in epistemology, 242, 244, 250; without identity, 120–21; identity theory of, 95, 125, 270n22; objective/subjective, 204–5; relevant, 116
simplicity, 40, 164–65, 252
Simson, Robert, 22–23, 264n13
social practice(s), 119, 217–19, 243–54
social values, 13, 14, 199, 219, 247, 251–54
spectator theory of knowledge, 12, 226
speech acts, 141, 143–44
spherical symmetry assumption (SSA), 74, 75–77, 183
SSA (spherical symmetry assumption), 74, 75–77, 183
SSV (structural semantic view), 137–40, 256, 275n14, 275n18, 276n20
star formation models, 70–77, *72*, 101, 108–9, 181–84
Stegmüller, Wolfgang, 133
stellar structure models, 70–77, *72*, 101, 108–9, 181–84
Stewart, Dugald, 24, 26, 264n11
Stokes, George Gabriel, 30, 180
structural identity, 98, 100
structural representation, 2, 99, 118, 125, 128–29, 133, 137, 139; without isomorphism, 123–24
structural semantic view (SSV), 137–40, 256, 275n14, 275n18, 276n20
substantive accounts of representation, 4, 90–103, 124–28, 140–41, 166, 168; in art, 202, 204; in epistemology, 226, 228, 232, 241, 244, 247–51, 253
Suppes, Patrick, 2, 54–55, 107, 121–22, 133, 137, 140

Swoyer, Chris, 123–24, 129, 272n16
symmetry, 9, 100, 110–12, 120–25, 129, 168, 204
syntactic view of theory, 1–2, 132–38, 140, 152–53, 273nn3–4, 275n19, 277n31; syntactic bare claim, 136–37

theoretical models, 46, 56–58, 65, 70, 97, 108–9, 258
theory, 2, 3, 50, 84, 132–38, 238–39, 242, 256, 258. *See also* pragmatic view of theory; semantic view of theory; syntactic view of theory
thermal equilibrium assumption (EA), 73–77, 183
Thomson, James, 21, 25
Thomson, William (Lord Kelvin), 2, 6, 20–32, 34, 265n31
Toulmin, Stephen, 3
transference, 86, 269n7
transitivity, 9, 110, 112, 125, 129, 168–69, 204
truth, 8, 14, 15–16, 88, 225–26, 231, 235, 238–47, 251
truth, theories of, 91–94; deflationary (*see* minimalism; redundancy theory [no-theory]; use-based theory); substantive (*see* coherence theory; correspondence theory; utility theory)
Tversky, Amos, 120–22, 125–27, 272n19
twofoldness, 189, 192, 204, 211, 213–15, 217, 219–23

UCA (uniform composition assumption), 74–77, 183–84
understanding (scientific), 14, 228, 254–59
uniform composition assumption (UCA), 74–77, 183–84
use-based theory, 91, 93, 143–44, 147, 151, 277n30
utility theory, 92, 270n16

values, 13, 14, 152, 199, 219, 243, 247, 251–54
van Fraassen, Bas, 12, 107, 114, 136, 140, 237–44
variety, argument from, 103–9, 118, 120, 122, 166
Velázquez, Diego, 110–13, *111*, 168–69, 193–94, 212
veridicality thesis, 164–65, 255–58
Vienna, 31, 32, 40, 134
visual fields, 205–6, 282n19
voluntarism, 228, 237, 286n19
vortices and idle wheels/vortex model, 27–29, *27*, 65, 105, 148, 150

warrant: causal, 230, 231, 233–35; experimental, 234–37, 285n16
Warwick, Andrew, 217, 263n4, 276n26, 276n28, 285n17
Weber, Wilhelm Eduard, 33, 34
weighted feature matching account (Tversky-Weisberg measure), 125–27
Weisberg, Michael, 46–49, 125–27, 141, 272n19
whales, evolution of, 57, 108
Whewell, William, 24, 26, 30, 263n1
Williams, Bernard, 13
Winther, Rasmus, 132, 140, 152–53, 274n8, 275n17, 278n38
Wollheim, Richard, 114, 191, 203, 204, 210–22, 283n33, 284n38
Wright, Crispin, 93, 143, 161, 262n14